Nerve-Muscle Cell Trophic Communication

Editor

Hugo L. Fernandez, Ph.D.

Director
Neuroscience Research Laboratory
Veterans Administration Medical Center
Kansas City, Missouri

Professor
Department of Physiology
University of Kansas Medical Center
Kansas City, Kansas

Associate Editor

J. Alejandro Donoso, Ph.D.

Research Physiologist
Neuroscience Research Laboratory
Veterans Administration Medical Center
Kansas City, Missouri

Research Associate Professor
Department of Neurology
University of Kansas Medical Center
Kansas City, Kansas

CRC Press, Inc.
Boca Raton, Florida

Library of Congress Cataloging-in-Publication Data

Nerve-muscle cell trophic communication.

Includes bibliographies and index.
1. Neuromuscular transmission. 2. Myoneural junction. 3. Muscle contraction—Regulation. I. Fernandez, Hugo L. II. Donoso, J. Alejandro. [DNLM: 1. Motor Neurons—physiology. 2. Muscles—innervation. 3. Neural Transmission. 4. Neuromuscular Junction—physiology. WL 102.7 N455]
QP364.5.N46 1988 599'.01852 87-23888
ISBN 0-8493-6660-7

Direct all inquiries to CRC Press, Inc., 2000 Corporate Blvd., N.W., Boca Raton, Florida, 33431.

International Standard Book Number 0-8493-6660-7

Library of Congress Card Number 87-23888
Printed in the United States

PREFACE

Research on nerve-target cell trophic interactions has a long history encompassing substantial contributions by scientists with a wide spectrum of interests extending from neurotransmission and axoplasmic transport to neurological diseases. It is now well established that aside from bioelectrical impulse conduction and other functions, nerve cells exert long-term trophic (nutritive) influences on themselves and on the target cells they innervate. It is also generally recognized that target cells play an important role in maintaining the functional state of their corresponding innervation. While the phenomenology of such life-supporting bidirectional influences has been extensively documented over the years, the underlying molecular mechanisms are only beginning to be appreciated.

Trophic mechanisms are apparently complex and traditionally there has been considerable controversy over which cellular factors are directly responsible for the maintenance of nerve and target cell characteristics. In an effort to contribute to a better understanding of trophic mechanisms, in this volume we attempt to illustrate the aforegoing controversy and, more importantly, to compile under a single cover some of the current views and hypotheses related to trophic phenomena. In view of obvious space limitations and for the sake of simplicity, we have restricted this work to discussions on the trophic interdependence between mammalian motor nerves and skeletal muscles. In this context, it must be kept in mind that the concepts behind trophic mechanisms in the neuromuscular system may not necessarily be applicable to other types of cellular arrangements.

Throughout this book analysis of nerve-muscle trophic effects has been centered mainly around the direct participation, physiological interrelationships, and relative importance of various "putative" trophic factors. Additional emphasis has been placed on those gaps remaining in our understanding of this subject matter, as well as on the relevance of trophic functions in the areas of nerve cell survival, regeneration, and neuromuscular plasticity.

We are indebted to all the contributors for their diligent efforts and cooperation in sharing their ideas and expertise. Finally, we hope that the readers find *Nerve-Muscle Cell Trophic Communication* of value when considering future directions in this challenging area of the Neurosciences.

Kansas City, 1986

Hugo L. Fernandez
J. Alejandro Donoso

DEDICATION

To Maritza, Stella & Milene Fernandez and Elena Lathrop

H. L. F.

To Vicki, Elena, Alexander and my parents

J. A. D.

THE EDITOR

Hugo L. Fernandez, Ph.D., is Director of the Neuroscience Research Laboratory, Veterans Administration Medical Center, Kansas City, Missouri and Professor of Physiology, College of Health Sciences, University of Kansas, Kansas City, Kansas.

Dr. Fernandez obtained his B.S. in Electrical Engineering at the Catholic University, Valparaiso, Chile (1965), M.S. in Biophysics at the Massachusetts Institute of Technology, Cambridge, Massachusetts (1969), and Ph.D. in Physiology and Cell Biology at the University of Kansas, Lawrence, Kansas (1971). He has served as Assistant Professor (1972-1973), Associate Professor (1973-1976), and Professor (1976-1977) of Neurobiology at the Catholic University, Santiago, Chile; and Assistant Professor (1977-1979) and Associate Professor (1981-1983) of Physiology at the University of Kansas College of Health Sciences. He assumed his current position as Director of the Neuroscience Research Laboratory in 1979 and was appointed Associate Research Career Scientist of the Veterans Administration in 1983. He has also served as Professor in the Department of Physiology at the University of Kansas College of Health Sciences since 1983.

Dr. Fernandez is a member of several scientific societies including the Neuroscience Society, American Society for Neurochemistry, International Brain Research Organization, International Society for Developmental Neuroscience, and Sigma Xi. His major research interests and publications are in the areas of axoplasmic transport, the biology of acetylcholinesterases, and nerve-target cell physiological interactions at the molecular level.

THE ASSOCIATE EDITOR

J. Alejandro Donoso, Ph.D., is Research Physiologist, Neuroscience Research Laboratory, Veterans Administration Medical Center, Kansas City, Missouri and Research Associate Professor of Neurology, College of Health Sciences, University of Kansas, Kansas City, Kansas.

Dr. Donoso obtained his B.S. in biology at the Catholic University, Santiago, Chile (1966), M.S. in Pharmacology (1976), and Ph.D. in Pharmacology at the University of Kansas College of Health Sciences, Kansas City, Kansas (1979). He has served as Assistant Professor (1970-1975) of Neurobiology at the Catholic University, Santiago, Chile; and Research Associate, Ralph L. Smith Research Center (1979-1983) at the University of Kansas College of Health Sciences. He assumed his current position as Research Physiologist at the Neuroscience Research Laboratory in 1983. He has also served as Research Associate Professor in the Department of Neurology at the University of Kansas College Health Sciences since 1983.

Dr. Donoso is a member of several scientific societies including the Neuroscience Society, American Society for Neurochemistry, International Brain Research Organization, Society for Toxicology, International Society for Developmental Neuroscience, and Sigma Xi. His major research interests and publications are in the areas of neurotoxicology, axoplasmic transport, and nerve-target cell interactions at the ultrastructural level.

CONTRIBUTORS

Stephen Brimijoin, Ph.D.
Professor
Department of Pharmacology
Mayo Clinic
Rochester, Minnesota

Heather L. Davis, Ph.D.
Assistant Professor
Department of Anatomy
Assistant Professor
School of Physical and Occupational Therapy
McGill University
Montreal, Canada

J. Alejandro Donoso, Ph.D.
Research Associate Professor
Department of Neurology
University of Kansas Medical Center
Kansas City, Kansas
Research Physiologist
Neuroscience Research Laboratory
Veterans Administration Medical Center
Kansas City, Missouri

Hugo L. Fernandez, Ph.D.
Professor
Department of Physiology
University of Kansas Medical Center
Kansas City, Kansas
Director
Neuroscience Research Laboratory
Veterans Administration Medical Center
Kansas City, Missouri

D. Bruce Gray, Ph.D.
Research Associate
Department of Physiology and Neurobiology
University of Connecticut
Storrs, Connecticut

Kristian Gundersen, M.Sc.
Research Assistant
Institute of Neurophysiology
University of Oslo
Oslo, Norway

Nibaldo C. Inestrosa, Ph.D.
Associate Professor
Department of Cell Biology
Catholic University
Santiago, Chile

Terje Lømo, M.D., Ph.D.
Professor
Institute of Neurophysiology
University of Oslo
Oslo, Norway

George J. Markelonis, Jr., Ph.D.
Associate Professor
Department of Anatomy
University of Maryland School of Medicine
Baltimore, Maryland

Stephen D. Meriney, Ph.D.
Associate
Jerry Lewis Neuromuscular Research Center
UCLA School of Medicine
Los Angeles, California

Sidney Ochs, Ph.D.
Professor
Department of Physiology and Biophysics
Indiana University School of Medicine
Indianapolis, Indiana

Tae H. Oh, Ph.D.
Professor
Department of Anatomy
University of Maryland School of Medicine
Baltimore, Maryland

Guillermo R. Pilar, M.D.
Professor and Head
Department of Physiology and Neurobiology
University of Connecticut
Storrs, Connecticut

Norman Robbins, M.D. Ph.D.
Department of Developmental Genetics and Anatomy
Case Western Reserve University School of Medicine
Cleveland, Ohio

Sang H. Shim, Ph.D.
Visiting Professor
Department of Anatomy
University of Maryland School of Medicine
Baltimore, Maryland

Linda H. Younkin, Ph.D.
Department of Pharmacology
Case Western Reserve University School of Medicine
Cleveland, Ohio

Steven G. Younkin, M.D., Ph.D.
Department of Pharmacology
Case Western Reserve University School of Medicine
Cleveland, Ohio

TABLE OF CONTENTS

Chapter 1

NERVE-MUSCLE CELL TROPHIC COMMUNICATION: INTRODUCTORY REMARKS

Hugo L. Fernandez and J. Alejandro Donoso

TABLE OF CONTENTS

I. INTRODUCTION

The importance of trophic (nutritive) interactions in the formation, maintenance, and function, of the nervous system cannot be overemphasized: elucidation of the underlying molecular mechanisms currently remains as one of the greatest challenges in neuroscience.

Despite the long history of research in this area (Chapter 2), it has not been possible to unambiguously determine which cellular activities and regulatory factors specifically mediate the mutual interdependence of nerve and target cells. Accordingly, in defining trophic functions, most investigators have appropriately avoided any allusion to such factors and mechanisms. In the late 1960s trophic effects were provisionally defined as "those interactions between nerves and other cells which initiate or control molecular modifications in the other cell. Such activities, being long-term in nature, would serve to regulate and/or modify some of the metabolic, physiological, and structural characteristics of the target tissues."[1] During the subsequent decade the bidirectional character of trophic phenomena was underscored by considering that "both the neuron and its target cell are interdependent, with molecular signals traveling between them in the maintenance of a dynamic balance".[2] Thus, broadening the foregoing definition it was stated that "nerve cells and target cells interact so as to influence the structure or function of either member of the pair".[3] In attempts to further distinguish between trophic and other aspects of neural function the term neurotrophic has often been restricted to include only those "long-term maintenance regulations not mediated by nerve impulses".[4,5] This latter statement, however, is somewhat distracting because it seems to dismiss nerve impulse-evoked target cell activity as one of the factors involved in trophic relationships. In this regard, it must be recognized that the mechanisms behind conceptually distinguishable nerve cell functions such as impulse conduction, axoplasmic transport, transmitter release, neurosecretion, and trophic effects are physiologically interrelated and may have certain factors in common. With this in mind and in order to take the broadest possible view of the subject matter, throughout this volume we have made every effort to dispense with the word neurotrophic.

II. PROPRIOTROPHIC AND EXTEROTROPHIC FUNCTIONS

More recently, trophic (from the Greek *trophikos,* meaning nutrition) phenomena in the nervous system have been considered to fall within two distinct categories designated as proprio- (from the Latin *proprius*, meaning own) and extero- (from the Latin *exter*, meaning from outside) trophic.[6,7] Propriotrophic functions refer to those long-term regulatory influences exerted by nerve cells in support of their own structural and functional integrity. Indeed, throughout life nerve cells maintain a dynamic equilibrium between anabolic and catabolic processes, thereby ensuring a continuous renewal of their molecular constituents.[8] The substances needed for such a renewal, as well as for neuronal growth and regeneration, are supplied through axoplasmic transport.[9] Exterotrophic functions refer to those long-term supportive and regulatory influences exerted by either neurons on target cells or vice versa. Such actions affecting target cells are dramatically illustrated by the loss of normal structure and operation of effector and receptor cells when their respective nerves are transected.[10] In turn, those influences related to the maintenance of nerve cells are best exemplified by the vital role played by target cells in the survival of motor neurons during development.[11] The particular direction of these junctions is often specified by adopting the same terminology used to describe axonal transport phenomena (Chapter 3), i.e., the terms anterograde and retrograde being applied to neuron-to-target cell and target cell-to-neuron exterotrophic influences, respectively. At this point it is pertinent to note that while both proprio- and exterotrophic interactions apparently take place throughout life, they may be subjected to physiologically important dynamic modifications with age.[12] In this regard, it must be conceded that the factors and mechanisms involved in trophic processes during cellular development may or may not be the same as those which are operative during adulthood and/or aging.[10]

III. NEUROMUSCULAR TROPHIC FACTORS (MEDIATORS)

As noted in the Preface we have limited the scope of this book to analyses of trophic interactions in the vertebrate neuromuscular system. Much of the present knowledge regarding exterotrophic influences of nerve on skeletal muscle (anterograde) has traditionally come from studies showing that denervation results in decreased resting membrane potentials,[13] increased extrajunctional sensitivity to acetylcholine (ACh),[14] loss of junctional acetylcholinesterase (AChE),[15] and various metabolic changes which eventually lead to complete muscle degeneration.[16] How much each of these alterations results from depletion of ACh, absence of other neural substances, muscle inactivity, and/or alterations in the structural organization of motor endplates is still a matter of controversy.[17] Part of the complication arises from the fact that a variety of different neuromuscular preparations have been studied, and often less than adequate molecular markers were chosen to evaluate the muscle's response to the simultaneous absence of several neural factors.[7] An additional problem is the lack of information on whether a single or several muscle characteristics are directly influenced by one or a combination of neural factors.[8] In an effort to illustrate and, at least to a certain extent, clarify the aforementioned controversy, important data and compelling arguments are presented for the relative importance of one or more of the following putative trophic factors (mediators): axoplasmic transport and ACh (Chapter 3); muscle electromechanical activity (Chapter 4); nerve-evoked muscle activity (Chapter 5); neurogenic substances other than the neurotransmitter (Chapters 6 and 7); and synaptic extracellular matrix components (Chapter 8). While convincing evidence is given for the individual preeminence of these factors, one must keep in mind the possible multifactorial nature of trophic mechanisms and their conceivable modulation by other agents (e.g., hormones).[3] Likewise, an analogous concept invoking the participation of multiple factors should perhaps be taken into account when analyzing exterotrophic effects of muscle on nerve (retrograde).[7] The latter are mainly considered within the general framework of trophic influences exerted by peripheral target cells on the growth and survival of developing motor neurons (Chapter 9). However, it is well established that adult motor neurons also depend on their target cells for trophic support.[4,18] The vital importance of these, as well as of anterograde trophic events during development, adulthood, and aging, is clearly emphasized in studies dealing with synaptic neuromuscular plasticity (Chapter 10).

IV. NEUROGENIC AND MYOGENIC SUBSTANCES

In recent years a certain degree of confusion has been introduced regarding the terminology used for cellular substances claimed to encode trophic information in the neuromuscular system, i.e., putative trophic substances. While substances extracted from nerves have indistinctively been referred to as neurotrophic or myotrophic, those derived from muscles are usually called neurotrophic.[19] In other words, the same terms are used in some cases to imply the cellular source of the substances, whereas in other instances they are applied to specify the particular type of cell they affect. For the sake of simplicity and not to dwell on semantic arguments, we will use here the terms neurogenic for nerve-derived substances affecting muscle cells and myogenic for muscle-derived substances influencing nerve cells. Although a number of these substances have been partially purified and tested for their effects on certain cellular features, there is still little agreement concerning their precise biochemical properties, distinction from growth factors or secondary hormones, and genuine trophic nature.[20,21] In attempts to facilitate the identification and evaluation of substances suspected of having trophic capabilities, the following criteria have been established.[7,22,23]

First, a neurogenic substance affecting skeletal muscle should (1) be present within motor neurons; (2) be synthesized in the neuronal cell body and conveyed by anterograde axoplasmic transport; (3) be released from the nerve terminal and supplied to the muscle; (4) interact with a molecular target or receptor in the postsynaptic membrane and/or within the muscle cell; and (5) exert a dose-dependent effect on one or several definable metabolic, functional, and/or structural properties of muscle. As a corollary to these guidelines one can expect such a substance to be absent, or reduced in concentration, in denervated muscle after complete nerve terminal degeneration; be depleted from nerve terminals upon selective inhibition of its synthesis or transport; and prevent or reverse those muscle alterations elicited by its absence. Second, a myogenic substance affecting motor neurons should (1) be synthesized by myofibers and supplied to the neuron; (2) interact with a molecular target or receptor in the presynaptic membrane and/or in the neuronal cell body after being carried by retrograde axoplasmic transport; and (3) exert a dose-dependent effect on one or several definable metabolic, functional, and/or structural properties of motor neurons. While these criteria may be modified as future knowledge becomes available, they currently provide a useful framework for research on trophic interactions.

V. TROPHIC MECHANISMS

During the past 3 decades an impressive amount of information has accumulated on the phenomenology of nerve-muscle cell interdependence, and the fundamental importance of trophic events in this system is no longer doubted. On the other hand, progress toward attaining an improved understanding of the molecular mechanisms involved in such events has certainly been minimal. Implicit in the literature, however, are important conceptual developments which will hopefully result in the formulation of newer research directions in this field (Chapter 11). For example, worthy of remark is the conceptual distinction between proprio- and exterotrophic functions. Under certain conditions neurons and myofibers separated from each other can temporarily survive by expressing their individual propriotrophic capabilities; however, their long-term maintenance critically depends on the continuous control and fine modulation of such capabilities by the other cell's exterotrophic actions.[7] Thus, in analyzing trophic mechanisms it becomes tempting to search for a final common pathway leading to regulation of the target cell's genetic machinery as related to preservation of a dynamic equilibrium between anabolic and catabolic processes. Also, not to be overlooked is the progressive widespread recognition now given to the multifactorial character of trophic mechanisms.[3] In fact, several studies are being actively pursued in connection with the relative contribution and possible physiological relationship of multiple mediators in the trophic control of one or more target cell features. Although unavoidably complex, the latter approach has led to the interesting idea that all such mediators may simultaneously or sequentially act upon a second messenger system which in turn would directly affect the target cell's genetic expression.[24] These and other possibilities are discussed in more detail in the following chapters where individual authors consider various future research strategies. Which of such strategies will eventually prove the most successful remains one of the most pressing challenges to students of trophic phenomena.

REFERENCES

1. **Guth, L.,** "Trophic" effects of vertebrate neurons, *Neurosci. Res. Prog. Bull.*, 7, 1, 1969.
2. **Smith, B. H. and Kreutzberg, G. W.,** Neuron-target cell interactions, *Neurosci. Res. Prog. Bull.*, 14, 222, 1976.
3. **Drachman, D. B.,** Trophic interactions between nerves and muscles: the role of cholinergic transmission (including usage) and other factors, in *Biology of Cholinergic Function,* Goldberg, A. M. and Hanin, I., Eds., Raven Press, New York, 1976, 161.
4. **Gutmann, E.,** Neurotrophic relations, *Ann. Rev. Physiol.*, 38, 177, 1976.
5. **Guth, L. and Albuquerque, E. X.,** The neurotrophic regulation of resting membrane potential and extrajunctional acetylcholine sensitivity in mammalian skeletal muscle, in *Muscle Regeneration,* Mauro, A., Ed., Raven Press, New York, 1979, 405.
6. **Fernandez, H. L.,** Neurotrophic effects in relation to axoplasmic transport, *Acta Fisiol. Latinoam.*, 23, 189, 1973.
7. **Fernandez, H. L. and Donoso, J. A.,** The relevance of axoplasmic transport in neurotrophic functions, in *Axoplasmic Transport,* Iqbal, Z., Ed., CRC Press, Boca Raton, Fla., 1986, 209.
8. **Gutmann, E. and Hník, P.,** Denervation studies in research of neurotrophic relationships, in *The Denervated Muscle,* Gutmann, E., Ed., Publishing House of the Czechoslovak Academy of Science, Prague, 1962, 13.
9. **Graftstein, B. and Forman, D. S.,** Intracellular transport in neurons, *Physiol. Rev.*, 60, 1167, 1980.
10. **Harris, A. J.,** Inductive functions of the nervous system, *Ann. Rev. Physiol.*, 36, 251, 1974.
11. **Landmesser, L.,** The generation of neuromuscular specificity, *Ann. Rev. Neurosci.*, 3, 279, 1980.
12. **Gutman, E., Hanzlikova, V., and Jakoubek, B.,** Changes in the neuromuscular system during old age, *Exp. Gerontol.*, 3, 141, 1968.
13. **Albuquerque, E. X., Schuh, F. T., and Kauffman, F. C.,** Early membrane depolarization of the fast mammalian muscle after denervation, *Pfluegers Arch.*, 328, 36, 1971.
14. **Axelsson, J. and Thesleff, S.,** A study of supersensitivity in denervated mammalian skeletal muscle, *J. Physiol. (London),* 147, 178, 1959.
15. **Guth, L., Albers, R. W., and Brown, W. C.,** Quantitative changes in cholinesterase activity of denervated muscle fibers and sole plates, *Exp. Neurol.*, 10, 236, 1964.
16. **Guth, L.,** "Trophic" influences of nerve on muscle, *Physiol. Rev.*, 48, 645, 1968.
17. **Gutmann, E.,** Problems in differentiating trophic relationships between nerves and muscle cells, in *Motor Innervation of Muscle,* Thesleff, S., Ed., Academic Press, New York, 1976, 323.
18. **Guth, L.,** Axonal regeneration and functional plasticity in the central nervous system, *Exp. Neurol.*, 45, 606, 1974.
19. **Davis, H. L., Heinicke, E. A., Cook, R. A., and Kiernan, J. A.,** Partial purification from mammalian peripheral nerve of a trophic factor that ameliorates atrophy of denervated muscle, *Exp. Neurol.*, 89, 159, 1985.
20. **Bradshaw, R. A.,** Nerve growth factor, *Ann. Rev. Biochem.*, 47, 191, 1978.
21. **Lentz, T. L., Addis, J. S., and Chester, J.,** Partial purification and characterization of a nerve trophic factor regulating muscle acetylcholinesterase activity, *Exp. Neurol.*, 73, 542, 1981.
22. **Younkin, S. G., Brett, R. S., Davey, B., and Younkin, L. H.,** Substances moved by axonal transport and released by nerve stimulation have an innervation-like effect on muscle, *Science,* 200, 1292, 1978.
23. **Beach, R. L., Popiela, H., and Festoff, B. W.,** Trophic factors and neuromuscular function: implications for ALS, in *Research Progress in Motor Neuron Diseases,* Rose, F. C., Ed., Pitman Press, London, 1984, 228.
24. **Thesleff, S.,** Studies on neurotrophic regulation of skeletal muscle, *Prog. Brain Res.*, 49, 385, 1979.

Chapter 2

AN HISTORICAL INTRODUCTION TO THE TROPHIC REGULATION OF SKELETAL MUSCLE

Sidney Ochs

TABLE OF CONTENTS

I. THE OLDER BACKGROUND

The term neurotrophic, derived from nerve (from the Greek *neuron*) and trophic (from the Greek *trophos*), expresses the idea that the muscle (and other target tissues) is supplied by the nerve with some substance needed for its maintenance.* The term atrophy is an ancient one used with respect to wasting of the body. Some indication that the term may have been used with respect to nerve and muscle may be inferred from the description given by Soranus (second century A.D.)[1] regarding chronic paralysis: "the affected part becomes pale and thin because of want of nourishment [atrophic] and there is a contraction of the affected parts." Nevertheless, a clear conception of earlier views of a trophic control of muscle by nerve is made difficult because of the uncertain distinction then made between nerves and tendons, a residue of the earlier confusion remaining in present-day dictionary definitions of the term nerve where both nerve and tendon (or sinew) are usually included. In any event, the relation of nerve to muscle was recognized in the second century by Galen,[7] who, drawing on the teachings of Erasistratus (4th century B.C.) and still earlier precursors, along with his own experimental studies, considered nerve as a conduit for a breathlike spirit (from the Latin *pneuma*) or humor passing down from the brain (see Chapter 1 in Reference 2). This model, further elaborated by Descartes[8] in the late 17th century, viewed a nerve substance flowing into the muscles as the cause of their swelling and contraction. The model could account for both the loss of the voluntary control of contractility and the atrophy which follows a nerve transection or crush. The concept was viewed favorably even late in the 18th century as may be seen from a quotation given by Prochaska:[9] "Tralles has lately again promulgated the doctrine denied by Haller and others, of a nutrient property in the nervous fluid, observing that if a nerve be tied, compressed or destroyed, not only are motion and sensation in the muscle destroyed, but also nutrition and atrophy comes on; whence he thinks, the conclusion is undoubted, that some fluid passes from the brain along the nerves to the muscles, by which not only the muscles but the nerves are nourished."

The plethymographic studies carried out by Swammerdam, Steno, and Glisson toward the end of the 17th century had already shown that muscles contract without a measurable increase in their volume, and this required other hypothesis to explain the nature of the nerve influence.[10] Galvani[11] believed it to be an imperceptible "animal electricity" produced in the brain flowing down within the nerves to cause the muscles to electrically discharge, much as a condenser is discharged. In the 19th century, the electrical nature of the nerve impulse became better understood, and improved microscopes and advances in neurobiology allowed an identification to be made of motor, sensory, and vasomotor nerve fibers. The question then arose as to whether there are specific nerve fibers which have a trophic influence on muscle.[12] Samuel[13] supported the existence of such a special class of trophic fibers. Others considered atrophy to be due to some interference with vasomotor nerve fibers or unmyelinated fibers responsible for both protopathic pain and a trophic influence, a concept maintained even fairly recently.[14]

The modern era in this field of study begins with the realization that all the various types of fibers have a trophic influence in addition to their main function. This was shown for sensory fibers by Von Vintschgau and Honigschmied[15] who found that taste buds in the dog tongue degenerated on transecting the hypoglossal nerve. The supply of a trophic substance by sensory nerve was further supported by Parker's[16] finding that the onset of degeneration of the sensory lateral line organs in fish following transection was related to the length of

* An overview of trophic regulation related to axoplasmic transport[2] has been drawn on in writing this chapter, which also is oriented toward the role of axoplasmic transport in the supply of trophic substances by the nerve. More general reviews of trophic control have been written by Tower,[3] Guth,[4] and Gutmann,[5] and a number of important papers can be found in the book edited by Harris.[7]

the sensory nerve stump remaining in connection with their sensory organ, the degeneration appearing earlier with shorter lengths of nerve remaining.

With respect to motor fibers, when it became established in the early 1950s that neuromuscular transmission is effected by the release of the neurotransmitter acetylcholine (ACh), this relatively brief synaptic event was separated from the much longer-lasting trophic influence which was defined as "the regulation of muscle and other tissues by nerve not mediated by activity".[4,5,17-19] Experiments designed to show that the regressive changes in the muscle on interruption of its motor nerve supply are due to elimination of a trophic influence are complicated by the resulting muscle inactivity (disuse) which in itself could cause some of the denervation changes observed. This possibility presented itself early on in experimental studies carried out to determine the nature of the atrophy seen following denervation or disuse and will be described in more recent studies of other consequences of denervation: fibrillation and the spread of ACh-receptors (AChR), depolarization, and alteration of the sodium channels of the muscle membrane.

II. DENERVATION EFFECTS

A. Atrophy

While the late appearance of muscle atrophy resulting from denervation was generally considered to be due to the loss of a trophic substance, muscle disuse was also considered likely.[20] That was the motive for the use of massage and electrical (galvanic) stimulation in attempts to reverse the consequences of the paralysis seen in a wide range of neurological conditions.[21] Based on the extensive pioneer studies of nerve wounds incurred in the American Civil War, Mitchell[22] advised that "for atrophy and muscular palsy, a treatment of electricity three or four times a week [is] part of a suitable treatment". Massage and electrical stimulation remain in use in cases of denervation to reduce muscle wasting and contractures.[23]

Animal studies have shown that electrical stimulation of denervated muscle prevents the weight loss seen following denervation,[24,25] evidence that at least some of the denervation-induced atrophy is caused by disuse. In order to differentiate the loss of a trophic substance from muscle inactivity, Tower[26] sectioned the spinal cord supplying the leg muscles from the rest of the cord and also cut the dorsal roots of those segments, resulting in reduced motoneuron discharge causing a disuse atrophy. Weight reductions in those muscles appeared in 3 weeks (50% of control), reductions comparable to those seen following denervation.[3] A similar decrease in muscle mass was observed following tenotomy[27] or placing a limb in a cast to prevent muscle movement.[28] Such changes were reversed by intermittent tetanization of the muscle,[28] with the restorative effect being greater in the anterior tibialis anticus and extensor digitorum longus than the gastrocnemius and soleus muscles.[29]

The loss of muscle substance seen following denervation (or disuse) would be due to metabolic changes in the muscles with a relative failure of synthesis of muscle tissue.[30] Metabolic changes were seen to have an onset earlier when the nerve supply was cut closer to, rather than farther away from, the muscle.[31]

Because of the complexity of the metabolic changes observed and the multiple factors involved, attention has more recently been focused on other muscle properties which change as a consequence of denervation or disuse as described in the following sections.

B. Fibrillation

Fibrillation, visible as a spontaneous irregular twitching or quivering of denervated muscle, was first reported by Schiff.[32] It is characterized by irregular, large-amplitude, spontaneous potentials present in the denervated muscles[33-35] with a rhythm of about 10/sec[25] and independent of nerve excitation.[35] Tower et al.[36] considered the possibility that the activity of fibrillation might counteract some of the disuse atrophy produced by cord isolation, tenotomy, or the placing of a limb in a cast. Langley and Hashimoto,[34] on the other hand, proposed that atrophy might even be the result of the fibrillation, namely, an exhaustion of the muscles. However, blocking the fibrillation with quinine did not prevent the appearance of atrophy.[35] The loss of a trophic influence as the cause of muscle fibrillation was supported by the studies of Luco and Eyzaguirre.[37] They showed that following nerve transection, fibrillatory potentials in the muscles appear spread out from the junctional regions with time to eventually extend over the whole length of the muscle. The time of onset of spontaneous fibrillation depended on the length of the nerve left connected with the muscle, the fibrillation occurring sooner when the nerve was cut closer to the muscle. This would point to the existence of a trophic substance which became exhausted from nerve sooner when a short, rather than a longer, length of nerve was left remaining in connection with the muscle. Such a hypothesis requires the continuation of axoplasmic transport for a time in the distal transected portion of the nerve after its interruption. In fact, axoplasmic transport is known to persist for up to several days in a portion of peripheral nerve separated from its cell bodies.[38]

C. Depolarization

The earliest change seen in muscle after sectioning its nerve is a depolarization which appears first in the endplate region and then spreads out into the muscle fibers until the resting membrane potential (RMP) level falls all along their length.[39] When the nerve was cut close to the muscle the decrease in RMP had an earlier onset than when cut farther away, a finding in accord with the loss of some trophic substance from the nerve rather than an absence of muscle activity as the cause of depolarization. A similar temporal difference in the onset of muscle depolarization was also reported by Redfern and Thesleff[40] and by Card,[41] though with somewhat different times of onset compared to that reported earlier.[39]

D. Sodium Channel Properties

The properties of sodium channels in the membrane of denervated muscle are altered by denervation. Normally, tetrodotoxin (TTX) blocks the sodium channel and thus prevents excitation of action potentials. The sodium channels of the denervated muscles, on the other hand, are insensitive to the blocking action of TTX.[40,42,43] Such TTX insensitivity begins a few days after denervation, appearing first in the endplate region and then spreading out in the muscle fiber over a period of about a week. The time at which TTX insensitivity begins depends on the length of the nerve left connected to the muscle, the effect appearing sooner when nerves are cut closer to (rather than farther from) the muscle, a length-related effect which again is in accord with the loss of trophic control over the synthesis of normal sodium channels.

E. Sensitivity to Acetylcholine

Denervated muscles become abnormally sensitive to ACh. The phenomenon is revealed by the injection of ACh into blood vessels close to the muscles they supply, such injections having little effect in normal muscles but causing large contractions in the denervated muscle.[44] Increased excitability appearing some time after denervation has been referred to as "denervation supersensitivity" by Cannon and Rosenblueth.[45] One possibility for the hypersensitivity advanced with respect to the adrenergic system was the loss of a restraining action by the neurotransmitter. This was indicated by a partial desensitization of the denervated nictitating membrane seen following repeated injections of adrenaline.[46] However, little evidence for a similar desensitization by ACh in skeletal muscles was found which could account for the effects of denervation. The explanation of the increased sensitivity to ACh became clear when it was shown that ACh, which normally acts only on the immediate vicinity of the endplate, extends its action into the extrajunctional membrane following transection of the muscle's nerve supply. Because of the interruption of scientific communication during World War II, the earlier work of Ginetzinski and Shamarma[47] showing this phenomenon was at first overlooked. The extrajunctional spread of ACh sensitivity was later shown by the depolarizations initiated by localized iontophoretic applications of ACh.[48,49]

The denervated muscle apparently reverts to its immature state at birth when, in most mammalian species, skeletal muscle fibers are receptive to ACh over their whole length. In the course of muscle maturation, a process which takes place over a period of a few weeks to approximately 1 month following birth, the region of the muscle fiber highly sensitive to ACh shrinks until it becomes centered on the endplate.[50]

The increase in ACh sensitivity in denervated muscles is due to an increase in the number of AChR in the extrajunctional membrane, as was shown by the use of ^{125}I-labeled α-bungarotoxin (α-BTX). The toxin binds irreversibly to AChR and serves as a tag of the receptor protein in the course of its isolation from the tissue.[51,52] Using this technique the rate of turnover of the receptor protein in the muscle membrane,[53] as well as its amount,[54] was shown to greatly increase in the denervated as compared to normal muscle.

F. Acetylcholinesterase

Another change seen in the denervated muscle is a decrease in the content of acetylcholinesterase (AChE), to approximately 30 to 50% of control levels within several days after transection of its nerve supply.[55,56] This would presumably be due to an effect on the synthesis of AChE by the muscle. However, AChE is also present in the motor nerve fibers where it is conveyed by fast transport[57] at a rate close to 410 mm/day.[58,59] The presence of fast-transported AChE in the motor fibers suggested the possibility that some of the enzyme may be released from the motor nerve terminals to become localized in the endplate region,[60] presumably in the basement membrane between the nerve terminal and the muscle membrane.[61,62] Some support for this possibility is the loss of AChE activity in rat diaphragm muscle seen following denervation, a loss which occurs with an apparent rate of 300 mm/day as determined by cutting nerves near to or far from their muscles.[63,64]

Multiple molecular forms of AChE isolated from nerve have been determined by their differing sedimentation rates in sucrose gradients.[65,66] Monomeric AChE is a glycoprotein of about 80,000 mol wt combining into dimeric and tetrameric globular (G) forms or with one, two, or three tetramers joining with three-stranded collagen-like tails to give rise to the asymmetric (A) forms[67] located in the junctional region.[62,68] The 16S species (A12) is transported fast,[69,70] and it shows the greatest loss from the endplate region following denervation,[69] a decrease which occurs sooner when the nerve is cut close to (rather than far from) the muscle.[66] In addition to the 16S form, the 10S (G4) form is also fast-transported.[70,71] While these findings would suggest that the 16S species (as well the 10S species) could be supplied by nerve to the endplate, it is also known that muscle itself is capable of synthesizing all these various AChE species, with differences in their intra- and extracellular location bearing on how the 16S species might be assembled in the muscle endplate region.[72] A possible origin of some of the AChE in the endplate region from the nerve still remains a possibility.[73] Estimating that 26% of the total amount of AChE is in the endplate region[62] and that A12 (16S) and G4 (10S), the fast-transported species of AChE supplied wholly by nerve and having a half-life of 1 week, the nerve could supply 40% of the AChE of the endplate region.[73] The question of the origin of all the various forms of AChE in the endplate region and how the fast transport of AChE species is related to that located in the endplate region of the muscle remains a challenging problem.

III. DIFFERENTIATION BETWEEN BLOCK OF MUSCLE ACTIVITY AND TROPHIC SUPPLY

A. Block of Nerve Activity

The question as to whether the decreased muscle activity or the loss of a trophic substance accounts for muscle fibrillation seen following nerve transection was addressed by Denny-Brown and Brenner.[74] A clip was used to produce a partial compression of cat sciatic nerves to the degree where transmission of nerve impulses through the compressed region was blocked without, however, interrupting the continuity of the nerve fibers and the supply of a trophic substance. The muscles innervated by the partially compressed nerves did not show the fibrillation typical of denervation, and the authors concluded that the normal state of the muscle was maintained by the continual supply of a trophic substance. In a refinement of that technique, Robert and Oester[75] constructed cuffs made of Silastic in which a local anesthetic had been incorporated and with a hole in the cuff the size of the sciatic nerve so that when placed around the nerve and left for several days, *in situ,* the anesthetic diffusing from the cuff maintained a block of nerve conduction. Signs of denervation did not appear in the muscles, and the authors also concluded that a block of nerve conduction, and the consequent muscle inactivity, was not the cause of the denervation changes seen in muscle following nerve transection.

On the other hand, Lømo and Rosenthal,[76] using Silastic cuffs impregnated with the local anesthetic Marcaine, obtained different results. Their cuffs, placed around rat sciatic nerves for 5 days, caused signs of fibrillation to appear in the soleus (SOL) and extensor digitalis longus (EDL) muscles. Neurotransmission was preserved below the site of cuffing, as indicated by the recording of miniature endplate potential (MEPP) activity from the muscle fibers. The MEPP were actually increased in amplitude, most likely as a result of the higher input impedance characteristic of denervated muscle fibers.[77] Futhermore, an increase of AChR appeared over the length of the muscle fibers, another sign of denervation (see Section II.E).

The results obtained by Lømo and Rosenthal[76] might be explained if Marcaine enters the nerve fibers to interfere with the axoplasmic transport of a trophic substance, the lack of which causes some of the signs of denervation to appear in the muscle. Marcaine was reported to block axoplasmic transport.[78] Bisby[79] showed that Marcaine and a number of other local anesthetics incorporated into silicone cuffs applied to rat sciatic nerves blocked axoplasmic transport is indicated either by the accumulation of catecholamines and of AChE at ligations placed below the cuffed regions or by the failure of transport of labeled proteins through the cuffed region. However, a complete block by Marcaine of all materials transported would not explain Lømo and Rosenthal's results. One would then see evidence of Wallerian degeneration in the fibers below the cuff with loss of MEPP activity and neurotransmission, rather than the remaining presence of function of the fibers as indicated by such activity. The evidence indicates a selective block of transport of a trophic substance(s) with other components still transported. It should be noted that evaluation of this point using the cuff technique is difficult. Differences in the degree of block have been seen when using colchicine and other mitotic blocking agents with the Silastic cuff technique. Differences are due to the size of the cuff hole with respect to the nerve, variations in the permeability of the perineurial sheath due to species or age of the animal, and the amount of the agent reaching the nerve fibers interior.

TTX offers an analytically simpler means to effect a selective block of nerve activity in that this agent does not enter the fibers and axoplasmic transport is not affected.[80] With TTX incorporated into Silastic cuffs placed around rat sciatic nerves and left *in situ* for times up to 7 days, the innervated muscles show an increased amount of AChR as indicated by an increased amount of ^{125}I-α-BTX binding, a finding associated with denervation.[81] However, a still greater degree of ^{125}I-α-BTX binding was seen in muscles following sectioning of their nerve supply. Thus, inactivity of muscle brought about by a block of its nerve activity with TTX does give rise to some of the signs of denervation, but a further increment of such denervation changes is seen following nerve transection, as would be expected of the loss of supply of a trophic substance. Essentially the same results were found by Pestronk and co-workers[82] using repeated subperineurial injections of TTX into sciatic nerves to maintain a chronic block of nerve conduction over a period of 7 days. Fast axoplasmic transport, as assessed by injecting either ^{3}H-leucine or ^{3}H-fucose into the ventral horn of the L5 lumbar cord and recording the passage of labeled proteins through the TTX-treated region of nerve, was not affected either in form or rate as compared to control nerves. Studies of nerve inactivation with TTX were further extended by Bray and colleagues[83] using a capillary implant technique to maintain a TTX block for longer times of up to 21 days. A fall in RMP was found in the SOL muscle when the nerve supplying the muscle was blocked by TTX with an additional decrease in potential seen following transection of the nerve, findings which are in agreement with the loss of a trophic factor in addition to a partial denervation-like effect due to muscle inactivity.

On compressing the nerve in the lower limb of baboons with narrow cuffs to affect the myelin but not disrupt the axons, a persistent block of nerve activity could be produced without the appearance of Wallerian degeneration.[84] After 28 days of such compression, a spread of extrajunctional ACh sensitivity, but not fibrillation, was found in the muscles. The authors concluded that the loss of muscle activity due to such compressions was responsible for the spread of AChR in the muscle fibers while the continued supply of a trophic substance(s) prevents the occurrence of fibrillation.

B. Block of Axoplasmic Transport

The inverse of the experiments described above, namely, a block of axoplasmic transport while nerve and muscle activity were left intact, was studied by Albuquerque et al.[43] Silastic cuffs containing either vinblastine or colchicine were applied *in situ* to the sciatic nerves of rats to block axoplasmic transport. This caused denervation-like changes to appear in the muscles innervated by those nerves, a decrease of RMP, and the development of TTX-insensitive action potentials with MEPP activity while neuromuscular transmission was seen to remain present. The MEPP were somewhat increased in their rate and amplitudes, changes related to an increased membrane impedance such as is seen in denervated muscle.[77] Albuquerque et al.[43] concluded that the denervation effects seen in the muscles were brought about by the lack of a trophic substance(s) insofar as nerve impulses and the activation of the muscles remained present. The finding of a maintained function of the axons as indicated by MEPP discharges indicated that a partial block of transport must have occurred. A more profound block of transport with Wallerian degeneration of the nerve fibers can occur following application of colchicine directly to nerves.[85,86] In those experiments the permeability of the nerves to colchicine was likely to have been greater and more of the agent taken up by the fibers.

A point to be considered in experiments where colchicine or Vinca alkaloids are applied to nerves, either by cuffs or by direct application, is that the agents may gain entry to the circulation and act directly on the muscles.[87] However, other studies showing a unilateral effect in the muscle with the agent applied to its nerve while the muscle on the other side remained unaffected[88,89] are not in accord with an indirect action through the circulation.

C. Direct Stimulation of Muscle

If inactivity can cause denervation changes in muscle, its direct stimulation following transection of its nerve supply should prevent or reverse the effects of denervation. The spread of AChR into the extrajunctional region typically seen in denervated muscles was blocked or reversed when the muscles were directly stimulated with a pattern of electrical pulses close to the normally occurring nerve discharges.[90] Such a restorative effect of stimulation was seen even when it was initiated as late as 5 days following denervation. Jones and Vrbova[91] also found that direct stimulation of denervated muscles prevented the spread of AChR into the extrajunctional region, though in their studies this only occurred when stimulation was initiated within the first 2 to 3 days following nerve transection. Direct stimulation of tissue-cultured muscles also decreased extrajunctional ACh sensitivity.[92,93]

Direct stimulation of muscle appears able to reverse other denervation changes, such as the ability of the muscle to accept foreign innervation,[94] fibrillation,[76,93] the decrease of RMP, and other changes in muscle membrane properties. However, all the effects of denervation cannot be explained simply by diminished muscle activity, e.g., the experiments cited above of the differing times of onset of the denervation effects seen when cutting the nerve near to or far from the muscle. Additionally, when the doubly innervated frog sartorius muscle is deprived of one of its nerve supplies, only the denervated portion of the muscle fiber becomes hypersensitive to ACh in spite of the fact that the whole length of the muscle contracts as a result of activation through its remaining nerve supply.[95]

Rather than the loss of a trophic substance(s), an agent arising from either degenerating nerve fibers or denervated muscle fibers has been proposed as a cause of the denervation effects. For example, Lømo and Westgaard[90] suggested that products of nerve degeneration could cause a transient ACh hypersensitivity in SOL muscles. This was also suggested by partially denervating EDL muscles (L5 root section) in the rat and later finding TTX-resistant action potentials in nearly all the fibers of the muscle, including those supplied by nerve branches from other roots intermingled with the denervated ones.[96] Conversely, some factor released from degenerating muscle fibers is required to explain the signal for regeneration of neurite sprouts from nearby normal nerve terminals which then reinnervate the denervated muscle fibers, as first shown by Van Harreveld.[97]

These results taken together suggest that muscle fibers exhibit denervation-like changes in response to several factors: diminished activity, loss of some trophic substance(s), and possibly also to some agent(s) arising from degenerated nerves or muscles. We can no longer, therefore, view all the changes seen in muscle after an interruption of its nerve supply as due wholly to a loss of a trophic substance or to a lack of activity. The nerves and the muscles they innervate constitute an interrelated system in which several control factors operate within the muscle fiber to maintain its function.

IV. CROSS-REINNERVATION

A control by the nerve over the characteristic fast or slow response of muscle was shown by cross-reinnervation. Slow and fast muscles are differentiated by their anatomical distribution, color, and functional properties;[98] differing by their speed of contraction and development of tension, rate of relaxation following a single contraction, and tetanic contractions on repetitive stimulation.[99-101] In the newborn kitten all the muscles are at first slow, the fast muscles taking on their adult characteristics by 3 to 4 weeks after birth.[99] In the newborn kitten typical contraction times of 80 msec and half-relaxation times of 89 msec in a fast muscle such as the flexor hallucis longus (FHL) shortened a month later to 25 msec and 16 msec, respectively, times characteristic of a fast muscle.[99]

To demonstrate the influence of nerve in effecting changes in muscle contraction properties, Buller et al.[99,100] cut the nerves supplying the fast FHL and the slow SOL muscles, crossing and reuniting these nerves so that their fibers could regenerate down into the distal nerve stumps. The cross-reinnervated muscles were then seen to take on the contraction properties characteristic of their new nerve supply, the slow muscle now giving rise to contractions typical of a fast muscle and vice versa.

Biochemical properties of the muscles are also changed by cross-innervation. The level of actomyosin ATPase activity (or the myosin ATPase which parallels it) of muscle is correlated with the speed of contraction.[102,103] The high level of ATPase activity characteristic of fast muscle falls to the lower level typical of slow muscle after its cross-reinnervation by a slow nerve. Conversely, the low level of myosin ATPase characteristic of slow muscle rises after cross-reinnervation by a fast nerve to the higher levels typical of a fast muscle.[104]

Buller and Lewis[105,106] reported that the slow cat SOL muscle cross-reinnervated by the nerve of the fast flexor digitorum longus (FDL) muscle showed only a 10 to 15% shortening of its contraction time, while a complete transformation was seen in the cross-reinnervated rat SOL muscle.[107] The relatively small change in the cat was traced to the large number of slow muscle fibers intermingled with the fast muscle fibers in its FDL muscle. When the nerve of the fast EDL cat muscle which contains only a few slow motor units was used, the cross-reinnervated SOL muscle was shifted completely in its responses to those typical of a fast muscle.[107]

The cross-reinnervation experiments have been explained in two ways. Buller et al.[99,100] suggested that differing trophic materials transported in the nerve fibers confer the special character of being a slow or a fast muscle on those they normally innervate or reinnervate after a cross-union. An alternative hypothesis is that the changes seen in muscles following their cross-reinnervation result from the different patterns and rates of nerve impulses imposed on them by their new motoneuron supply. In support of the latter hypothesis, direct stimulation of fast muscles at a slow rate was shown to convert them to the longer-duration contractions typical of slow muscles.[108] The effect of fast and slow patterns of muscle stimulation was further studied by Lømo et al.[109] using brief trains of 100 pulses/sec stimulation to induce fast, and 10 pulses/sec stimulation to induce slow, muscle changes in denervated muscles. The slow SOL muscle of the rat was shifted toward a faster contraction response following a fast pattern of stimulation, whereas it was not affected when stimulated at a slow rate. Other properties associated with slow and fast muscles were similarly shifted by the imposition of these two different patterns of stimulation.[110] The higher level of alkaline-resistant actomyosin ATPase typical of fast muscles appeared in slow muscles either cross-reinnervated with fast nerves or when directly activated by a fast pattern of stimulation.[111]

V. REGENERATION OF MINCED MUSCLE

Studitsky[112] discovered that minced muscle can reform *in situ* into more or less normal-appearing functional muscle. He removed the gastrocnemius, plantaris, and SOL muscles from the back of the hindlimb of adult rats, minced them, and then placed the mince back into the hollowed out region in the back of the leg. The muscle fragments were found to gradually become transformed first into myoblasts and then into myotubes, and eventually to form muscle fibers. The nerve must be present in the regenerating field for the minced graft to progress beyond the first stages of development.[113] If the limb has been denervated, regeneration stops,[114] leaving mostly narrow bands of connective tissue remaining with some cross-striated myotubes also present, indicating that the failure of regeneration occurs just after the myotube stage.[115] Motor nerve fibers are required for the muscle fibers to develop, with sensory or autonomic fibers being ineffective in this regard. Further, fast nerves, when present, organized the minced muscle preparation into fast muscles, and slow nerves produced slow muscles.[116]

VI. NATURE OF TROPHIC FACTORS

A. Acetylcholine as a Trophic Factor

Clostridium botulinum toxin (BT) acts in very low concentrations to prevent the release of ACh from motor nerve terminals, and it also causes signs of denervation, i.e., the spread of increased ACh sensitivity outward in the muscle fibers.[117] Other signs of denervation produced by BT are muscle fibrillation and the acceptance of foreign innervation in extra-junctional regions with formation of new endplates.[118,119] The possibility was raised that ACh itself could be a trophic substance (see Chapter 3), its block of release by BT producing denervation changes. However, the increased spread of ACh sensitivity indicative of denervation was not blocked in tissue-cultured muscles when ACh was added in high concentrations.[120] An alternative explanation for the denervation-like effects seen with BT is that the agent prevents not only the release of ACh from the nerve terminal but that of another trophic substance as well.[121] BT and α-BTX could additionally interfere with the release of a trophic substance, either one coupled with ACh release or one independently released from the terminal. Alternatively, ACh could increase the permeability of the muscle membrane to the action of a trophic substance.[122]

B. Characterization of Neurogenic Trophic Substances

To identify and characterize trophic substances Lentz[123] used organ-cultured newt triceps muscle as an assay. On cutting its nerve supply, the axon terminals in the muscle fibers were seen to rapidly degenerate and then, more slowly, the junctional folds until eventually the endplate as a whole disappeared. Along with these morphological changes a gradual fall in the content of AChE in the denervated muscles was found. When a sensory ganglion explant was placed near the organ-cultured muscles, these regressive changes in the muscle were mostly prevented or reversed. The junctional folds maintained their normal appearance, and AChE levels, while they did not revert to control levels, were higher than those seen without the ganglion present. Newt brain homogenates added to the muscle culture also had a similar trophic effect, as did cyclic AMP plus theophylline and adenosine.

Trophic substances were implicated in the studies of tissue-cultured muscles carried out by Peterson and Crain.[124,125] Isolated fragments of 18-day fetal rodent skeletal muscle when explanted to a culture medium showed, after a time, the formation of myotubes with cross-striations. The cross-striations remained present in some of the fibers after the second week, but after the third week the muscles underwent atrophy.[126] When a spinal cord ganglion fragment was implanted near the cultured muscles, the fibers went on to further stages of maturation and were maintained for months without atrophy. Furthermore, those muscle fibers could become innervated. Even muscle fibers which had become atrophic took on a more normal appearance under the influence of the nearby neural implants, and they also were able to become innervated.

When organ-cultured muscles are used to assay for possible neurogenic trophic substances, an appropriate stage of development of the muscle must be selected. If taken too late after denervation the muscle may have developed response requirements other than those present at an earlier stage. The use of an undefined culture medium also introduces considerable variability into the response characteristics of the tissue. All these factors make it difficult to completely predict the action of a given substance suspected of being a trophic agent unless adequate precautions are taken.[127] With these provisos, tissue culture is a valuable tool for the identification and isolation of possible neurogenic trophic substances,[128] especially as well-defined media are now available (see Chapters 4 and 6).

Oh et al.[129] and Oh[130] used chick embryo skeletal muscle cells in culture as an assay and found that extracts of chick brain and spinal cord stimulated morphological development, protein synthesis, and an increased level of AChE activity in the cultured muscle fibers. Soluble extracts of adult chick sciatic nerve were fractionated, and a 10,000- to 50,000-dalton protein added to the culture medium stimulated synthesis, increased the level of AChE activity, and supported a long-term maintenance of the muscle cells. The neurogenic substance is a glycoprotein,[131] the most active fraction isolated from chicken nerve having a molecular weight of 84,000 daltons. Its protein nature was shown by its inactivation with trypsin or by proteolytic hydrolysis.[132] In transected adult chicken nerves the trophic material was found in the distal stumps of transected nerves, possibly sequestered by the Schwann cells following its release from degenerated axons.[133]

Davey et al.[134] used as an assay the EDL of the rat in which a decreased level of AChE was measured 5 days after denervation. The assay schedule used was to remove the muscles from the animal 3 days after denervation and keep them in culture for a further 2 days, during which time substances were added to the culture medium to test their trophic influence. Brain, spinal cord, or muscle extracts added to the culture were found able to reverse some of the decrease of AChE seen in the controls.

Evidence that neurogenic substances are transported in nerve fibers was strengthened by the finding that they accumulate proximal to nerve transections.[132] A similar result was seen by Fernandez et al.,[69,135] who found that a soluble component of nerve extracts was able to augment the level of 16S AChE assayed in organ-cultured denervated rat anterior gracilis muscle. A less well-defined protein released from nerve on its stimulation was also found to act as a neurogenic trophic substance.[136] Oh and Markelonis[137] found that chicken serum transferrin duplicated the effects of the neurogenic substance they previously called sciatin, though some differences were noted. Differences in the effectiveness of transferrins were also found in various species.[138] One dissimilarity between sciatin and transferrin seen was that transferrin did not reverse atrophy following denervation, while the substance obtained from nerve did.[139]

Another approach is to use the denervated EDL muscles of the adult rat as an assay. Some of the denervation effects were shown to be reversed by injection of nerve extracts into the animal.[140] In this muscle the cross-sectional area of type IIb muscle fibers decreases to about 40% of control levels following denervation, and about half of this decrease was reversed after 7 days of intramuscular injection with a protein isolated from nerve.[141] How proteins with their relatively large molecular weights are able to enter muscle fibers to reverse some of the deficiencies of denervation when given in vivo remains to be determined. Presumably an endocytosis or special transfer mechanism would be involved (see Chapter 7).

Nerve growth factor (NGF) remains the one completely identified trophic (and tropic) substance. It is taken up by neurites and carried by retrograde transport to the cell bodies of adrenergic and sensory neurons where it acts in very small concentrations to augment synthesis.[142,143] The materials synthesized are directed to the appropriate set of neurites by an intra-axonal process of routing.[144] Other trophic substances[144] most likely exist which act on motoneurons. When these are completely characterized the powerful methods so successful in the study of NGF can be applied to them. We look to the new advances reported in this book and to future studies for the identification and further understanding of the nature of trophic agents and eventually to their beneficial application in medicine.

VII. SUMMARY AND CONCLUDING REMARKS

This brief survey shows that the earlier dialectic opposition between explanations for the regressive changes in muscle seen following denervation as due either to the loss of a trophic substance or to the resultant inactivity could not be so simply resolved.

A trophic regulation of muscle has been established, but it was also shown that a considerable portion of the regressive changes following denervation can be reversed by activating the muscle (see Chapter 5). The analysis of the muscle effects of differential block of the propagated action potential or of axoplasmic transport has been instrumental in leading to our present insight that both a trophic substance(s) (see Chapter 6 and 7) and activity (see Chapter 4) are necessary for normal muscle function. While the analysis is dependent on our better understanding of axoplasmic transport (see Chapter 3), those experiments also indicated that the transport of trophic components may be interfered with by exposure to mitotic blocking agents while other components needed for the function of the fiber and for neurotransmission are not blocked. Such a differential blocking action demands further study, one which would be facilitated by the isolation of the trophic substance(s) from nerve. The use of such substances should also reveal how muscles are normally regulated by their nerve supply and how the activity of the muscle cooperates with trophic controls to maintain its normal function. Finally, such trophic materials may have therapeutic usefulness.

ACKNOWLEDGMENTS

Supported by NSF BNS 82-17727 and NIH PHS ROI NS 8706-16.

REFERENCES

1. **Drabkin, I. E., Trans.,** *Caelius Aurelianus and Acute Diseases and on Chronic Diseases,* University of Chicago Press, Chicago, 1950. (The work of Soranus was transcribed by Caelius Aurelianus.)
2. **Ochs, S.,** *Axoplasmic Transport and Its Relation to Other Nerve Functions,* John Wiley & Sons, New York, 1982, 1.
3. **Tower, S.,** The reaction of muscle to denervation, *Physiol. Rev.,* 19, 1, 1939.
4. **Guth, L.,** "Trophic" influences of nerve on muscle, *Physiol. Rev.,* 48, 645, 1968.
5. **Gutmann, E.,** Neurotrophic relations, *Ann. Rev. Physiol.,* 38, 177, 1976.
6. **Gutmann, E.,** *The Denervated Muscle,* Publishing House of the Czechoslovak Academy of Sciences, Prague, 1962.
7. **Harris, A. J.,** Trophic effort of nerve on muscle, in *The Physiology of Peripheral Nerve Disease,* Sumner, A. J., Ed., W. B. Saunders, Philadelphia, 1980, 195.
8. **Descartes, R.,** *Treatise of Man,* Hall, T. S., Trans., Harvard University Press, Cambridge, Mass., 1972.
9. **Prochaska, G.,** *A Dissertation on Functions of the Nervous System,* Laycock, T., Trans., Sydenham Society, London, 1851, 425.
10. **Bastholm, E.,** *The History of Muscle Physiology,* Munksgaard, Copenhagen, 1950, 1.
11. **Galvani, L.,** *Commentary on the Effects of Electricity on Muscular Motion,* Foley, M. G., Trans., Burndy Library, Norwalk, Conn., 1954, 1.
12. **Jones, H.,** Are there special trophic nerves?, *St. George's Hosp. Rep.,* 3, 89, 1868.
13. **Samuel, S.,** *Trophschen Nerven,* Wigund, Leipzig, 1860, 1.
14. **Wyburn-Mason, R.,** *Trophic Nerves,* Kimpton, London, 1950, 1.
15. **Von Vintshgau, M. and Honigschmied, J.,** Nervus Glossopharyngeus und Schmeckbecher, *Arch. Gesammete Physiol.,* 14, 443, 1877.
16. **Parker, G. H.,** On the trophic impulse so-called, its rate and nature, *Am. Nat.,* 66, 147, 1932.
17. **Miledi, R.,** An influence of nerve not mediated by impulses, in *The Effect of Use and Disuse of Neuromuscular Functions,* Gutmann, E. and Hník, P., Eds., Publishing House of the Czechoslovak Academy of Sciences, Prague, 1963, 35.
18. **Cannon, W. B. and Rosenblueth, A.,** *The Supersensitivity of Denervated Structures,* Macmillan, New York, 1949, 1.
19. **Harris, A. J.,** Inductive functions of the nervous system, *Ann. Rev. Physiol.,* 36, 251, 1974.
20. **Brown-Sequard, C. E.,** *Course of Lectures on the Physiology and Pathology of the Central Nervous System,* Collins, Philadelphia, 1860, 1.
21. **Remak, R.,** *Galvanotherapie der Nerven — und Muskelkrankheiten,* Hirschwald, Berlin, 1858, 1.
22. **Mitchell, S. W.,** *Injuries of Nerves and their Consequences,* Dover, New York, 1965, 1.
23. **Stookey, B.,** *Surgical and Mechanical Treatment of Peripheral Nerves,* W. B. Saunders, Philadelphia, 1922, 1.
24. **Grodins, F. S., Osbourne, S. L., Johnson, F. R., Arana, S., and Ivy, A. C.,** The effect of appropriate electrical stimulation on atrophy of denervated skeletal muscle in the rat, *Am. J. Physiol.,* 142, 222, 1944.
25. **Eccles, J. C.,** Changes in muscle produced by nerve degeneration, *Med. J. Aust.,* 1, 573, 1941.
26. **Tower, S.,** Trophic control of non-nervous tissues by the nervous system: a study of muscle and bone innervated from an isolated and quiescent region of the spinal cord, *J. Comp. Neurol.,* 67, 241, 1937.
27. **Lipschultz, A. and Audova, A.,** The comparative atrophy of the skeletal muscle after cutting the nerve and after cutting the tendon, *J. Physiol. (London),* 55, 300, 1921.
28. **Chor, H. and Dolkeart, R. E.,** A study of "simple disuse atrophy" in the monkey, *Am. J. Physiol.,* 117, 626, 1936.
29. **Eccles, J. C.,** Disuse atrophy of skeletal muscle, *Med. J. Aust.,* 2, 160, 1941.
30. **Bass, A.,** Energy metabolism in denervated muscle, in *The Denervated Muscle,* Gutmann, E., Ed., Publishing House of the Czechoslovak Academy of Sciences, Prague, 1962, chap. 6.

31. **Gutmann, E. and Hník, P.,** Denervation studies in research of neurotrophic relationships, in *The Denervated Muscle,* Gutmann, E., Ed., Publishing House of the Czechoslovak Academy of Sciences, Prague, 1962, chap. 1.
32. **Schiff, J. M.,** Muskel — und nervenphysiologie, in *Cyclus Part 9,* Schauenburg, L., Ed., Lahr, 1885, 1.
33. **Langley, J. N.,** Observations on denervated muscle, *J. Physiol. (London),* 50, 335, 1916.
34. **Langley, J. N. and Hashimoto, M.,** Observations on the atrophy of denervated muscle, *J. Physiol. (London),* 52, 15, 1919.
35. **Solandt, D. Y. and Magladery, J. W.,** The relation of atrophy to fibrillation in denervation muscle, *Brain,* 63, 255, 1940.
36. **Tower, S. H., Howe, H., and Bodian, D.,** Fibrillation in skeletal muscle in relation to denervation and to inactivation without denervation, *J. Neurophysiol.,* 4, 398, 1941.
37. **Luco, J. V. and Eyzaguirre, C.,** Fibrillation and hypersensitivity to ACh in denervated muscle: effect of length of degenerating nerve fibers, *J. Neurophysiol.,* 18, 65, 1955.
38. **Ochs, S.,** Retention and redistribution of proteins in mammalian nerve fibers by axoplasmic transport, *J. Physiol. (London),* 253, 459, 1975.
39. **Albuquerque, E. X., Schuh, F. T., and Kauffman, F. C.,** Early membrane depolarization of the fast mammalian muscle after denervation, *Pfluegers Arch.,* 328, 36, 1971.
40. **Redfern, P. and Thesleff, S.,** Action potential generation in denervated rat skeletal muscle. II. The action of tetrodotoxin, *Acta Physiol. Scand.,* 82, 70, 1971.
41. **Card, D. J.,** Denervation: sequence of neuromuscular degenerative changes in rats and the effect of stimulation, *Exp. Neurol.,* 54, 251, 1977.
42. **Harris, J. B. and Thesleff, S.,** Nerve stump length and membrane changes in denervated skeletal muscle, *Nature (London) New Biol.,* 236, 60, 1972.
43. **Albuquerque, E. X., Warnick, J. E., Sansone, F. M., and Onur, R.,** The effects of vinblastine and colchicine on neural regulation of muscle, *Ann. N. Y. Acad. Sci.,* 228, 224, 1974.
44. **Brown, G. L.,** The actions of acetylcholine on denervated mammalian and frog's muscle, *J. Physiol. (London),* 89, 438, 1937.
45. **Cannon, W. B. and Rosenblueth, A.,** *The Supersensitivity of Denervated Structures,* Macmillan, New York, 1949, 1.
46. **Simeone, F. A.,** The effect of previous stimulation on the responsiveness of the cat's nictitating membrane sensitized by denervation, *Am. J. Physiol.,* 122, 650, 1938.
47. **Ginetzinski, A. G. and Shamarma, N. M.,** The tonomoter phenomenon in denervated muscles, *Usp. Sovrem. Biol.,* 15, 283, 1942.
48. **Axelsson, J. and Thesleff, S.,** A study of supersensitivity in denervated mammalian skeletal muscle, *J. Physiol. (London),* 147, 178, 1959.
49. **Miledi, R.,** Properties of regenerating neuromuscular synpases in the frog, *J. Physiol. (London),* 154, 190, 1960.
50. **Diamond, J. and Miledi, R.,** A study of foetal and newborn rat muscle fibers, *J. Physiol. (London),* 162, 393, 1962.
51. **Miledi, R., Molinoff, P., and Potter, L. T.,** Isolation of the cholinergic receptor protein of *Torpedo* electric tissue, *Nature (London),* 229, 554, 1971.
52. **Potter, L. T. and Molinoff, P. B.,** Isolation of cholinergic receptor proteins, in *Perspectives in Neuropharmacology: A Tribute to Julius Axelrod,* Snyder, S. H., Ed., Oxford University Press, New York, 1972, 1.
53. **Devreotes, P. N. and Fambrough, D. M.,** Turnover of acetylcholine receptors in skeletal muscle, *Cold Spring Harbor Symp. Quant. Biol.,* 40, 237, 1975.
54. **Berg, D. K. and Hall, Z. W.,** Fate of α-bungarotoxin bound to acetylcholine receptors of normal and denervated muscle, *Science,* 184, 473, 1974.
55. **Guth, L., Albers, R. W., and Brown, W. C.,** Quantitative changes in cholinesterase activity of denervated muscle fibers and sole plates, *Exp. Neurol.,* 10, 236, 1964.
56. **Guth, L., Brown, W. C., and Watson, P. K.,** Studies on the role of nerve impulses and acetylcholine release in the regulation of the cholinesterase activity of muscle, *Exp. Neurol.,* 18, 443, 1967.
57. **Lubinska, L. and Niemierko, S.,** Velocity and intensity of bidirectional migration of acetylcholinesterase in transected nerves, *Brain Res.,* 27, 329, 1971.
58. **Ranish, N. and Ochs, S.,** Fast axoplasmic transport of acetylcholinesterase in mammalian nerve fibers, *J. Neurochem.,* 19, 2641, 1972.
59. **Brimijoin, S. and Wiermaa, M. J.,** Rapid orthograde and retrograde axonal transport of acetylcholinesterase as characterized by the stop-flow technique, *J. Physiol. (London),* 285, 129, 1978.
60. **Ochs, S.,** Fast axoplasmic transport of materials in mammalian nerve and its integrative role, *Ann. N.Y. Acad. Sci.,* 193, 43, 1972.
61. **McMahan, U. J., Sanes, J. R., and Marshall, L. M.,** Cholinesterase is associated with the basal lamina at the neuromuscular junction, *Nature (London),* 271, 172, 1978.

62. **Younkin, S. G., Rosenstein, C., Collins, P. L., and Rosenberry, T. L.,** Cellular localization of the molecular forms of acetylcholinesterase in rat diaphragm, *J. Biol. Chem.*, 257, 13630, 1982.
63. **Ranish, N. A., Dettbarn, W. D, and Wecker, L.,** Nerve stump length-dependent loss of acetylcholinesterase activity in endplate regions of rat diaphragm, *Brain Res.*, 191, 379, 1980.
64. **Davey, B. and Younkin, S. G.,** Effect of nerve stump length on cholinesterase in denervated rat diaphragm, *Exp. Neurol.*, 59, 168, 1978.
65. **Hall, Z. W.,** Multiple forms of acetylcholinesterase and their distribution in endplate and non-endplate regions of rate diaphragm muscle, *J. Neurobiol.*, 4, 343, 1973.
66. **Fernandez, H. L., Duell, M. J., and Festoff, B. W.,** Neurotrophic control of 165 acetylcholinesterase at the vertebrate neuromuscular junction, *J. Neurobiol.*, 10, 441, 1979.
67. **Brimijoin, S.,** Molecular forms of acetylcholinesterase in brain, nerve, and muscle: nature, localization, and dynamics, *Prog. Neurobiol.*, 21, 291, 1983.
68. **Dreyfus, P. and Rieger, F.,** Isolation of rat and mouse neuromuscular junctions: tailed asymmetric acetylcholinesterase and junctional-enriched muscle basal lamina, in *Neuromuscular Diseases*, Serratrice, G., Ed., Raven Press, New York, 1984, 429.
69. **Fernandez, H. L., Patterson, M. R., and Duell, M. J.,** Neurotrophic control of 16S acetylcholinesterase from mammalian skeletal muscle in organ culture, *J. Neurobiol.*, 11, 557, 1980.
70. **DiGiamberardino, L. and Couraud, J. Y.,** Rapid accumulation of high molecular weight acetylcholinesterase in transected sciatic nerve, *Nature (London)*, 271, 170, 1978.
71. **Couraud, J. Y. and Di Giamberardino, L.,** Axonal transport of the molecular forms of acetylcholinesterase in chick sciatic nerve, *J. Neurochem.*, 35, 1053, 1980.
72. **Fernandez, H. L., Inestrosa, N. C., and Stiles, J. R.,** Subcellular localization of acetylcholinesterase molecular forms in endplate regions of adult mammalian skeletal muscle, *Neurochem. Res.*, 9, 1211, 1984.
73. **DiGiamberardino, L. and Couraud, J. Y.,** Axonal transport as a potential source of synaptic acetylcholinesterase, in *Neuromuscular Diseases*, Serratrice, G., Ed., Raven Press, New York, 1984, 451.
74. **Denny-Brown, D. and Brenner, C.,** Lesion in peripheral nerve resulting from compression by spring clip, *Arch. Neurol. Psychiatry (Chicago)*, 52, 1, 1944.
75. **Robert, E. D. and Oester, Y. T.,** Absence of supersensitivity to acetylcholine in innervated muscle subjected to a prolonged pharmacologic nerve block, *J. Pharmacol. Exp. Ther.*, 174, 133, 1970.
76. **Lømo, T. and Rosenthal, J.,** Control of ACh sensitivity by muscle activity in the rat, *J. Physiol. (London)*, 221, 493, 1972.
77. **Hubbard, S. J.,** The electrical constants and the component conductances of frog skeletal muscle after denervation, *J. Physiol. (London)*, 165, 443, 1963.
78. **Fink, B. R., Kennedy, R. D., Hendrickson, A. E., and Middaugh, M. E.,** Lidocaine inhibition of rapid axonal transport, *Anesthesiology*, 36, 422, 1972.
79. **Bisby, M. A.,** Inhibition of axonal transport in nerves chronically treated with local anesthetics, *Exp. Neurol.*, 47, 481, 1975.
80. **Ochs, S. and Hollingsworth, D.,** Dependence of fast axoplasmic transport in nerve on oxidative metabolism, *J. Neurochem.*, 18, 107, 1971.
81. **Lavoie, P. A., Collier, B., and Tenenhouse, A.,** Comparison of α-bungarotoxin binding to skeletal muscles after inactivity or denervation, *Nature (London)*, 260, 349, 1976.
82. **Pestronk, A., Drachman, D. B., and Griffin, J. W.,** Effect of muscle disuse on acetylcholine receptors, *Nature (London)*, 260, 352, 1976.
83. **Bray, J. J., Hubbard, J. I., and Mills, R. G.,** The trophic influence of tetrodotoxin-inactive nerves on normal and reinnervated rat skeletal muscles, *J. Physiol. (London)*, 297, 479, 1979.
84. **Gilliatt, R. W., Westgaard, R. H., and Williams, I. R.,** Extrajunctional acetylcholine sensitivity of inactive muscle fibers in the baboon during prolonged nerve pressure block, *J. Physiol. (London)*, 280, 499, 1978.
85. **Perísíe, M. and Cuénod, M.,** Synaptic transmission depressed by colchicine blockage of axoplasmic flow, *Science*, 175, 1140, 1972.
86. **Pilar, G. and Landmesser, L.,** Axotomy mimicked by localized colchicine application, *Science*, 177, 1116, 1972.
87. **Cangiano, A. and Fried, J. A.,** The production of denervation-like changes in rat muscle by colchicine, without interference with axonal transport or muscle activity, *J. Physiol. (London)*, 265, 63, 1977.
88. **Camerino, D. and Bryant, S. H.,** Effects of denervation and colchicine treatment on the chloride conductance of rat skeletal muscle fibers, *J. Neurobiol.*, 7, 221, 1976.
89. **DeCoursey, T. E., Younkin, S. G., and Bryant, S. H.,** Neural control of chloride conductance in rat extensor digitorum longus muscle, *Exp. Neurol.*, 61, 705, 1978.
90. **Lømo, T. and Westgaard, R. H.,** Control of ACh sensitivity in rat muscle fibers, *Cold Spring Harbor Symp. Quant. Biol.*, 40, 263, 1975.
91. **Jones, R. and Vrbova, G.,** Two factors responsible for the development of denervation hypersensitivity, *J. Physiol. (London)*, 236, 517, 1974.

92. **Fischbach, G. D., Cohen, S. A., and Henkart, M. P.,** Some observations on trophic interactions between neurons and muscle fibers in cell culture, *Ann. N.Y. Acad. Sci.*, 228, 35, 1974.
93. **Purves, D. and Sakmann, B.,** The effect of contractile activity on fibrillation and extra-junctional acetylcholine-sensitivity in rat muscle maintained in organ culture, *J. Physiol. (London)*, 237, 157, 1974.
94. **Jansen, J. K. S., Lømo, T., Nicolaysen, K., and Westgaard, R. H.,** Hyperinnervation of skeletal muscle fibers: dependence on muscle activity, *Science*, 181, 559, 1973.
95. **Katz, B. and Miledi, R.,** The development of acetylcholine sensitivity in nerve-free segments of skeletal muscle, *J. Physiol. (London)*, 170, 389, 1964.
96. **Cangiano, A. and Lutzemberger, L.,** Partial denervation affects both denervated and innervated fibers in the mammalian skeletal muscle, *Science*, 196, 542, 1977.
97. **Van Harreveld, A.,** Re-innervation of denervated muscle fibers by adjacent functioning motor units, *Am. J. Physiol.*, 144, 477, 1945.
98. **Denny-Brown, D.,** The histological features of striped muscle in relation to its functional activity, *Proc. R. Soc. London Ser. B*, 104, 371, 1929.
99. **Buller, A. J., Eccles, J. C., and Eccles, R. M.,** Differentiation of fast and slow muscles in the cat hind limb, *J. Physiol. (London)*, 150, 399, 1960.
100. **Buller, A. J., Eccles, J. C., and Eccles, R. M.,** Interactions between motoneurons and muscles in respect of the characteristic speeds of their responses, *J. Physiol. (London)*, 150, 417, 1960.
101. **Close, R.,** Properties of motor units in fast and slow skeletal muscles of the rat, *J. Physiol. (London)*, 193, 45, 1967.
102. **Mommaerts, W. F. H. M.,** Energetics of muscular contraction, *Physiol. Rev.*, 49, 427, 1969.
103. **Barany, M.,** ATPase activity of myosin correlated with speed of muscle shortening, *J. Gen. Physiol.*, 50, 197, 1967.
104. **Buller, A. J., Mommaerts, W. F. H. M., and Seraydarian, K.,** Enzymic properties of myosin in fast and slow twitch muscles of the cat following cross-innervation, *J. Physiol. (London)*, 205, 581, 1969.
105. **Buller, A. J. and Lewis, D. M.,** The rate of tension development is isometric tetanic contractions of mammalian fast and slow skeletal muscle, *J. Physiol. (London)*, 176, 337, 1965.
106. **Buller, A. J. and Lewis, D. M.,** Further observations on mammalian cross-innervated skeletal muscle, *J. Physiol. (London)*, 178, 343, 1965.
107. **Luff, A. R.,** Dynamic properties of fast and slow skeletal muscles in the cat and rat following cross-reinnervation, *J. Physiol. (London)*, 248, 83, 1975.
108. **Salmons, S. and Vrbova, G.,** The influence of activity on some contractile characteristics of mammalian fast and slow muscles, *J. Physiol. (London)*, 201, 535, 1969.
109. **Lømo, T., Westgaard, R. H., and Dahl, H. A.,** Contractile properties of muscle: control by pattern of muscle activity in the rat, *Proc. R. Soc. London Ser. B*, 187, 99, 1974.
110. **Lømo, T., Westgaard, R. H., and Engebretsen, L.,** Different stimulation patterns affect contractile properties of denervated rat soleus muscles, in *Plasticity of Muscle*, Pette, D., Ed., Walter de Gruyter, New York, 1980, 297.
111. **Guth, L., Samaha, F. J., and Albers, R. W.,** The neural regulation of some phenotypic differences between the fiber types of mammalian skeletal muscle, *Exp. Neurol.*, 26, 126, 1970.
112. **Studitsky, A. N.,** Free auto- and homografts of muscle tissue in experiments on animals, *Ann. N.Y. Acad. Sci.*, 120, 789, 1964.
113. **Carlson, B. M.,** The regeneration of skeletal muscle — a review, *Am. J. Anat.*, 137, 119, 1973.
114. **Zhenevskaya, R. P.,** The influence of de-efferentation on the regeneration of skeletal muscle, *Arkh. Anat. Gist. Embriol.*, 39, 42, 1960; as cited in **Carlson, B. M.,** *Am. J. Anat.*, 137, 119, 1973.
115. **Zhenevskaya, R. P.,** Restoration of muscle by the method of transplantation of minced muscle tissue under conditions of sensory deprivation, *Arkh. Anat. Gistol. Embriol.*, 40, 46, 1961; as cited in **Carlson, B. M.,** *Am. J. Anat.*, 137, 119, 1973.
116. **Hsu, L.,** The Role of Nerves in the Regeneration of Minced Muscle in Adult Anurans, Ph.D. thesis, University of Michigan, 1971; as cited in **Carlson, B. M.,** *Am. J. Anat.*, 137, 119, 1973.
117. **Thesleff, S.,** Supersensitivity of skeletal muscle produced by botulinum toxin, *J. Physiol. (London)*, 151, 598, 1960.
118. **Purves, D. and Lichtmann, J. W.,** Formation and maintenance of synaptic connections in autonomic ganglia, *Physiol. Rev.*, 58, 821, 1978.
119. **Purves, D. and Lichtman, J. W.,** Elimination of synapses in the developing nervous system, *Science*, 210, 153, 1980.
120. **Miledi, R.,** The acetylcholine sensitivity of frog muscle fibers after complete or partial denervation, *J. Physiol. (London)*, 151, 1, 1960.
121. **Thesleff, S.,** Physiological effects of denervation of muscle, *Ann. N.Y. Acad. Sci.*, 228, 89, 1974.
122. **Drachman, D. B., Stanley, E. F., and Pestronk, A.,** Neural regulation of muscle properties, in *Neuromuscular Diseases*, Serratrice, G., Ed., Raven Press, New York, 1984, 415.

123. **Lentz, T. L.,** Neurotrophic regulation at the neuromuscular junction, *Ann. N.Y. Acad. Sci.*, 228, 323, 1974.
124. **Peterson, E. R. and Crain, S. M.,** Innervation in cultures of fetal rodent skeletal muscle by organotypic explants of spinal cord from different animals, *Z. Zellforsch.*, 106, 1, 1970.
125. **Peterson, E. R. and Crain, S. M.,** Regeneration and innervation in cultures of adult mammalian skeletal muscle coupled with fetal rodent spinal cord, *Exp. Neurol.*, 36, 136, 1972.
126. **Crain, S. and Peterson, E. R.,** Development of neural connections in culture, *Ann. N.Y. Acad. Sci.*, 228, 6, 1974.
127. **Peterson, E. R. and Crain, S. M.,** Factors involved in muscle atrophy after surgical denervation of long-term cultures of cord-innervated fibers, in *Muscle Regeneration*, Mauro, A. et al., Eds. Raven Press, New York, 1979, 305.
128. **Crain, S. M.,** Neurophysiological studies, in *Tissue Cultures*, Raven Press, New York, 1976, 1.
129. **Oh, T. H., Johnson, D. D., and Kim, S. U.,** Neurotrophic effect on isolated chick embryo muscle in culture, *Science*, 178, 1298, 1972.
130. **Oh, T. H.,** Neurotrophic effects: characterization of the nerve extract that stimulates muscle development in culture, *Exp. Neurol.*, 46, 432, 1975.
131. **Markelonis, G. J. and Oh, T. H.,** A protein fraction from peripheral nerve having neurotrophic effects on skeletal muscle cells in culture, *Exp. Neurol.*, 58, 285, 1978.
132. **Markelonis, G. J. and Oh, T. H.,** A sciatic nerve protein has a trophic effect on development and maintenance of skeletal muscle cells in culture, *Proc. Natl. Acad. Sci. U.S.A.*, 76, 2470, 1979.
133. **Oh, T. H., Markelonis, G. J., Reier, P. J., and Zalewski, A. A.,** Persistence in degenerating sciatic nerve of substances having a trophic influence upon cultured muscle, *Exp. Neurol.*, 67, 646, 1980.
134. **Davey, B., Younkin, L. H., and Younkin, S. G.,** Neural control of skeletal muscle cholinesterase: a study using organ-cultured rat muscle, *J. Physiol. (London)*, 289, 501, 1979.
135. **Fernandez, H. L. and Inestrosa, N. C.,** Role of axoplasmic transport in neurotrophic regulation of muscle endplate acetylcholinesterase, *Nature (London)*, 262, 55, 1976.
136. **Musick, J. and Hubbard, J. I.,** Release of protein from mouse motor nerve terminals, *Nature (London)*, 237, 279, 1972.
137. **Oh, T. H. and Markelonis, G. J.,** Chicken serum transferrin duplicates the myotrophic effects of sciatin on cultured muscle cells, *J. Neurosci. Res.*, 8, 535, 1982.
138. **Beach, R. L., Popiela, H., and Festoff, B. W.,** The identification of neurotrophic factor as a transferrin, *FEBS Lett.*, 156, 151, 1983.
139. **Davis, H. L. and Heinicke, E. A.,** Prevention of denervation atrophy in muscle: mammalian neurotrophic factor is not transferrin, *Brain Res.*, 309, 293, 1984.
140. **Davis, H. L. and Kiernan, J. A.,** Neurotrophic effects of sciatic nerve extract on denervated extensor digitorium longus muscle in the rat, *Exp. Neurol.*, 69, 124, 1980.
141. **Davis, H. L. and Kiernan, J. A.,** Effect of nerve extract on atrophy of denervated or immobilized muscles, *Exp. Neurol.*, 72, 582, 1981.
142. **Levi-Montalcini, R. and Angeletti, P. V.,** Nerve growth factor, *Physiol. Rev.*, 48, 534, 1968.
143. **Mobley, W. C., Server, A. C., Ishii, D. N., Riopelle, R. J., and Shooter, E. M.,** Nerve growth factor, *N. Engl. J. Med.*, 297, 1096, 1977.
144. **Ochs, S.,** Regeneration of nerve in relation to axoplasmic transport: an analysis, in *Posttraumatic Peripheral Nerve Regeneration: Experimental Basis and Clinical Implications*, Gorio, A. et al., Eds., Raven Press, New York, 1981, 115.
145. **Barde, Y. A., Edgar, D., and Thoenen, H.,** New neurotrophic factors, *Ann. Rev. Physiol.*, 45, 601, 1983.

Chapter 3

TROPHIC ROLES OF AXOPLASMIC TRANSPORT

Stephen Brimijoin

TABLE OF CONTENTS

I. AXOPLASMIC TRANSPORT

Axoplasmic transport is the bidirectional process that conveys intraneuronal particles, macromolecules, and other axoplasmic constituents from one region of the neuron to another. This process has a clear trophic role in neuronal maintenance and renewal (propriotrophic) and may influence the relationship between the neuron and its effector cell (exterotrophic). To evaluate these roles, it may help to begin by considering some of the current information about the nature and mechanism of transport. Several recent monographs and reviews provide an excellent overview of this topic.[1-4] Here we will deal only with information which is most pertinent to an understanding of the involvement of axonal transport in trophic effects.

A. General Properties of Transport

Materials are known to be transported in both directions along axons by an active, temperature-dependent, energy-requiring process.[5-8] Up to five transport rates have been identified, but with rare exceptions any given material moves at only one rate. This has led to a structural hypothesis which accounts for the different phases of transport in terms of separate constellations of organelles.[3]

Fast transport involves primarily membrane-limited organelles such as synaptic vesicles and their precursors. The maximal velocity of fast transport is about 400 mm/day in the anterograde direction (toward the nerve terminal) and 300 mm/day in the retrograde direction (toward the cell body), at 37°C, and is independent of fiber diameter.[9-11] There is good evidence that the average velocity of fast transport is appreciably slower than the maximum, owing to a tendency for transported particles to become stationary for variable periods of time.[12,13] The exhaustion of fast-transported material from any given length of nerve stump, therefore, presumably occurs at rates well below the maximum.

At the other extreme of the velocity spectrum is slow transport, which may be divided into two components. Slow component a (SCa, 0.2 to 1.0 mm/day) consists almost entirely of the neurofilament triplet proteins and tubulin.[14] A persuasive case has been made that SCa represents the coordinate movement of the axonal cytoskeleton.[3] Slow component b (SCb, 2 to 8 mm/day) comprises an additional component of tubulin, possibly of a different form.[15] In addition, actin, myosin, spectrin, clathrin, cytoplasmic enzymes such as nerve-specific enolase, pyruvate kinase, aldolase, calmodulin, and a host of unidentified polypeptides move in this phase of transport.[3] The extent to which these constituents move in register is striking, but the underlying structure responsible for SCb is still unknown. Ellisman[16] has drawn attention to a "microtrabecular meshwork" in the axon. Because the constituents of SCb are partly or fully soluble in physiologic buffers, their association with this organizing structure, or any other, must be transient and reversible. Even less well understood from a structural point of view is intermediate transport (3 to 20 mm/day), a phenomenon that has been observed in certain types of nerves.[17]

Axoplasmic transport is only marginally influenced by electrical activity. It is quite independent of the direction of impulse conduction. Transport velocity is also largely independent of the firing frequency,[18] although the amount and nature of the material being transported may be altered. A number of local anesthetics are capable of interrupting transport, but this effect requires doses higher than necessary for blockade of impulse conduction,[19-21] and it probably reflects interference with the local mechanisms of generation of high-energy phosphates. Tetrodotoxin (TTX) is even more selective, causing no measurable change in fast axoplasmic transport even after total blockade of action potentials.[5,22]

B. Mechanism of Transport

It is widely believed that a single mechanism accounts for all of the different phases and velocities of axonal transport. The mechanism of fast anterograde transport has been investigated most thoroughly. Though major questions remain, important details are beginning to emerge. Fast transport is primarily under local control and will continue for hours in axons isolated from their cell bodies by transection or ligation.[23,24] Because transport is sensitive to agents that block oxidative phosphorylation or glycolysis, it was suggested that the process must require ATP.[5,25] Other pharmacological information pointed to an involvement of microtubules, since antimitotic agents such as colchicine and vinblastine block transport roughly in parallel with their disaggregation of microtubules.[26-28]

Optical studies with permeabilized cells in vitro have now established that fast transport is truly ATP dependent.[29,30] Recently, the advent of video-enhanced and differential interference contrast microscopy[31] has greatly aided analysis of the interaction of moving particles with other elements of the axon. By this method, several groups have demonstrated that particles in neurons and other cells of ectodermal origin move along isolated microtubules or bundles of microtubules.[32-34] Studies with vesicles isolated from the squid giant axon show that particles can even move rapidly in both directions along the same microtubule.[35]

Progress in identifying the ATP-dependent transport motor is now being made with the aid of motility assays in cell-free systems. A promising candidate is kinesin, a 600-kdalton ATPase complex isolated from squid optic lobe.[36,37] This complex, consisting of 110- to 120- and 60- to 70-kdalton polypeptides, is not dynein-like, but it promotes the transport of latex beads along purified microtubules. Further work is required to determine whether the same protein is involved in the transport of physiological particles in squid axons and other nerves.

II. PROPRIOTROPHIC ROLES OF TRANSPORT

A. Anterograde Transport

The role of anterograde axonal transport in maintaining neuronal structure and function is related to the paucity of ribosomes in axons, dendrites, or nerve terminals. With the exception of some proteins produced within the semiautonomous mitochondria, nearly all neuronal polypeptides are synthesized in the cell body and transported to their ultimate destinations. In particular, anterograde axonal transport is responsible for distal translocation of the enzymes involved in intermediary metabolism and neurotransmitter biosynthesis,[38-41] as well as of neuropeptides and biogenic amines produced in the proximal part of the cell.[42-45] In addition, anterograde transport must supply the axon with mitochondria, cytoskeletal elements,[14] and many complex phospholipids and glycolipids.[46-48] There is good evidence that anterograde transport is involved not only in axonal elongation and maintenance of axonal caliber,[49,50] but also in renewal of the axolemma and presynaptic membranes.[51-53]

Interesting evidence for the propriotrophic role of anterograde axonal transport was obtained by Lubinska[54] in studies on the early course of Wallerian degeneration after nerve transection. Her results suggested that degeneration of the distal nerve stump proceeded in a wavelike fashion from the point of transection, at a rate of 50 to 250 mm/day. This phenomenon presumably underlies the well-known relationship between nerve stump length and the time of onset of degenerative changes in nerve terminals.[55] A reasonable interpretation of both phenomena is that nerve terminal degeneration follows the exhaustion of some rapidly transported material. Lubinska[54] found that the velocity of the degeneration wave was dependent on fiber size. Since the velocity of axonal transport is independent of the fiber size,[9] this observation implies that the size and turnover of local pools of critical materials are also important.

B. Retrograde Transport

Retrograde transport probably plays a subsidiary propriotrophic role in neuronal housecleaning. Axons are notably poor in lysosomes[56] and are therefore likely to be inefficient in metabolizing worn-out macromolecules. Accordingly, retrograde transport may be necessary to carry such materials back to the cell body for ultimate disposal. Ultrastructural studies support the view that many of the organelles undergoing retrograde transport are in the form of multivesicular bodies and other prelysosomal structures.[57,58]

III. EXTEROTROPHIC ROLES OF TRANSPORT

The remainder of this chapter will focus on the exterotrophic roles of axonal transport, with emphasis on the motor nerve and muscle cell. Exterotrophic functions are bidirectional, since neurons and effector cells mutually influence one another; both anterograde and retrograde transport may therefore be involved. The extent to which anterograde transport contributes to the dramatic trophic effects of neurons upon their effector cells is a major issue. Perhaps equally important is the question of how retrograde transport mediates the influences of effector cells upon their neuronal input. As the latter issue is less controversial, we consider it first.

A. Retrograde Transport

Postsynaptic substances that affect neuronal survival or the production of neuronal macromolecules must exert their actions in the perinuclear zone where transcription and translation occur. Three routes of access are possible: (1) through the circulation; (2) through the extracellular fluid of the endoneurium; and (3) through the axon (by retrograde transport). Only the last route offers the kind of specificity that can account for developmental effects like cell death among supernumerary axons (see Chapter 9).

Many trophic substances affecting neurons have recently been defined in part.[59] Perhaps there may be one for each class of neuron. To date, however, nerve growth factor (NGF) remains the only well-characterized example. The physiology of NGF has been reviewed by Thoenen and Barde.[60] For present purposes it suffices to note that this small basic protein may be secreted by most cells receiving adrenergic innervation, is essential for the early development of sensory as well as sympathetic nerve fibers,[61] and has a major effect on the production of neurotransmitter-related proteins in adrenergic neurons.[62] Exogenous NGF is captured by sensory and adrenergic nerve terminals via a specific, saturable uptake mechanism and is then carried back to cell bodies by retrograde transport.[63,64] Studies on the kinetics of NGF delivery, localization of action, and the effects of nerve ligation[65,66] underscore the biological importance of retrograde transport in NGF trophism. Additional strong support comes from recent evidence that endogenous NGF is retrogradely transported in sympathetic axons of the rat[67] and the guinea pig.[68]

Taking NGF as a paradigm, it seems that the identification of trophic substances will depend on demonstrating the presence of tissue components that promote the survival, growth, and normal function of the neurons in question. Demonstrating the physiological significance of any such substances will require proof that they are actually captured in vivo and somehow conveyed to cell bodies, e.g., by retrograde axonal transport. Crude extracts of muscle do contain a substance(s) that promotes the survival of motor neurons in vitro,[69-71] and the potency of these extracts is enhanced by prior denervation of the muscle.[72,73] Similar substances have been demonstrated in peripheral nerve,[74] possibly en route to the spinal cord. Unfortunately, their low concentration in tissues has greatly hindered attempts to isolate any of these substances.

The *in situ* dependency of motor nerves on tissue substances is not well established. One suspects that competition for such substances during early development is at least partly responsible for the cell death that occurs among supernumerary motor nerves[75,76] (see Chapter 9). Additional relevant information has recently been obtained in studies on nerve and muscle acetylcholinesterase (AChE) (see Chapter 4). This enzyme occurs in a variety of molecular forms, whose relative concentrations are characteristic of a given tissue.[77,78] It has been found that slow and fast muscle fibers differ in the proportion of certain high-molecular-weight forms of AChE,[79,80] and corresponding differences have been identified in the nerve supplies.[80] Cross-innervation techniques have now been used to show that the pattern of AChE forms in nerve is sensitive to the type of muscle innervated.[81] At present, one cannot rule out the possibility that a portion of this effect is due to AChE released by the muscle and recaptured by the nerve. However, since nerve cells produce, transport, and release their own AChE,[24,82-84] the operation of a trophic substance is more plausible. More work is needed to define the extent to which trophic substances of muscle origin can control the biochemical properties of motor neurons.

B. Anterograde Transport

1. General Considerations

The view that anterograde axonal transport delivers crucial trophic substances to muscle and other tissues was formed early (see Chapter 2) and is still strongly held.[85] Some of this conviction, however, is based on teleology rather than evidence. Apparently, the machinery exists; for example, motor nerves are known to carry proteins to their terminals and release them in response to action potentials.[86] Even AChE is subject to this effect,[84] which might suggest a trophic role for this enzyme at least as a contributor to the synaptic enzyme assembly. On the other hand, the amount of AChE transported in nerve and eventually released is minuscule in relation to the normal turnover of the enzyme in muscle.[87] Furthermore, no one has demonstrated a trophic role for AChE or any of the other identified proteins whose neuronal localization and axonal transport have been well characterized.

Most investigations of the trophic role of anterograde axonal transport have not dealt with particular proteins. Instead, efforts have been directed at defining the contribution of axonal transport in general to particular exterotrophic phenomena, such as the effect of nerve on resting membrane potential in muscle. This has proved a difficult task. Attempts to isolate effects on transport from effects on acetylcholine (ACh) release and synaptic activation are beset by formidable problems that are inherent in the physiology of the system. At best, the isolation can only be transient, since synaptic function eventually fails without axonal transport and the composition of transported material is related to nerve activity. There are also semantic difficulties, like the question of whether ACh might ever be considered a trophic factor. How should one classify this substance if it did support muscle function and structure, but only by inducing postsynaptic action potentials? Also, in such a case, how should one view the role of axonal transport?

Though ACh is axonally transported,[43] most of the transmitter released from motor nerve terminals is probably synthesized locally.[88] In a broad sense axonal transport contributes to the delivery of this substance by supplying the choline acetyltransferase necessary for its biosynthesis,[41] as well as the synaptic vesicles for its storage and release.[89] Still, this role is rather indirect.

There are other potential exterotrophic mechanisms to which axonal transport could contribute indirectly. These mechanisms include direct cell-to-cell contact (see Chapters 8 and 10), which would probably involve specialized macromolecules conveyed from the perikaryon to the terminal axolemma. What arouses more interest and controversy is the idea of specific trophic substances elaborated in the nerve cell body and delivered by axonal transport for release onto muscle cells. Some evidence in favor of this idea comes from observations that the effects of nerve transection can be delayed in proportion to the length of nerve left with the muscle.

2. Significance of Length-Dependent Effects of Nerve Transection

The existence of transported trophic substances was suspected early on, especially after it was shown that the onset of certain denervation-related changes in the muscle depends on the length of the distal nerve stump. As reviewed by Ochs in Chapter 2 of this volume, there is a length dependency in the times of onset of fibrillation,[90] decreased resting membrane potential,[91,92] TTX-resistant action potentials,[91,93,94] and depletion of muscle endplate AChE.[95-97] All of these changes tend to occur later when the nerve stump is longer, as though something in the pipeline sustained normal function until exhausted. Thus, most studies have found that the onset of denervation-related changes in muscle is delayed 1 hr for each 2 to 6 mm of nerve stump length.[90,93,98,99] The corresponding velocities, 2 to 6 mm/hr or 48 to 144 mm/day, are within the range considered to represent fast transport. In experiments on the loss of endplate AChE activity[95] even higher velocities have been inferred, on the order of 300 mm/day (i.e., approaching the maximal rate possible for transport).

The length dependency data are clearly consistent with the existence of a rapidly transported trophic substance, which might well be located in a membrane-limited organelle and be subject to exocytotic release. However, one can explain length-dependent denervation effects without invoking the concept of specific, axonally transported trophic substances. The major alternative mechanisms for the changes occurring in denervated muscle are loss of contact with nerve, direct effects of materials released from degenerating nerve fibers, and altered neurotransmitter release. The first two mechanisms would be length dependent because the degeneration of the nerve terminal is itself length dependent, with characteristics implying the migration of some essential materials at a rate of 360 mm/day.[55,100] It has been pointed out that some biochemical changes in muscle precede complete nerve terminal degeneration,[97] but most denervation-related effects do tend to parallel structural disintegration of the nerve terminals.[55] As for mechanisms involving altered transmitter release, one must recognize that action potentials are eliminated soon after nerve transection. Yet there is good evidence that spontaneously released ACh can have important effects on muscle. Since the local stores of transmitter depend at least partly on axonal transport, it is not unreasonable to expect a length dependency in this mechanism. Such considerations emphasize the importance of experiments to isolate the exterotrophic contributions of transport from other neural mechanisms.

3. Experimental Attempts to Dissect the Exterotrophic Influences of Nerve

Much ingenuity has been applied to dissection of the exterotrophic influences of nerve and to account for length dependency and other phenomena of denervation. Experiments have been particularly aimed at distinguishing the effects of muscle activity, axonal transport, and nerve degeneration. These experiments fall into four groups: (1) introduction of degenerating nerve into innervated muscle; (2) attempts to block axonal transport without blocking impulse conduction or causing nerve degeneration; (3) attempts to block impulse conduction or muscle activation without blocking axonal transport or causing nerve degeneration; and (4) attempts to prevent denervation-induced changes in muscles by direct activation. The results of these experiments are remarkably consistent, with a few exceptions, but there has been strong diagreement regarding their interpretation.

a. Products of Nerve Degeneration

Evidence that products of nerve terminal degeneration contribute to the denervation effects on muscle was first obtained by Jones and Vrbova[101] and Vrbova.[102] These investigators showed that pieces of peripheral nerve placed on the surface of innervated soleus or extensor digitorum longus muscles induced local drops in resting membrane potential and hypersensitivity to ACh. However, even a graft of suture thread produced similar effects, suggesting that an inflammatory response was involved. More recently, Guth et al.[103] made a careful histological examination of the nerve autograft model and found extensive muscle degeneration, including sarcoplasmic vacuolization. Since such changes are radically different from those normally accompanying denervation, the pertinence of this model is questionable.

Cangiano and Lutzemberger's[104] study of partial denervation is sometimes cited in support of the view that functional changes in denervated muscle are due to agents liberated from degenerating nerves. This study involved a lesion of the L5 ventral root, with the result that innervated fibers intermingled with denervated ones in the extensor digitorum longus muscle. Cangiano and Lutzemberger observed TTX-resistant action potentials in both kinds of fibers and concluded that a diffusible substance from degenerating terminals was responsible. On the other hand, Tiedt et al.[105] pointed out technical difficulties and, in a similar experiment, found no such effect nor any change in resting membrane potential or extrajunctional ACh sensitivity. At present, therefore, the evidence for countertrophic factors as a cause of denervation-like changes in muscle is at best weak.

b. Selective Block of Axoplasmic Transport

In attempts to block axoplasmic transport without causing nerve degeneration, failure of impulse conduction, or direct effects on muscle, the antimicrotubule agents colchicine or vinblastine have been applied to nerve via subperineurial injections or drug-impregnated Silastic cuffs.[106-109] Such treatments have uniformly induced many of the characteristic signs of denervation in the corresponding muscles, including a reduction of resting membrane potential, appearance of extrajunctional ACh sensitivity,[106-108] onset of fibrillation,[108] in vitro release of AChE,[110] and decrease of AChE activity at the motor endplate.[111]

Two key criticisms have been leveled at these experiments: the treatments may have caused partial denervation or impaired impulse conduction, and the drug may have escaped from the site of application and acted directly on the muscle membrane. As evidence against partial denervation, the treated systems retained miniature endplate potentials of normal or increased frequency, but increased amplitude.[106] As evidence against impaired impulse conduction, the muscles could be activated by stimulation of the nerve above the colchicine cuff.[106,112] However, Cangiano and Fried[109] observed transient hemiparesis after applying comparatively loose, colchicine-impregnated cuffs to rat sciatic nerve. They point out that colchicine or any other means of blocking axonal transport is likely to induce a localized nerve swelling (as is commonly observed). The resulting nerve compression could account for the effectiveness of drug-containing cuffs as compared with control cuffs. Also, even a transient block of nerve conduction might have lasting effects on the muscle. This point is demonstrated by Lømo and Slater's[113] finding that interruption of muscle activity for 48 hr can cause generalized supersensitivity. It is therefore important to document the incidence of supersensitive fibers in muscle of colchicine-cuffed nerves. Unless supersensitivity is known to be uniform, one must allow for the possibility that the supersensitive fibers have actually suffered from transient, compression-induced blockade of impulse conduction.

As for the second criticism, there is evidence against artifacts that are due to colchicine leaking out of nerve cuffs or injection sites and acting directly on muscle. For example, Kauffman et al.[114] showed that little radiolabeled drug escapes from the site of application. Nonetheless, concern has again been raised by Cangiano and Fried,[109] who found bilateral muscle supersensitivity after injecting high concentrations of colchicine into nerve. Deliberately applying colchicine to innervated muscle tends to block neuromuscular transmission and does not lower membrane potential.[106] Therefore, colchicine does not directly cause the postsynaptic effects of denervation. However, if failure of neuromuscular transmission were the normal trigger for these effects, colchicine might be able to induce them indirectly, in ways unrelated to a blockade of transport.

Most studies of colchicine treatment used low doses that produced ipsilateral, rather than systemic, effects. Nevertheless, it remains possible that small but effective amounts of drug reached the ipsilateral neuromuscular junction by diffusion or bulk flow through the endoneurial fluid. For this reason, one cannot evaluate experiments with locally applied transport blockers without knowing the relative sensitivity of axonal transport and neurotransmitter release to the drug being used and the actual effect of the experimental treatment on total ACh release at the neuromuscular junctions where function is being assessed. Since much of this release is not reflected in either endplate or miniature endplate potentials, direct biochemical measurements are called for.

In summary, there is room to doubt whether a perfectly selective block of transport has been or can be achieved without impairing impulse conduction or directly affecting the muscle membrane. Certainly there is as yet no compelling evidence that an isolated deficit of transport can cause any of the prominent effects associated with muscle denervation, such as decreased resting membrane potential and appearance of extrajunctional acetylcholine receptors (AChR). Even if such evidence were in hand, it would not be safe to infer a trophic substance other than ACh unless all forms of neurotransmitter release were shown to be normal.

c. Selective Block of Impulse Conduction or Muscle Activation

One means of blocking muscle activation is with botulinum toxin (BT). It has been known for some time that BT produces signs of denervation, including the induction of collateral sprouting from nearby motor nerve fibers.[115-117] Since BT selectively interferes with exocytosis in cholinergic terminals, it is tempting to infer that the denervation-like syndrome simply reflects the loss of ACh. This inference is not necessarily justified, because cholinergic vesicles may well contain other substances with important trophic effects. A more popular experimental paradigm, therefore, has been to block either the propagation of action potentials along motor axons or the activation of AChR in the muscle.

Modern attempts to selectively block impulse conduction have used pharmacological means. Local anesthetics such as lidocaine or ion channel blockers such as TTX have been injected through the perineurium or placed onto the nerve as drug-impregnated silicone cuffs.[22,118-122] With one unexplained exception,[118] all such experiments led to definite denervation-like changes in the muscles, including fall of resting membrane potential and increased AChR concentration as measured by α-bungarotoxin (α-BTX) binding. As a rule, these changes were less severe than those following true denervation, and they could be enhanced by nerve transection.[122] Taken at face value these findings imply that the normal impulse traffic has a meaningful role in the trophic effect of nerve, but there must be at least one additional component.

Some of the experiments just described are open to the objection that the anesthetic or channel blocker could have affected nerve function in ways besides simply blocking impulse conduction. The primary alternative effect would be an interruption of axonal transport, either mechanically or pharmacologically induced. Mechanical compression is surely possible when drugs are applied locally to nerve, as pointed out by Cangiano and Fried.[109] On the other hand, numerous control experiments with drug-free cuffs or endoneurial injections of saline solution have shown that this kind of artifact is avoidable and can hardly account for the experimental results.[22,110,120]

Drug effects on axonal transport are a more serious possibility since the doses of local anesthetic required to stop transport are often not much greater than those required for sustained electrical silence.[19,21] However, this objection does not apply to the experiments with TTX.[22,121,122] TTX does not block axonal transport when added to bathing media.[5] Furthermore, when rapid axonal transport of radiolabeled protein was measured in the toxin-treated nerves used to test the trophic role of impulse conduction, it was found to be unaffected.[22,121,122] Conceivably the transport of one crucial trophic substance was blocked selectively. However, this suggestion does not fit the structural view of Lasek et al.,[3] which holds that transport operates on entire organelle classes, not protein by protein.

Some particularly interesting experiments have compared the blockade of nerve impulses by TTX with that of muscle activation by the long-acting receptor antagonist, α-BTX. Initial work indicated that α-BTX was more effective than TTX in inducing denervation-like changes in the muscle.[121] This issue was carefully reexamined by Drachman et al.[22] The latter study used an implanted osmotic pump to apply α-BTX continuously to the soleus muscle while TTX was repeatedly injected into the sciatic nerve at 48-hr intervals. Quantitative measurements of resting membrane potential and AChR content of the muscle were made frequently. The effects of the two toxins were qualitatively similar, but those of α-BTX were substantially more pronounced. TTX plus nerve transection had greater effects than TTX alone. However, the effects of α-BTX treatment were identical in timing and severity to the changes induced by nerve transection, and there was no additive effect of transection.

These results strongly suggest that denervation-effects on resting membrane potential and receptor density are triggered by failure to activate the postsynaptic receptors (but see Chapter 2). It seems most likely that the failure of TTX to induce full denervation-like changes in the muscle is due to continuing spontaneous release of ACh. Nonstimulated release of ACh from nerve is both quantal and nonquantal. The former is responsible for miniature endplate potentials, but there are reasons to suspect that the latter is more important: (1) motor nerves release most of their ACh nonquantally; (2) spontaneous quantal release persists indefinitely after blockade of nerve impulses, but signs of denervation appear in the muscle anyway; and (3) spontaneous quantal release persists for many hours after nerve section, but receptor blockade does not hasten the signs of denervation.[22]

d. Direct Activation of Denervated Muscle

Since the nicotinic AChR is primarily an ionophore that serves to trigger muscle action potentials, one can bypass the receptor by stimulating the muscle directly. This provides an alternative means of testing the role of ACh release and muscle activation in maintaining muscle function and biochemistry. In fact a large body of work demonstrates that chronic stimulation of denervated muscle can prevent or reverse many physiological changes.[119,123-127] This is particularly true with regard to extrajunctional ACh sensitivity (i.e., spread of receptors into extrajunctional regions) and fibrillation.[102,119,128]

In considering the potentially trophic effects of muscle activation it is worth noting that the pattern of AChE forms in innervated muscle may be physiologically regulated by the rate or intensity of impulse traffic. When compared with fast muscle, slow muscles like the rat soleus normally contain less AChE overall, much less G_4 AChE (a tetrameric globular form), but proportionately more A_8 AChE (an asymmetric form of intermediate size).[19,80,129,130] There are indications that these differences can be reversed by appropriate physiological maneuvers, for example, by imposing different patterns (fast and slow) of electrical stimulation (see Chapter 5).

e. Summary of Experimental Attempts to Isolate Transport, Activation, and Degeneration

Loss of contact with nerve and the presence of nerve degeneration products are unsatisfactory explanations for the changes that take place in denervated muscle. The evidence against these explanations is twofold: (1) drugs can bring about denervation-like changes without inducing either degeneration of the nerve terminals or loss of myoneural contact; and (2) denervation-like changes can be minimized by electrical stimulation of the muscle or application of neural extracts.

The idea that denervation-induced changes in muscle result from the loss of a special axonally transported trophic substance has slightly better experimental support. It is true that interrupting axonal transport reliably causes the classic signs of muscle denervation, including fall of resting membrane potential, appearance of extrajunctional ACh sensitivity, and muscle atrophy. However, no one has produced these effects in a setting that eliminates the possibility of a partial blockade of impulse conduction or transmitter release. It is unfortunate that none of the attempts to impair transport selectively has used biochemical methods to assess total ACh release, particularly the large nonquantal component.

It is striking how closely one can mimic denervation by preventing the release or action of ACh, at least as shown by resting membrane potential and receptor density. Graded effects are obtained with treatments that produce graded degrees of interference with receptor occupancy. TTX (blocking-evoked quantal ACh release) is least effective. BT (blocking-evoked and spontaneous quantal release) is somewhat more effective. α-BTX (block of receptor activation by ACh from all sources) is most effective and is indistinguishable from surgical denervation in certain tests. These observations suggest that all of the ACh released by nerve can have an important trophic role.

The fact that released ACh can be partly supplanted by direct electrical stimulation of the muscle implies that the AChR participates in trophic events only by mediating muscle activation. It remains possible, even likely, that axonally transported substances do contribute to the trophic effects of motor nerves. Indirect evidence for this view stems from findings that muscle activation cannot replace all of the trophic support from the nerve since it does not fully eliminate muscle atrophy after nerve section.[131] Direct evidence for this view derives largely from studies on the effects of nerve extracts upon muscle in vitro and in vivo.

IV. BIOCHEMICAL STUDIES ON NEUROGENIC TROPHIC SUBSTANCES

Early studies of trophic substances in nerve used tissue-cultured embryonic myotubes.[132] Trophic potency was assessed in terms of morphological development, protein synthesis, production or maintenance of AChE activity, and number or clustering of AChR. These studies are reviewed extensively in Chapters 4 and 6. Here it is worth remarking that soluble extracts from various kinds of neural tissue, including spinal cord and nerve, have been found capable of promoting properties normally associated with maturing, innervated muscle.[133-137] In one case the responsible agent was characterized as a transferrin-like 86.4-kdalton glycoprotein termed sciatin.[138-140] Sciatin is believed to be axonally transported (see Chapter 6), since it, or a related substance, accumulates proximal to a nerve ligation.[141] The velocity of transport, and the relative importance of transport and local sources are unknown. It is also unknown whether sciatin is truly trophic, in the sense of maintaining normal muscle properties, or is more of a stimulus to muscle differentiation. Significantly, transferrins have not been found capable of reducing the characteristic muscle atrophy that accompanies denervation.[142]

A number of studies have dealt with the effects of nerve extracts on muscle in explant or organ culture. Among these are experiments from Younkin's group demonstrating that substances enriched in ligated nerve (i.e., presumably transported) cause a generalized increase in AChE activity and an increase in junctional ACh sensitivity.[143] Material released from curarized, electrically stimulated neuromuscular preparations had a similar effect.[143,144]

Fernandez et al.[145] have refined the analysis of effects of nerve extracts on muscle esterase by focusing on the 16S form of AChE. This enzyme form, characteristic of motor endplates,[146] is poorly represented in nerve and absent from most other tissues. Cross-contamination and other experimental artifacts are therefore unlikely to account for the observation that nerve extracts retard the loss of 16S AChE from cultured muscle. More recent work from the same group has not been published in detail, but was summarized in a recent review.[147] Extracts of ligated obturator nerve were found significantly more potent in retarding the decay of endplate 16S AChE in gracilis muscles, which suggests but does not prove that an axonally transported trophic substance was responsible. Gel electrophoresis showed many protein bands, several of which were removed by affinity chromatography with antirat serum antibodies. The remaining clean extracts not only prevented the decrease in 16S AChE in muscle explant cultures, but also promoted a net gain in 16S AChE activity of predenervated explants. This important finding implies that the nerve substance(s) were not merely acting to prevent enzyme degradation but may have been truly trophic.

Another line of evidence for trophic substances in nerve was obtained by Davis and Kiernan,[148] who showed that neural extracts would reduce in vivo muscle wasting if injected on a daily basis after nerve transection (see Chapter 7). In fact, quantitative measurements suggested that the extracts completely prevented the wasting that was not attributable to disuse per se.[148] A more recent report by the same authors[149] showed that the responsible substance is heat labile and binds to concanavalin A. Upon gel filtration the apparent molecular weight of this presumptive glycoprotein was 90 to 130 kdaltons, which suggests that the trophic substance is not sciatin or another transferrin. Unfortunately, attempts at further purification by ion exchange chromatography caused a marked loss of biological activity. Since the cleanest preparation had at least 15 bands on SDS gel electrophoresis, the characterization is still far from complete (but see Chapter 7).

It is remarkable that Davis and Kiernan's nerve substances, like the ones studied by Younkin's group, induced increases in the number or density of muscle AChR.[150] In other words, the effect was superficially in the same direction as that caused by denervation, rather than opposite, as might be expected. One suspects, however, that junctional AChRs rather than extrajunctional ones were being induced or preserved. This is a testable hypothesis, since the two types of receptor can be distinguished electrophysiologically as well as anatomically. Meanwhile, the possibility emerges that trophic substances from nerve are more directly related to junctional properties than to extrajunctional properties and are especially relevant to synaptic biochemistry (see Chapters 4, 5, and 11).

Further support for a synaptic role of neuronal substances comes from recent experiments by Busis et al.[151] with mutant cholinergic neuroblastoma-glioma hybrid cell lines that have lost the capacity to form synapses with co-cultured myotubes. Certain of these lines appear normal except for being unable to release ACh, in some cases because of defects in voltage-sensitive Ca^{2+} channels.[152] It may be inferred that ACh release and, presumably, receptor activation are required for synaptogenesis. Other mutant cell lines have functional Ca^{2+} channels and release ACh into the medium but still cannot form functional synapses. These cells do not induce aggregation of AChR on myotubes and are structurally distinguished by their lack of large, dense core vesicles, which may normally contain a receptor-aggregating substance.[151]

In an ingenious experiment it was shown that synapses do form when myotubes are cultured simultaneously with both kinds of deficient hybrids, namely, cells lacking dense-cored vesicles and cells unable to release ACh.[151] This finding unequivocally establishes a synaptogenic role for trophic substances other than ACh and strongly suggests a vesicular location. Since large dense core vesicles in neurons are prominent among the class of organelles that are rapidly transported from cell bodies to presynaptic terminals, it is overwhelmingly likely that the trophic substances they contain are axonally transported. These substances must now be characterized biochemically.

V. SUMMARY AND CONCLUDING REMARKS

It is unlikely that all of the trophic effects of motor nerves are mediated by a single mechanism or factor. Indeed, as the just cited study of synaptogenesis shows, it is even likely that certain of these effects represent interactions between two or more factors. However, it is intellectually defeatist to end by concluding that every one of the proposed mechanisms contributes to any particular manifestation of trophism (but see Chapter 11). At present ACh appears to be the most important factor controlling resting membrane potential and the spread of extrajunctional AChR. All sources of ACh are relevant, though evoked quantal release and spontaneous nonquantal release seem most crucial. It is also clear that nerves synthesize, transport, and release proteinaceous materials that have powerful effects on muscle differentiation and growth, as well as on synapse formation (see Chapters 8 to 10).

One of the most promising directions for future research is to exploit the advantages of cell culture systems, especially with established cell lines. As dramatically demonstrated in the studies with glioma-neuroblastoma hybrids, mutant cells provide natural experiments in which a few cytoplasmic constituents, even a single protein, may be altered with a selectivity that cannot be approximated in other ways. An obvious next step with regard to the vesicle-associated substance reported by Busis et al.[151] is to clarify its relationship, if any, to the trophic substances investigated by Younkin, Fernandez, Davis and Kiernan, and others.

Identification, purification, and characterization of trophic proteins and related substances will allow increasingly sophisticated studies of their roles. For example, instead of blocking the axonal transport of all neuronal materials, it should be possible to selectively interfere with the production or delivery of specific substances. One way to achieve this would be to introduce specific monoclonal antibodies into synaptic regions (either in culture or *in situ*). Monoclonal antibodies are especially appropriate at the present stage of ignorance, since their production does not require full purification of the trophic substances or any knowledge of their chemical nature.

Eventually it may be possible to perturb particular motor units even more selectively by injecting neurons with messenger RNA from hybridomas secreting antibody to a given trophic substance. The expected result would be intraneuronal synthesis of immunoglobulins against the substance, with consequent inactivation. Assuming that the trophic substances prove to be proteins, it will be important to establish their amino acid sequences. After this has been accomplished, one can envision causing a selective blockade of trophic substance synthesis by using antisense messenger RNA to prevent translation of the endogenous message. Such methods could make it possible to work out the physiological consequence of an isolated deficiency of any given molecule, without searching for a suitable mutant cell line that is naturally devoid of it.

The profound importance of trophic mechanisms and their implications for development and regeneration, as well as normal neuromuscular function, have stimulated interest in this problem for decades. The pace of investigation is now accelerating. We can be confident that modern approaches will lead to a detailed picture of trophic mechanisms in general and axonally transported trophic substances in particular.

REFERENCES

1. **Ochs, S.,** *Axoplasmic Transport and Its Relation to Other Nerve Functions,* John Wiley & Sons, New York, 1982, 1.
2. **Grafstein, B. and Forman, D. S.,** Intracellular transport in neurons, *Physiol. Rev.,* 60, 1167, 1980.
3. **Lasek, R. J., Garner, J. A., and Brady, S. T.,** Axonal transport of the cytoplasmic matrix, *J. Cell Biol.,* 99, 212s, 1984.
4. **Weiss, D. G., Ed.,** *Axoplasmic Transport,* Springer-Verlag, Berlin, 1982, 1.
5. **Ochs, S. and Hollingsworth, D.,** Dependence of fast axoplasmic transport in nerve on oxidative metabolism, *J. Neurochem.,* 18, 107, 1971.
6. **Garcia, A. G., Kirpekar, S. M., Prat, J. C., and Wakade, A. R.,** Metabolic and ionic requirements for the axoplasmic transport of dopamine beta-hydroxylase, *J. Physiol. (London),* 241, 809, 1974.
7. **Edström, A. and Hanson, M.,** Temperature effects on fast axonal transport of proteins *in vitro* in frog sciatic nerves, *Brain Res.,* 58, 345, 1973.
8. **Ochs, S. and Smith, C. B.** Low temperature slowing and cold-block of fast axoplasmic transport in mammalian nerves *in vitro, J. Neurobiol.,* 6, 85, 1975.
9. **Ochs, S.,** Rate of fast axoplasmic transport in mammalian nerve Fibers, *J. Physiol. (London),* 227, 627, 1972.

10. **Forman, D. S., Padjen, A. L., and Siggins, G. R.,** Effect of temperature on the rapid retrograde transport of microscopically visible intra-axonal organelles, *Brain Res.*, 136, 215, 1977.
11. **Brimijoin, S. and Helland, L.,** Rapid retrograde transport of dopamine-β-hydroxylase as examined by the stop-flow technique, *Brain Res.*, 102, 217, 1976.
12. **Ochs, S.,** Retention and redistribution of proteins in mammalian nerve fibers by axoplasmic transport, *J. Physiol. (London)*, 253, 459, 1975.
13. **Brimijoin, S.,** Axonal Transport in autonomic nerves: views on its kinetics, in *Trends in Autonomic Pharmacology*, Vol. 2, Kalsner, S., Ed., Urban & Schwartzenberg, Baltimore, 1982, 17.
14. **Hoffman, P. N. and Lasek, R. J.,** The slow component of axonal transport. Identification of major structural polypeptides of the axon and their generality among mammalian neurons, *J. Cell Biol.*, 66, 351, 1975.
15. **Tashiro, T., Kurokawa, M., and Komiya, Y.,** Two populations of axonally transported tubulin differentiated by their interactions with neurofilaments, *J. Neurochem.*, 43, 1220, 1984.
16. **Ellisman, M. and Porter, K. R.,** Microtrabecular structure of the axoplasmic matrix: visualization of cross-linking structures and their distribution, *J. Cell Biol.*, 87, 464, 1980.
17. **Willard, M., Cowan, W. M., and Vagelos, P. R.,** The polypeptide composition of intra-axonally transported proteins: Evidence for four transport velocities, *Proc. Natl. Acad. Sci. U.S.A.*, 71, 2183, 1974.
18. **Jankowska, E., Lubinska, L., and Niemierko, S.,** Translocation of AChE-containing particles in the axoplasm during nerve activity, *Comp. Biochem. Physiol.*, 28, 907, 1969.
19. **Bisby, M. A.,** Inhibition of axonal transport in nerves chronically treated with local anesthetics, *Exp. Neurol.*, 47, 481, 1975.
20. **Lavoie, P.-A.,** Block of fast axonal transport *in vitro* by the local anesthetics dibucaine and etidocaine, *J. Pharmacol. Exp. Ther.*, 223, 251, 1982.
21. **Lavoie, P.-A.,** Inhibition of fast axonal transport *in vitro* by the local anesthetics prilocaine, mepivacaine, and bupivacaine, *Can. J. Physiol. Pharmacol.*, 61, 1478, 1983.
22. **Drachman, D. B., Stanley, E. F., Pestronk, A., Griffin, J. W., and Price, D. L.,** Neurotrophic regulation of two properties of skeletal muscle by impulse-dependent and spontaneous acetylcholine transmission, *J. Neurosci.*, 2, 232, 1982.
23. **Ochs, S. and Ranish, N.,** Characteristics of the fast transport system in mammalian nerve fibers, *J. Neurobiol.*, 1, 247, 1969.
24. **Lubinska, L. and Niemierko, S.,** Velocity and intensity of bidirectional migration of acetylcholinesterase in transected nerves, *Brain Res.*, 27, 329, 1971.
25. **Ochs, S. and Smith, C. B.** Fast axoplasmic transport in mammalian nerve *in vitro* after block of glycolysis with iodoacetic acid, *J. Neurochem.*, 18, 833, 1971.
26. **Dahlström, A.,** Effect of colchicine on transport of amine storage granules in sympathetic nerves of the rabbit, *Eur. J. Pharmacol.*, 5, 111, 1968.
27. **Kreutzberg, G. W.,** Neuronal dynamics and axonal flow. IV. Blockage of intra-axonal enzyme transport by colchicine, *Proc. Natl. Acad. Sci. U.S.A.*, 62, 722, 1969.
28. **Brimijoin, S.,** Microtubules and the "capacity" of the system for rapid axonal transport, *Fed. Proc.*, 41, 2312, 1982.
29. **Forman, D. S.,** New approaches to the study of the mechanism of fast axonal transport, *Trends Neurosci.*, 7, 112, 1984.
30. **Forman, D. S., Brown, K., Promersberger, M., and Adelman, M.,** Nucleotide specificity for reactivation of organelle movements in permeabilized axons, *Cell Motil.*, 4, 121, 1984.
31. **Allen, R. D., Allen, N. S., and Travis, J. L.,** Video-enhanced contrast, differential interference contrast (AVEC-DIC) microscopy: a new method capable of analyzing microtubule-related motility in the reticulopodial network of *Allogromie laticollaris, Cell Motil.*, 1, 291, 1981.
32. **Allen, R. D., Metuzals, J., Taskai, I., Brady, S. T., and Gilbert, S. P.,** Fast axonal transport in squid giant axon, *Science*, 218, 1127, 1982.
33. **Gilbert, S. P., Allen, R. D., and Sloboda, R. D.** Translocation of vesicles from squid axoplasm on flagellar microtubules, *Nature (London)*, 315, 245, 1985.
34. **Vale, R. D., Schnapp, B. J., Reese, T. S., and Sheetz, M. P.,** Movement of organelles along filaments dissociated from axoplasm of the squid giant axon, *Cell*, 40, 449, 1985.
35. **Schnapp, B. J., Vale, R. D., Sheetz, M. P., and Reese, T. S.,** Single microtubules from squid axoplasm support bidirectional movement of organelles, *Cell*, 40, 455, 1985.
36. **Vale, R. D., Reese, T. S., and Sheetz, M. P.,** Identification of a novel force-generating protein, kinesin, involved in microtubule-based motility, *Cell*, 42, 39, 1985.
37. **Schnapp, B., Vale, R. D., Mitchison, T., Reese, T. S., and Sheetz, M. P.,** Translocator protein (kinesin) for rapid axonal transport mediates anterograde movement of beads along purified microtubules, *Soc. Neurosci. Abstr.*, 11, 1133, 1985.
38. **Brady, S. T. and Lasek, R. J.,** Nerve-specific enolase and creatine phosphokinase in axonal transport: soluble proteins and the axoplasmic matrix, *Cell*, 23, 515, 1981.

39. **Laduron, P. and Belpaire, F.,** Transport of noradrenaline and dopamine-β-hydroxylase in sympathetic nerves, *Life Sci.*, 7, 1, 1968.
40. **Wooten, G. F. and Coyle, J. T.,** Axonal transport of catecholamine synthesizing and metabolizing enzymes, *J. Neurochem.*, 20, 1361, 1973.
41. **Frizell, M., Hasselgren, P. O., and Sjöstrand, J.,** Axoplasmic transport of acetylcholinesterase and choline acetyltransferase in the vagus and hypoglossal nerve of the rabbit, *Exp. Brain Res.*, 10, 526, 1970.
42. **Dahlström, A. and Fuxe, K.,** A method for the demonstration of adrenergic nerve fibers in peripheral nerves, *Z. Zellforsch. Mikrosk. Anat.*, 62, 602, 1964.
43. **Dahlström, A., Evans, C. A. N., Häggendal, C. J., Heiwall, P.-O., and Saunders, N. R.,** Rapid transport of acetylcholine in rat sciatic nerve proximal and distal to a lesion, *J. Neural. Transm.*, 35, 1, 1974.
44. **Brimijoin, S., Lundberg, J. M., Brodin, E., Hökfelt, T., and Nilsson, G.,** Axonal transport of substance P in the vagus and sciatic nerves of the guinea pig, *Brain Res.*, 191, 443, 1980.
45. **Gilbert, R. F. T., Emson, P. C., Fahrenkrug, J., Lee, C. M., Penman, E., and Wass, J.,** Axonal transport of neuropeptides in the cervical vagus nerve of the rat, *J. Neurochem.*, 34, 108, 1980.
46. **Sherbany, A. A., Ambron, R. T., and Schwartz, J. H.,** Membrane glycolipids: regional synthesis and axonal transport in a single identified neuron of *Aplysia californica, Science*, 203, 78, 1979.
47. **Toews, A. D., Goodrum, J. F., and Morell, P.,** Axonal transport of phospholipids in rat visual system, *J. Neurochem.*, 32, 1165, 1979.
48. **Gould, R. M., Spivack, W. D., Sinatra, R. S., Lindquist, T. D., and Ingoglia, N. A.,** Axonal transport of choline lipids in normal and regenerating rat sciatic nerve, *J. Neurochem.*, 39, 1569, 1982.
49. **Hoffman, P. N. and Lasek, R. J.,** Axonal transport of the cytoskeleton in regenerating motor neurons: constancy and change, *Brain Res.*, 202, 317, 1980.
50. **Hoffman, P. N., Griffin, J. W., and Price, D. L.,** Changes in the axonal transport of the cytoskeleton during development, aging and regeneration, *Soc. Neurosci. Abstr.*, 7, 743, 1981.
51. **DiGiamberardino, L. D., Bennett, G., Koenig, H. L., and Droz, B.,** Axonal migration of protein and glycoprotein to nerve endings. III. Cell fraction analysis of chicken ciliary ganglion after intracerebral injection of labeled precursors of proteins and glycoproteins, *Brain Res.*, 60, 147, 1973.
52. **Lorenz, T. and Willard, M.,** Subcellular fractionation of intraaxonally transported polypeptides in the rabbit visual system., *Proc. Natl. Acad. Sci. U.S.A.*, 75, 505, 1978.
53. **Brimijoin, S., Skau, K. A., and Wiermaa, M. J.,** On the origin and fate of external acetylcholinesterase in peripheral nerve, *J. Physiol. (London)*, 285, 143, 1978.
54. **Lubinska, L.,** Early course of Wallerian degeneration in myelinated fibers of the rat phrenic nerve, *Brain Res.*, 130, 47, 1977.
55. **Miledi, R. and Slater, C. R.,** On the degeneration of rat neuromuscular junctions after nerve section, *J. Physiol. (London)*, 207, 507, 1970.
56. **Holtzman, E.,** *Lysosomes: A Survey*, Springer, New York, 1976, 1.
57. **Geffen, L. B. and Ostberg, A.,** Distribution of granular vesicles in normal and constricted sympathetic neurons, *J. Physiol. (London)*, 204, 583, 1969.
58. **Tsukita, S. and Ishikawa, H.,** The movement of membranous organelles in axons; electronmicroscopic identification of anterogradely and retrogradely transported organelles, *J. Cell Biol.*, 84, 513, 1980.
59. **Barde, Y.-A., Edgar, D., and Thoenen, H.,** New neurotrophic factors, *Ann. Rev. Physiol.*, 45, 601, 1983.
60. **Thoenen, H. and Barde, Y.-A.,** Physiology of nerve growth factor, *Physiol. Rev.*, 60, 1284, 1980.
61. **Levi-Montalcini, R. and Angeletti, P. V.,** Nerve growth factor, *Physiol. Rev.*, 48, 534, 1968.
62. **Thoenen, H., Angeletti, P. V., Levi-Montalcini, R., and Kettler, R.,** Selective induction by nerve growth factor of tyrosine hydroxylase and dopamine-β-hydroxylase in the rat superior cervical ganglion, *Proc. Natl. Acad. Sci. U.S.A.*, 68, 1598, 1971.
63. **Stoekel, K., Paravicini, U., and Thoenen, H.,** Specificity of the retrograde axonal transport of nerve growth factor, *Brain Res.*, 76, 413, 1974.
64. **Hendry, I. A., Stach, R., and Herrup, D.,** Characteristics of the retrograde axonal transport system for nerve growth factor in the sympathetic nervous system, *Brain Res.*, 82, 117, 1974.
65. **Paravicini, U., Stoeckel, K. G., and Thoenen, H.,** Biological importance of retrograde axonal transport of nerve growth factor in adrenergic neurons, *Brain Res.*, 84, 279, 1975.
66. **Hendry, I. A.,** The effect of the retrograde axonal transport of nerve growth factor on the morphology of adrenergic neurons, *Brain Res.*, 134, 213, 1977.
67. **Korsching, S. and Thoenen, H.,** Quantitative demonstration of the retrograde axonal transport of endogenous nerve growth factor, *Neurosci. Lett.*, 39, 1, 1983.
68. **Palmatier, M. A., Hartman, B. K., and Johnson, E. M., Jr.,** Demonstration of retrogradely transported endogenous nerve growth factor in axons of sympathetic neurons, *J. Neurosci.*, 4, 751, 1984.
69. **Bennett, M. R., Lai, K., and Nurcombe, V.,** Identification of embryonic motoneurons *in vitro:* their survival is dependent on skeletal muscle, *Brain Res.*, 190, 537, 1980.

70. **Henderson, C. E., Huchet, M., and Changeux, J.-P.,** Neurite outgrowth from embryonic chicken spinal neurons is promoted by media conditioned by muscle cells, *Proc. Natl. Acad. Sci. U.S.A.*, 78, 2625, 1981.
71. **Smith, R. G. and Appel, S. H.,** Extracts of skeletal muscle increase neurite outgrowth and cholinergic activity of fetal rat spinal motor neurons, *Science*, 219, 1079, 1983.
72. **Henderson, C. E., Huchet, M., and Changeux, J.-P.,** Denervation increases a neurite-promoting activity in extracts of skeletal muscle, *Nature (London)*, 302, 609, 1983.
73. **Slack, J. R. and Pockett, S.,** Motor neurotrophic factor in denervated adult skeletal muscle, *Brain Res.*, 247, 138, 1982.
74. **Popiela, H., Porter, T., Beach, R. L., and Festoff, B. W.,** Peripheral nerve extract promotes long-term survival and neurite outgrowth in cultured spinal cord neurons, *Cell Mol. Neurobiol.*, 4, 67, 1984.
75. **O'Connor, T. M. and Wyttenbach, C. R.,** Cell death in the embryonic chick spinal cord, *J. Cell Biol.*, 60, 448, 1974.
76. **Hamburger, V.,** Cell death in the development of the lateral motor column of the chick embryo, *J. Comp. Neurol.*, 160, 535, 1975.
77. **Massoulié, J. and Bon, S.,** The molecular forms of cholinesterase and acetylcholinesterase in vertebrates, *Ann. Rev. Neurosci.*, 5, 57, 1982.
78. **Brimijoin, S.,** Molecular forms of acetylcholinesterase in brain, nerve, and muscle: nature, localization, and dynamics, *Prog. Neurobiol.*, 21, 291, 1983.
79. **Bacou, F., Vigneron, P., and Massoulié, J.,** Acetylcholinesterase forms in fast and slow rabbit muscle, *Nature (London)*, 296, 661, 1982.
80. **Gisiger, V. and Stephens, H.,** Acetylcholinesterase content in both motor nerve and muscle is correlated with twitch properties, *Neurosci. Lett.*, 31, 301, 1982.
81. **Bacou, F., Vigneron, P., and Couraud, J.-Y.,** Retrograde effect of muscle on forms of acetylcholinesterase in peripheral nerves, *J. Neurochem.*, 45, 1178, 1985.
82. **Ranish, N. and Ochs, S.,** Fast axoplasmic transport of acetylcholinesterase in mammalian nerve fibers, *J. Neurochem.*, 19, 2641, 1972.
83. **Brimijoin, S. and Wiermaa, M. J.,** Rapid orthograde and retrograde axonal transport of acetylcholinesterase as characterized by the stop-flow technique, *J. Physiol. (London)*, 285, 129, 1978.
84. **Skau, K. A. and Brimijoin, S.,** Release of acetylcholinesterase from rat hemidiaphragm preparations stimulated through the phrenic nerve, *Nature (London)*, 275, 224, 1978.
85. **Ochs, S.,** An historical introduction to the trophic regulation of skeletal muscle, this volume chap. 2.
86. **Musick, J. and Hubbard, J. I.,** Release of protein from mouse motor nerve terminals, *Nature (London)*, 237, 279, 1972.
87. **Brimijoin, S. and Carter, J.,** Turnover of the molecular forms to acetylcholinesterase in the rat diaphragm, *J. Neurochem.*, 38, 588, 1982.
88. **Hebb, C. O.,** Biosynthesis of acetylcholine in nervous tissue, *Physiol. Rev.*, 52, 918, 1972.
89. **Holtzman, E.,** The origin and fate of secretory packages, especially synaptic vesicles, *Neuroscience*, 2, 327, 1977.
90. **Luco, J. V. and Eyzaguirre, C.,** Fibrillation and hypersensitivity ot ACh in denervated muscle: effect of length of degenerating nerve fibers, *J. Neurophysiol.*, 18, 65, 1955.
91. **Redfern, P. and Thesleff, S.,** Action potential generation in denervated rat skeletal muscle. II. The action of tetrodotoxin, *Acta Physiol. Scand.*, 82, 70, 1971.
92. **Card, D. J.,** Denervation: sequence of neuromuscular degenerative changes in rats and the effect of stimulation, *Exp. Neurol.*, 54, 251, 1977.
93. **Harris, J. B. and Thesleff, S.,** Nerve stump length and membrane changes in denervated skeletal muscle, *Nature (London) New Biol.*, 236, 60, 1972.
94. **Albuquerque, E. X., Warnick, J. E., Sansone, F. M., and Onur, R.,** The effects of vinblastine and colchicine on neural regulation of muscle, *Ann. N.Y. Acad. Sci.*, 228, 224, 1974.
95. **Ranish, N. A., Dettbarn, W. D., and Wecker, L.,** Nerve stump length-dependent loss of acetylcholinesterase activity in endplate regions of rat diaphragm, *Brain Res.*, 191, 379, 1980.
96. **Davey, B. and Younkin, S. G.** Effect of nerve stump length on cholinesterase in denervated rat diaphragm, *Exp. Neurol.*, 59, 168, 1978.
97. **Fernandez, H. L., Duell, M. J., and Festoff, B. W.,** Neurotrophic control of 16S acetylcholinesterase at the vertebrate neuromuscular junction, *J. Neurobiol.*, 10, 441, 1979.
98. **Carlsen, R. C.,** The possible role of cyclic AMP in the neurotrophic control of skeletal muscle, *J. Physiol. (London)*, 247, 343, 1975.
99. **Uchitel, O. and Robins, N.,** On the appearance of acetylcholine receptors in denervated rat diaphragm, and its dependence on nerve stump length, *Brain Res.*, 153, 539, 1978.
100. **Slater, C. R.,** Time course of failure of neuromuscular transmission after motor nerve section, *Nature (London)*, 209, 305, 1966.
101. **Jones, R. and Vrbova, G.,** Two factors responsible for the development of denervation hypersensitivity, *J. Physiol. (London)*, 236, 517, 1974.

102. **Vrbova, G.,** Induction of an extra-junctional chemosensitive area in intact innervated muscle fibers, *J. Physiol. (London),* 191, 20P, 1967.
103. **Guth, L., Richman, E., Barrett, C., Warnick, J. E., and Albuquerque, E. X.,** The mechanism by which degenerating peripheral nerve produces extrajunctional acetylcholine sensitivity in mammalian muscle, *Exp. Neurol.,* 68, 465, 1980.
104. **Cangiano, A. and Lutzemberger, L.,** Partial denervation affects both denervated and innervated fibers in the mammalian skeletal muscle, *Science,* 196, 542, 1977.
105. **Tiedt, T. N., Albuquerque, E. X., and Guth, L.,** Degenerating fiber products do not alter physiological properties of adjacent innervated skeletal muscle fibers, *Science,* 198, 839, 1977.
106. **Albuquerque, E. X., Warnick, J. E., Tasse, J. R., and Sansone, F. M.,** Effects of vinblastine and colchicine on neural regulation of the fast and slow skeletal muscles of the rat, *Exp. Neurol.,* 37, 607, 1972.
107. **Hoffmann, W. W. and Thesleff, S.,** Studies of the trophic influence of nerve on skeletal muscle, *Eur. J. Pharmacol.,* 20, 256, 1972.
108. **Fernandez, H. L. and Ramirez, B. U.,** Muscle fibrillation induced by blockade of axoplasmic transport in motor nerves, *Brain Res.,* 79, 385, 1974.
109. **Cangiano, A. and Fried, J. A.,** The production of denervation-like changes in rat muscle by colchicine, without interference with axonal transport or muscle activity, *J. Physiol. (London),* 265, 63, 1977.
110. **Inestrosa, N. C., Ramirez, B. U., and Fernandez, H. L.,** Effects of denervation and of axoplasmic transport blockage on the *in vitro* release of muscle endplate acetylcholinesterase, *J. Neurochem.,* 28, 941, 1977.
111. **Fernandez, H. L. and Inestrosa, N. C.,** Role of axoplasmic transport in neurotrophic regulation of muscle endplate acetylcholinesterase, *Nature (London),* 262, 55, 1976.
112. **Hofmann, W. W. and Peacock, J. H.,** Post-junctional changes induced by partial interruption of axoplasmic flow in motor nerves, *Exp. Neurol.,* 41, 345, 1973.
113. **Lømo, T. and Slater, C. R.,** Control of acetylcholine sensitivity and synapse formation by muscle activity, *J. Physiol. (London),* 275, 391, 1978.
114. **Kauffman, F. C., Warnick, J. E., and Albuquerque, E. X.,** Uptake of [^{3}H]-colchicine from Silastic implants by mammalian nerves and muscles, *Exp. Neurol.,* 44, 404, 1974.
115. **Thesleff, S.,** Supersensitivity of skeletal muscle produced by botulinum toxin, *J. Physiol. (London),* 151, 598, 1960.
116. **Thesleff, S.,** Physiological effects of denervation of muscle, *Ann. N.Y. Acad. Sci.,* 228, 89, 1974.
117. **Purves, D. and Lichtman, J. W.,** Elimination of synapses in the developing nervous system, *Science,* 210, 153, 1980.
118. **Robert, E. D. and Oester, Y. T.,** Absence of supersensitivity to acetylcholine in innervated muscle subjected to a prolonged pharmacologic nerve block, *J. Pharmacol. Exp. Ther.,* 174, 133, 1970.
119. **Lømo, T. and Rosenthal, J.,** Control of ACh sensitivity by muscle activity in the rat, *J. Physiol. (London),* 221, 493, 1972.
120. **Lavoie, P. A., Collier, B., and Tenenhouse, A.,** Comparison of α-bungarotoxin binding to skeletal muscles after inactivity or denervation, *Nature, (London),* 260, 349, 1976.
121. **Pestronk, A., Drachman, D. B., Stanley, E. F., Price, D. L., and Griffin, J. W.,** Cholinergic transmission regulates extrajunctional acetylcholine receptors, *Exp. Neurol.,* 70, 690, 1980.
122. **Bray, J. J., Hubbard, J. I., and Mills, R. G.,** The trophic influence of tetrodotoxin-inactive nerves on normal and reinnervated rat skeletal muscles, *J. Physiol. (London),* 297, 479, 1979.
123. **Jones, R. and Vrbova, G.,** Effect of muscle activity on denervation hypersensitivity, *J. Physiol. (London),* 210, 144P, 1970.
124. **Jones, R. and Vrbova, G.,** Can denervation hypersensitivity be prevented?, *J. Physiol. (London),* 217, 67P, 1971.
125. **Drachman, D. B. and Witzke, F.,** Trophic regulation of acetylcholine sensitivity of muscle: effect of electrical stimulation, *Science,* 176, 514, 1972.
126. **Cohen, S. A. and Fischbach, C. D.,** Regulation of muscle acetylcholine sensitivity by muscle activity in cell culture, *Science,* 181, 76, 1973.
127. **Purves, D. and Sakmann, B.,** The effect of contractile activity on fibrillation and extra-junctional acetylcholine-sensitivity in rat muscle maintained in organ culture, *J. Physiol. (London),* 237, 157, 1974.
128. **Lømo, T. and Westgaard, R. H.,** Control of ACh sensitivity in rat muscle fibers, *Cold Spring Harbor Symp. Quant. Biol.,* 40, 263, 1975.
129. **Gisiger, V. and Stephens, H.,** Correlation between the acetylcholinesterase content in motor nerves and their muscles, *J. Physiol. (Paris),* 78, 720, 1983.
130. **Groswald, D. E. and Dettbarn, W.-D.,** Characterization of acetylcholinesterase molecular forms in slow and fast muscle of the rat, *Neurochem. Res.,* 8, 983, 1983.
131. **Cole, B. G. and Gardiner, P. F.,** Does electrical stimulation of denervated muscle, continued after reinnervation, influence recovery of contractile function?, *Exp. Neurol.,* 85, 52, 1984.

132. **Younkin, S. G. and Younkin, L. H.,** Trophic regulation of skeletal muscle acetylcholinesterase, this volume, chap. 4.

133. **Oh, T. H., Johnson, D. D., and Kim, S. U.,** Neurotrophic effect on isolated chick embryo muscle in culture, *Science,* 178, 1298, 1972.

134. **Oh, T. H.,** Neurotrophic effect: characterization of the nerve extract that stimulates muscle development in culture, *Exp. Neurol.,* 46, 432, 1975.

135. **Oh, T. H.,** Neurotrophic effect of sciatic nerve extract on muscle development in culture, *Exp. Neurol.,* 50, 376, 1976.

136. **Markelonis, G. J. and Oh, T. H.,** A protein fraction from peripheral nerve having neurotrophic effects on skeletal muscle cells in culture, *Exp. Neurol.,* 58, 285, 1978.

137. **Stevens, W. F., Hooisma, J., Slaaf, D. W., and Meeter, E.,** Neurotrophic influence in tissue culture, *Prog. Brain Res.,* 48, 31, 1978.

138. **Markelonis, G. J., Oh, T. H., Eldefrawi, M. E., and Guth, L.,** Sciatin: a myotrophic protein increases the number of acetylcholine receptors and receptor clusters in cultured skeletal muscles, *Dev. Biol.,* 89, 353, 1982.

139. **Markelonis, G. J. and Oh, T. H.,** Purification of sciatin using affinity chromatography on concanavalin-A agarose, *J. Neurochem.,* 37, 95, 1981.

140. **Markelonis, G. J., Bradshaw, R. A., Oh, T. H., Johnson, J. L., and Bates, O. J.,** Sciatin is a transferrin-like polypeptide, *J. Neurochem.,* 39, 315, 1982.

141. **Markelonis, G. J. and Oh, T. H.,** A sciatic nerve has a trophic effect on development and maintenance of skeletal muscle cells in culture, *Proc. Natl. Acad. Sci. U.S.A.,* 76, 2470, 1979.

142. **Davis, H. L. and Heinicke, E. A.,** Prevention of denervation atrophy in muscle: mammalian neurotrophic factor is not transferrin, *Brain Res.,* 309, 293, 1984.

143. **Younkin, S. G., Brett, R. S., Davey, B., and Younkin, L. H.,** Substances moved by axonal transport and released by nerve stimulation have an innervation-like effect on muscle, *Science,* 200, 1292, 1978.

144. **Davey, B., Younkin, L. H., and Younkin, S. G.,** Neural control of skeletal muscle cholinesterase: a study using organ-cultured rat muscle, *J. Physiol. (London),* 289, 501, 1979.

145. **Fernandez, H. L., Patterson, M. R., and Duell, M. J.,** Neurotrophic control of 16S acetylcholinesterase from mammalian skeletal muscle in organ culture, *J. Neurobiol.,* 11, 557, 1980.

146. **Hall, Z.,** Multiple forms of acetylcholinesterase and their distribution in endplate and non-endplate regions of rat diaphragm muscle, *J. Neurobiol.,* 4, 343, 1973.

147. **Fernandez, H. L. and Donoso, J. A.,** The relevance of axoplasmic transport in neurotrophic functions, in *Axoplasmic Transport,* Iqbal, Z., Ed., CRC Press, Boca Raton, Fla., 1986, 209.

148. **Davis, H. L. and Kiernan, J. A.,** Neurotrophic effects of sciatic nerve extract on denervated extensor digitorum longus muscle in the rat, *Exp. Neurol.,* 69, 124, 1980.

149. **Davis, H. L., Heinicke, E. A., Cook, R. A., and Kiernan, J. A.,** Partial purification from mammalian peripheral nerve of a trophic factor that ameliorates atrophy of denervated muscle, *Exp. Neurol.,* 89, 159, 1985.

150. **Davis, H. L. and Kiernan, J. A.,** Effect of nerve extract on number of acetylcholine receptors in denervated muscles of rats, *Exp. Neurol.,* 83, 108, 1984.

151. **Busis, N. A., Daniels, M. P., Bauer, H. C., Pudimat, P. A., Sonderegger, P., Schaffner, A. E., and Nirenberg, M.,** Three cholinergic neuroblastoma hybrid cell lines that form few synapses on myotubes are deficient in acetylcholine receptor aggregation molecules and large dense core vesicles, *Brain Res.,* 324, 201, 1984.

152. **Nirenberg, M., Wilson, S., Higashida, H., Rotter, A., Krueger, K., Busis, N., Ray, R., Kenimer, J. G., and Adler, M.,** Modulation of synapse formation by cyclic adenosine monophosphate, *Science,* 222, 794, 1983.

Chapter 4

TROPHIC REGULATION OF SKELETAL MUSCLE ACETYLCHOLINESTERASE

Steven G. Younkin and Linda H. Younkin

TABLE OF CONTENTS

I. INTRODUCTION

It has long been known that motor nerve, in addition to initiating action potentials and contraction in skeletal muscle, also has a long-term trophic influence on many functionally important muscle proteins. The trophic influence of nerve on skeletal muscle is, in fact, one of the most intensively studied examples of the general phenomenon of trophism, wherein neurons cause long-term molecular modifications in the cells they innervate (see Chapters 1 and 2). One enzyme that is clearly influenced by innervation is acetylcholinesterase (AChE). In embryonic skeletal muscle, local accumulations of AChE are not present prior to innervation, but develop at newly formed synapses as the muscle becomes innervated.[1-5] In adult mammalian skeletal muscle, denervation causes a rapid and dramatic decrease in both the AChE specifically associated with endplates and that found outside endplates.[6-10] Reinnervation of adult mammalian skeletal muscle at either the original or at an ectopic site causes a reappearance of both endplate and non-endplate AChE.[11,12]

In studying trophism, it is appropriate to focus on AChE for several reasons:

1. AChE is a functionally important enzyme. It is crucial for normal transmission at cholinergic synapses, where it rapidly hydrolyzes released acetylcholine (ACh), thereby terminating transmission.
2. AChE is present in high concentration at endplates, but in terms of absolute values the majority of the enzyme is in the non-endplate region. Since innervation profoundly influences both endplate and non-endplate AChE, AChE is a suitable molecular marker for studying neural regulation of junctional and extrajunctional proteins and for gaining insight into the relationship between these two pools.
3. AChE is not a single protein. The enzyme exists in several molecular forms, and all forms are neurally regulated. There are globular forms, which are secreted; sarcolemmal globular forms; and asymmetric forms, which appear to be associated with the synaptic basal lamina that ensheathes each muscle fiber. Thus, one can gain insight into the neural regulation of several major classes of proteins by systematically studying the various forms of AChE.

The influence of nerve on muscle AChE must begin with signals that pass between nerve and muscle. These signals must (through what may be a very complicated series of molecular events) alter the metabolism of AChE somewhere between the production of mRNA and the degradation (or secretion) of the enzyme so that ultimately the pattern of AChE molecular forms found in innervated muscle is generated. In our laboratory, we have investigated the signals that mediate the influence of innervation on skeletal muscle AChE and the changes in AChE metabolism that these signals produce. At present, little is known about the intermediary events that are involved in the regulation of AChE metabolism by neural signals (see Chapter 11).

In this chapter, we discuss the results of our experimentation and focus, in particular, on recent studies in cultured embryonic rat myotubes which have proven to be a good model system for evaluating the influence of electromechanical activity on AChE metabolism. Much of the related literature is cited, but we do not attempt to cover it comprehensively.

II. SIGNALS THAT MEDIATE THE INFLUENCE OF MOTOR NERVE ON SKELETAL MUSCLE ACETYLCHOLINESTERASE

Several lines of evidence indicate that the influence of nerve on non-endplate AChE is mediated primarily by the activity (electrical and/or mechanical) exerted by nerve on muscle and that the influence of nerve on endplate-specific AChE is mediated in part by activity and by a mechanism which is activity independent (see Chapter 5). Leg immobilization, which causes an 85 to 95% decrease in electromyographic activity, produces a significant decrease in non-endplate AChE,[13] but has little effect on endplate AChE.[14] Blockade of muscle activity with botulinum toxin (BT) causes a decrease in non-endplate AChE equivalent to that caused by denervation, but causes a decrease in endplate AChE which is only 50% of that seen after denervation.[8] Direct electrical stimulation of denervated rat soleus muscle markedly increases AChE.[15] The molecular forms induced in this preparation depend on the pattern of stimulation, fast patterns producing AChE forms similar to those in fast muscles and slow patterns producing AChE forms similar to those in slow muscles (see Chapter 5).[15] In developing neuromuscular systems both in vitro[16] and in vivo,[17] direct electrical stimulation can at least partially substitute for the effects of innervation on the development of endplate AChE. In cultured myotubes derived from 18-day-old rat embryos, Rieger et al.[18] have shown that the 16S (A12) form of AChE appears shortly after fibrillation begins and that blockade of fibrillatory activity with tetrodotoxin (TTX) reversibly inhibits the appearance of this form. Studies in our laboratory,[10] and in others,[19,20] examining the effect of nerve stump length on AChE in denervated muscles have shown that nerve stump length has a significant effect on endplate AChE (endplate AChE decreases more rapidly in muscle with a short nerve stump), but has no effect on non-endplate AChE. The influence of nerve stump length on endplate AChE cannot be attributed to a change in muscle activity because muscles become inactive at the time of nerve section regardless of nerve stump length.

To account for such activity-independent influence of nerve on muscle AChE, several authors proposed that AChE might be influenced by trophic substances(s) moved in nerve by axoplasmic transport and delivered to muscle (see Chapters 2 and 3).[21-25] A number of early studies showed that the AChE in cultured muscle could be influenced by a variety of neural explants and extracts.[22,23,26-30] Those studies raised the possibility that cultured muscle could be used to assay the trophic substance(s) normally delivered to muscle by nerve to maintain AChE, but they were difficult to interpret because there was little evidence showing that the changes induced were due to a physiologically relevant trophic substance(s) and not to a nonspecific substance(s) (see Chapter 6).

Physiologically relevant trophic substances(s) influencing skeletal muscle AChE should be (1) present and moved by axoplasmic transport in nerve; (2) released from nerve; and (3) present in innervated muscle (within intramuscular nerve and possibly within the muscle itself). In addition, one would expect the effect(s) of these substances to be exerted primarily on endplate as opposed to non-endplate AChE (see Chapter 1).

In the late 1970s[31,32] we reported the results of an extensive series of experiments in which rat muscles denervated in vivo for several days and then placed in organ culture were used to assay the substance(s) influencing muscle AChE. These experiments provided evidence that the substance(s) which cause an innervation-like increase in muscle AChE is (1) present in nerve; (2) moved by axoplasmic transport; (3) present in innervated muscle, but undetectable in denervated muscle; and (4) released from nerve upon stimulation. Such substance(s) continued to be released when muscle contraction was blocked by adding *d*-tubocurarine (*d*-TB) to the conditioned medium, but release of the substance(s) was significantly reduced upon supplementation of the medium with magnesium. Additional experimentation showed that the substance(s) released from nerve by stimulation increase AChE in the region of the diaphragm muscle where endplates are found, but not in the endplate-free region. These experiments showed that the AChE in organ-cultured rat muscles could be increased by a substance(s) meeting criteria for physiologic relevance. Taken together they provided strong evidence that junctional AChE is at least partially maintained by a substance(s) which is normally conveyed by axoplasmic transport and delivered to muscle. This concept was confirmed and extended in a subsequent study by Fernandez et al.[33] on rat anterior gracilis muscles maintained for 48 hr in organ culture. The latter study showed that a nerve extract could partially prevent the denervation-induced decrease in the 16S form of AChE that is found in the endplate region of those muscles.

The chemical nature of the substance(s) influencing skeletal muscle AChE is currently uncertain (see Chapters 6 and 7). Our attempts to purify and characterize the substance(s) involved have been only partially successful despite considerable effort.[34] The major difficulty has been the bioassay, which was well suited for our initial analysis, but cumbersome as a basis for purification.

III. INFLUENCE OF INNERVATION OF ACETYLCHOLINESTERASE METABOLISM

Our hypotheses concerning the influence of innervation on AChE metabolism were generated initially form quantitative data on the effect of denervation on the molecular forms of AChE in rat diaphragm. There are at least six structurally distinct forms of AChE.[35] Three AChE forms are globular proteins, and three are asymmetric. The three globular forms are monomers, dimers, and tetramers of the basic catalytic subunit, and are referred to as G1, G2, and G4, respectively. The three asymmetric forms consist of an elongated collagen-like tail with one, two, or three attached tetramers, and are referred to as A4, A8, and A12, respectively. Globular forms can be solubilized in low-ionic-strength buffer, but asymmetric forms require high-ionic-strength conditions. The structures of the various forms of mammalian AChE have been evaluated most carefully using enzyme derived from bovine superior cervical ganglion.[36,37] In adult rat skeletal muscle, the three forms that can be solubilized in low-ionic-strength buffer containing detergent have sedimentation coefficients of 4.0S, 6.5S, and 10.0S, which correspond to the G1, G2, and G4 forms in bovine superior ganglion. In turn, the three forms that require high-ionic-strength buffer for solubilization have sedimentation coefficients of 8.5S, 12.5S, and 16.0S, which correspond to the A4, A8, and A12 forms.

To evaluate the forms of AChE quantitatively,[38] diaphragms were dissected to separate the muscle into endplate-containing and endplate-free regions. These regions were then subjected to sequential extraction to isolate globular and asymmetric forms, and the extracts were applied to 5 to 25% sucrose gradients to evaluate individual asymmetric and globular forms. The nonextractable AChE that remained in the pellet after extraction of globular and asymmetric forms was assayed routinely and turned out to have interesting properties, as described below. In the same study,[38] the cellular localization of the forms of AChE was evaluated by using echothiophate (ECHO), a cationic AChE inhibitor that penetrates muscle fibers very slowly at low temperature, to selectively inactivate external AChE. ECHO diethylphosphorylates a serine in the active site of the enzyme, thereby causing its inactivation. Since the diethylphosphorylated enzyme is quite stable, the inhibition is essentially irreversible, but it can be rapidly and completely reactivated with 1-methyl-2-hydroxyiminomethylpyridinium (2-PAM). Under these conditions, the amount of free (intracellular), inactivated (external), and total enzyme in each extract was readily determined. By combining ECHO inactivation with our sequential extraction procedure, it was possible to obtain comprehensive quantitative data on the forms comprising endplate and non-endplate AChE and to gather information on their cellular localization.

A. Non-Endplate Acetylcholinesterase

In innervated diaphragm, non-endplate AChE is predominantly (82%) composed of globular forms. The 4S (G1) and 10S (G4) forms are the most abundant forms, but there is measurable 6.5S (G2) AChE. Asymmetric forms are also found in the non-endplate region. In fact, our analysis indicates that 48% of the total asymmetric AChE in the rat diaphragm is non-endplate enzyme. Control experiments, carried out using ^{125}I-α-bungarotoxin (α-

BTX) to label ACh receptors (AChR) clearly showed that the asymmetric enzyme in the non-endplate region of the diaphragm is not associated with stray endplates.

Our experiments using ECHO to selectively inactivate external AChE showed that 50% of non-endplate AChE is intracellular. The 4S form is predominantly (74%) intracellular, and this form comprises most of the intracellular non-endplate AChE, but 34% of the 10S form and 38% of the asymmetric AChE in the non-endplate region also remain active after segments of diaphragm are exposed to ECHO for 1 hr at 0°C.

To develop a working hypothesis concerning the effect of innervation on the metabolism of AChE, we denervated rat diaphragms and used the techniques described above to examine AChE forms 1, 2, 3, 7, and 14 days later.[39] In non-endplate regions, after 2 days of denervation, the 4S form decreased rapidly to a minimum of 38% of control, whereas the 10S form decreased more slowly, reaching 67% in the same time and a minimum of 33% after 7 days. Three mechanisms that could account for such early decrease in non-endplate AChE activity are decreased synthesis, increased secretion, and /or accelerated internalization and degradation of sarcolemmal AChE. The rapid decrease in the predominantly intracellular 4S form suggested that denervation was causing a rapid change in AChE intracellular metabolism and made it unlikely for increased internalization and degradation of sarcolemmal enzyme to be the mechanism involved. Several laboratories had reported [40,41] that AChE secretion, expressed as a percentage of total muscle AChE, increases in denervated mammalian muscle; thus, the possibility that the decrease in 4S enzyme was due to increased secretion was seriously considered. The predominant form of AChE secreted from denervated muscle is the 10S form; thus, the 4S (G1) AChE lost through secretion presumably would first be assembled into the 10S (G4) form. We examined organ-cultured diaphragm segments undergoing denervation-induced loss of AChE activity and determined that the released enzyme accounted for only a fraction of that lost from the muscle. This result suggested that increased secretion was not causing the loss of non-endplate AChE after denervation and favored decreased synthesis as the mechanism responsible. There is good evidence that the influence of innervation on non-endplate AChE is mediated by activity; therefore, these results led us to postulate that nerve-evoked muscle activity increases AChE synthesis.

To pursue this possibility, a method for evaluating AChE metabolism was required. Study of the metabolism of AChR had been aided enormously by the availability of ^{125}I-α-BTX, a toxin that binds selectively and essentially irreversibly to the AChR; thus, we sought a similar label for AChE. To be effective, the labeling compound had to (1) be specific for AChE; (2) be easily measured; (3) form a stable complex with AChE; and (4) have minimal effect on the metabolism of AChE. ECHO met these criteria reasonably well.

ECHO is not a radiolabel. As mentioned earlier, this compound diethylphosphorylates a serine in the active site of the enzyme, thereby inactivating it. Diethylphosphorylated enzyme is stable (see below), but it can be rapidly and quantitatively reactivated by treatment with 2-PAM. To measure the amount of labeled (diethylphosphorylated) AChE, one measures active (unlabeled) enzyme, reactivates the diethylphosphorylated (labeled) enzyme with 2-PAM, measures total AChE activity, and then subtracts unlabeled from total AChE activity to obtain labeled AChE activity. One is, in effect, measuring the AChE activity revealed by 2-PAM reactivation; thus, the tag is as specific as the AChE assay employed. One major problem encountered in any attempt to label the AChE in skeletal muscle is that there are very few of these molecules present; therefore, a tag like radioactively labeled diisopropylfluorophosphate (DFP) is unattractive not only because it is not specific, but also because it is difficult to detect. By using 2-PAM reactivated enzyme to evaluate labeling, one overcomes both of these difficulties. We have evaluated the stability of diethylphosphorylated enzyme by labeling muscle homogenate with ECHO, dialyzing overnight to remove free ECHO, and measuring the rate of reappearance of AChE.[42] In vitro at 37°C, AChE regenerates with a half-time of 27 hr. In vivo, regeneration probably occurs at this rate or more slowly. The complex is not as stable as one would like, but it is stable enough to be useful in evaluating any AChE metabolism that occurs at an appreciable rate. Any significant effect of ECHO on the metabolism of AChE should cause a change in total AChE activity. Total AChE is always measured after 2-PAM reactivation and, at the doses employed in the studies described below, we have observed no effect of ECHO in either the endplate or the non-endplate regions of diaphragm muscle.

AChE turnover was compared in 18 denervated (left) and innervated (right) rat diaphragms by injecting 0.20 mℓ of 0.24 mm ECHO intravenously through the tail vein. The loss of labeled (diethylphosphorylated) AChE was followed at various times up to 24 hr after labeling and measuring labeled enzyme using 2-PAM as described above. In the non-endplate region of the innervated right diaphragm, label was lost exponentially with a half-time of 13 hr. This represents a maximal rate for the turnover of non-endplate enzyme. If one corrects for the loss of label on the basis of the rate of reappearance of activity in homogenates, then non-endplate enzyme is found to turn over with a half-time of 26 hr. In the interval between 18 and 42 hr after denervation, the denervated left diaphragm lost considerable non-endplate AChE, but the rate of loss of labeled enzyme was identical in innervated and denervated muscles. The fact that non-endplate AChE decreased without an increase in turnover strongly suggested that denervation decreased AChE synthesis and provided firm support for our hypothesis that the activity setup in muscle by nerve increases AChE synthesis.

B. Endplate-Specific Acetylcholinesterase

In rat diaphragm muscle, the AChE specifically associated with endplates comprises 28% globular forms, 45% asymmetric forms, and 27% nonextractable AChE.[38] It has long been known that the 16S (A12) form of the enzyme is concentrated in the endplate region.[9] Our data show that the 12.5S form and nonextractable AChE are also concentrated in this region. Analysis of the AChE released from the nonextractable fraction by collagenase showed that it is composed of asymmetric forms.[38] Nonextractable AChE is almost completely inactivated by ECHO and thus appears to be a largely asymmetric enzyme that is tightly associated with the junctional basal lamina. Protease inhibitors were normally included in the buffers used to extract AChE. When these inhibitors were omitted a substantial fraction of the endplate-specific asymmetric forms was converted to globular forms. Since we cannot be certain that proteases are completely inhibited by the buffers we normally employ, it is possible that the globular forms which appear to be specifically associated with the endplate region are generated artifactually at the time of homogenization.

The functionally important AChE in the diaphragm is the external junctional enzyme that hydrolyzes released transmitter. Our analysis indicates that virtually all (84%) of the external endplate-specific AChE is composed of asymmetric forms (45% readily extractable and 39% nonextractable). It has been shown[43,44] that treatment of skeletal muscle with collagenase releases AChE and prolongs the falling phase of miniature endplate potentials. These observations have been interpreted to mean that the collagen-like tail of the asymmetric form anchors this enzyme in the basal lamina at the neuromuscular junction, where it is important for degrading the ACh released from nerve terminals. McMahan's group has developed a procedure in which all of the cells and the cell membranes in the frog cutaneous pectoris muscle are destroyed so that only basal lamina remains. Histochemical examination of the basal lamina shells that remain after this procedure has established[45] that there is AChE associated with the junctional basal lamina. Taken together, these data strongly suggest that the functionally important external AChE at endplates is virtually all asymmetric AChE that is associated with the basal lamina located between the motor nerve terminal and the post-junctional membrane (see Chapter 8).

In our hands, endplate-specific AChE is unaffected for about 24 hr after denervation and then decreases very rapidly, reaching 27% of normal 3 days after denervation. Our analysis of the individual asymmetric forms in the endplate region showed that the 16S (A12) form decreases very rapidly, the 12.5S (A8) form decreases more slowly, and the 8.5S (A4) form actually increases transiently as the larger forms decrease. This pattern suggests that denervation accelerates the degradation of endplate-specific asymmetric forms, causing the production of smaller asymmetric forms from the larger ones as well as the destruction of enzyme activity. On this basis, we postulated that innervation may decrease the rate at which asymmetric forms are degraded at the endplate.

This possibility was further pursued in the experimentation described above in which ECHO was used to evaluate AChE turnover in innervated and 18 hr-denervated rat diaphragm. There was considerable scatter in the labeled endplate-specific AChE measured in innervated muscle. Overall, the rate of loss was significantly slower than that of non-endplate label and occurred at about the rate at which label is lost in homogenates in vitro. From these results it appears that endplate-specific AChE is more stable than non-endplate enzyme, but in innervated muscle the turnover of endplate-specific enzyme is too slow to evaluate quantitatively by following diethylphosphorylated AChE. In the interval between 18 and 24 hr after denervation the rate of loss of labeled endplate-specific AChE was significantly faster in denervated than in innervated muscle.

The observation that endplate-specific AChE is more rapidly degraded in 18 hr-denervated than in innervated diaphragm supports, but does not prove, the hypothesis that innervation decreases the rate at which asymmetric forms are degraded at the endplate. Nerve terminal degeneration and the decrease in endplate-specific AChE occur at about the same time after denervation, and the external endplate-specific AChE that was labeled is located just below the nerve terminal. It is possible, therefore, that the accelerated degradation of endplate-specific AChE seen 1 to 2 days after denervation does not represent an ongoing increase in the turnover of endplate-specific AChE caused by removing the trophic influence of the nerve, but is a transient event caused by increased proteolysis associated with nerve terminal degeneration.

IV. ANALYSIS OF CULTURED CELLS

Analysis of several different culture systems has yielded valuable information about AChE metabolism. In this section, we briefly review the major studies that have been performed on AChE metabolism and then discuss our examination of cultured embryonic rat myotubes as a model system to study the influence of electromechanical activity on AChE metabolism.

A. Acetylcholinesterase Metabolism

Rotundo and Fambrough[46] have reported data from a series of experiments in which the metabolism of AChE in cultured embryonic chick myotubes was examined. In this system, most of the AChE that is produced is secreted as dimers. To measure cell surface AChE, these authors washed the myotubes to remove secreted enzyme, inhibited further secretion (low temperature), and then added [^{14}C]-ACh. AChE generates acetate at a rapid rate at 0°C, and ACh does not enter the myotubes at this temperature; thus, it was possible to measure surface AChE activity by simply measuring the rate at which the AChE on the surface of the intact myotubes generated labeled acetate. Using this and a second method in which surface enzyme was selectively protected by BW284c51 dibromide (BW) when cells were exposed to DFP, the authors showed that one third of the enzyme in chick myotubes is surface AChE and two thirds is intracellular. Chick myotubes contain no asymmetric AChE. The surface enzyme is a mixture of 7.1S (G_2) and 11.5S (G_4) AChE and behaves like an integral membrane protein. The intracellular enzyme is mostly the 7.1S (G_2) form.

Three different procedures[46] were used to evaluate AChE metabolism: (1) the recovery of total AChE, surface AChE, and secreted AChE was followed for up to 2 days after blocking virtually all AChE activity with DFP; (2) the loss of total AChE, surface AChE, and secreted AChE was followed for 8 to 12 hr after blocking protein synthesis with puromycin; and (3) surface and total AChE was labeled with [^{3}H]-DFP, and the loss of label was followed. Results from these experiments indicate that, in this system, (1) AChE is synthesized at a rate of about 20% of the total cell-associated enzyme per hour with most of the enzyme destined for secretion; (2) the intracellular pool fills (or empties) in 2 to 3 hr; (3) the time from synthesis to secretion and from synthesis to the appearance of surface enzyme is 2 to 3 hr; (4) surface enzyme accumulates at a rate of 2 to 3% of total surface AChE per hour; (5) surface AChE is degraded by a first-order process with a half-life of about 50 hr; and (6) there is a quantitative transfer of secreted AChE from the intracellular to the extracellular compartment without any detectable residence time on the plasma membrane. These results suggest that there is an assembly line type of process operating within chick myotubes which transports secreted and integral membrane forms of AChE to the surface over 2 to 3 hr. The half-life of the plasma membrane AChE is noteworthy because it is considerably longer than the 19-hr half-life of the AChR. It would seem, therefore, that a number of different glycoproteins in the plasma membrane can be regulated by different rates of degradation, as well as by different rates of synthesis.

AChR and AChE are transported to the cell surface with similar kinetics.[47] In a second series of experiments, Rotundo and Fambrough[48] showed that temperature reduction and a variety of drugs known to affect secretory processes had identical effects on AChE secretion and on the rate of accumulation of AChR. This result suggests that there is a single transport pathway in muscle cell responsible for the externalization of both secreted and integral membrane proteins.

More recently, Rotundo[49] has analyzed AChE synthesis in chick myotubes by radiolabeling newly synthesized peptides and characterizing them by immunoprecipitation and electrophoresis. He found that most of the newly synthesized polypeptides recognized by antibody were enzymatically inactive forms that were rapidly degraded. Rotundo[50] has also analyzed AChE assembly in primary cultures of quail myotubes after DFP inactivation using lectins that recognize sugars attached in the rough endoplasmic reticulum and Golgi apparatus, and has concluded that asymmetric forms of AChE are assembled in the Golgi apparatus.

Lazar et al.[51] evaluated AChE metabolism in T28 cells by labeling the enzyme with heavy amino acids. The T28 cells, G_1, G_4, and to a lesser extent G_2 forms are secreted. Of the cell-associated AChE, the G_1 form is largely intracellular, and the G_4 form is largely associated with the plasmalemma. In this system, 70 to 80% of the enzyme synthesized appears to enter a G_1 pool and is rapidly turned over, presumably through intracellular degradation. Less than 10% of the enzyme produced is secreted, and only a small fraction enters the cell-associated G_4 pool. The data obtained in that study did not permit the authors to determine the metabolic relationship between the G_1 and G_4 pool, but they did conclude that G_4 was produced either from inactive precursors or from a rapidly turning over G_1 pool through G_2.

There is considerable indirect evidence suggesting that the simpler forms of AChE may be precursors of the more complex forms. During embryonic development, the simpler forms of AChE appear before the more complex forms.[5,52-54] After irreversible inactivation of AChE with organophosphates, there is, in all systems studied to date, a rapid recovery of the smaller globular forms of AChE and a slower recovery of G_4 and asymmetric forms.[46,52,54,55] Secretion of G_4 AChE from rat diaphragm occurs at a rate too rapid to be explained by the slow turnover of cellular G_4 enzyme.[40,56] This observation suggests that secreted G_4 AChE may be assembled from the rapidly turning over G_1 enzyme.

Recently, we have performed experiments to examine directly the metabolic relationship between G_1 AChE and the larger G_4 and asymmetric forms in cultured embryonic rat myotubes.[57] In these cells, 85% of the 4S, 69% of the 10S, 59% of the 16S, and 56% of the 12.5S forms are intracellular.[58] Precursor-product relationships among the various forms of AChE are best examined through pulse-chase experimentation in which the putative precursor forms of AChE are selectively labeled. Traditional methods using radiolabeled amino acids cannot be employed to label AChE in rat myotubes because the enzyme is present only in trace amounts. To overcome this problem, we employed a procedure in which AChE was labeled by diethylphosphorylating the enzyme at its active site with paraoxon (PAX). As mentioned earlier, the rate of spontaneous reactivation of diethylphosphorylated AChE is 0.026 hr^{-1}; thus, the labeled enzyme is quite stable, but the diethylphosphate group is rapidly and quantitatively removed, and the enzyme reactivated by treatment with 2-PAM. Labeled AChE could, therefore, be followed as AChE that regains its enzymatic activity after 2-PAM treatment.

To perform pulse-chase experimentation, it was first necessary to define a discrete pool of the hypothesized precursor form (G_1) that could be selectively labeled with PAX and followed as G_4 forms were assembled. To do this, the AChE in myotube cultures was inactivated with 5 m*M* methanesulfonylfluoride (MSF), MSF was washed out, and the cells were allowed to recover. MSF inactivated over 97% of total cellular AChE, leaving a small pool of active enzyme that was 92% G_1 AChE. The majority of this small pool of G_1 enzyme probably represents enzyme synthesized during the short period of time required for MSF washout. One hour after MSF treatment a sizable pool of AChE had been synthesized that consisted almost entirely of G_1 enzyme. By 3 hr after washout 38% of total AChE had recovered as additional G_1 AChE was synthesized and appreciable G_4 and A12 forms appeared. On the basis of these data, it was decided to pulse-label the enzyme that appears in the first 40 to 55 min after MSF-treatment (virtually all G_1 AChE) and to chase for about 2 hr before looking for the incorporation of labeled G_1 AChE into G_4 and asymmetric forms.

Cultures that had been inactivated with MSF were exposed to medium containing 0.1 μ*M* PAX for 40 to 55 min and then washed. Half of the cultures were harvested at this point to evaluate labeled AChE, and half were incubated an additional 120 to 130 min to evaluate the incorporation of labeled G_1 AChE into the G_4 and asymmetric forms. To avoid continued labeling by the low concentration of PAX remaining during the chase period, we deliberately used a concentration of PAX that labeled only 75% of the newly synthesized AChE. Labeled AChE is evaluated as the difference in enzyme activity before and after 2-PAM reactivation. Thus, unlabeled activity represents a kind of background noise above which the labeled enzyme signal is measured. To improve the signal-to-noise ratio, we reduced background noise by treating both the zero-time and 120 to 130 min cultures with MSF to inactivate unlabeled AChE.

In our first experimental series, four identical pulse-chase tests were performed. In each experiment, cultures were treated with MSF to inactivate AChE, washed, pulse-labeled for 55 min, washed again, and either evaluated immediately or chased for 120 min. Fractions (S1, S2, and S3) containing globular forms and the remaining asymmetric and nonextractable AChE (H4) were prepared utilizing the sequential extraction procedure previously described. Individual globular forms were then separated by velocity sedimentation on sucrose gradients. All fractions (S1, S2, S3, and H4) were assayed before and after 2-PAM reactivation, as were the globular and the asymmetric forms.

In our initial series of four experiments, there was a large pool of G_1 AChE labeled during the 55-min labeling period, a significant ($p < 0.01$) 55% decrease in labeled G_1 AchE during the 120-min chase, a significant ($p < 0.05$) 36% increase in labeled G_4 AChE during the chase, and a significant ($p < 0.025$) 36% increase in labeled asymmetric/nonextractable AChE during the chase. Figure 1 shows the globular forms present immediately after labeling (Figure 1A) and following a 120-min chase (Figure 1B). Each plot shows the activity both before (△) and after (x) 2-PAM reactivation. The gradients shown are the average of the gradients obtained in the four separate experiments. Labeled globular AChE, calculated from Figure 1 as the difference in AChE activity before and after 2-PAM reactivation, is shown

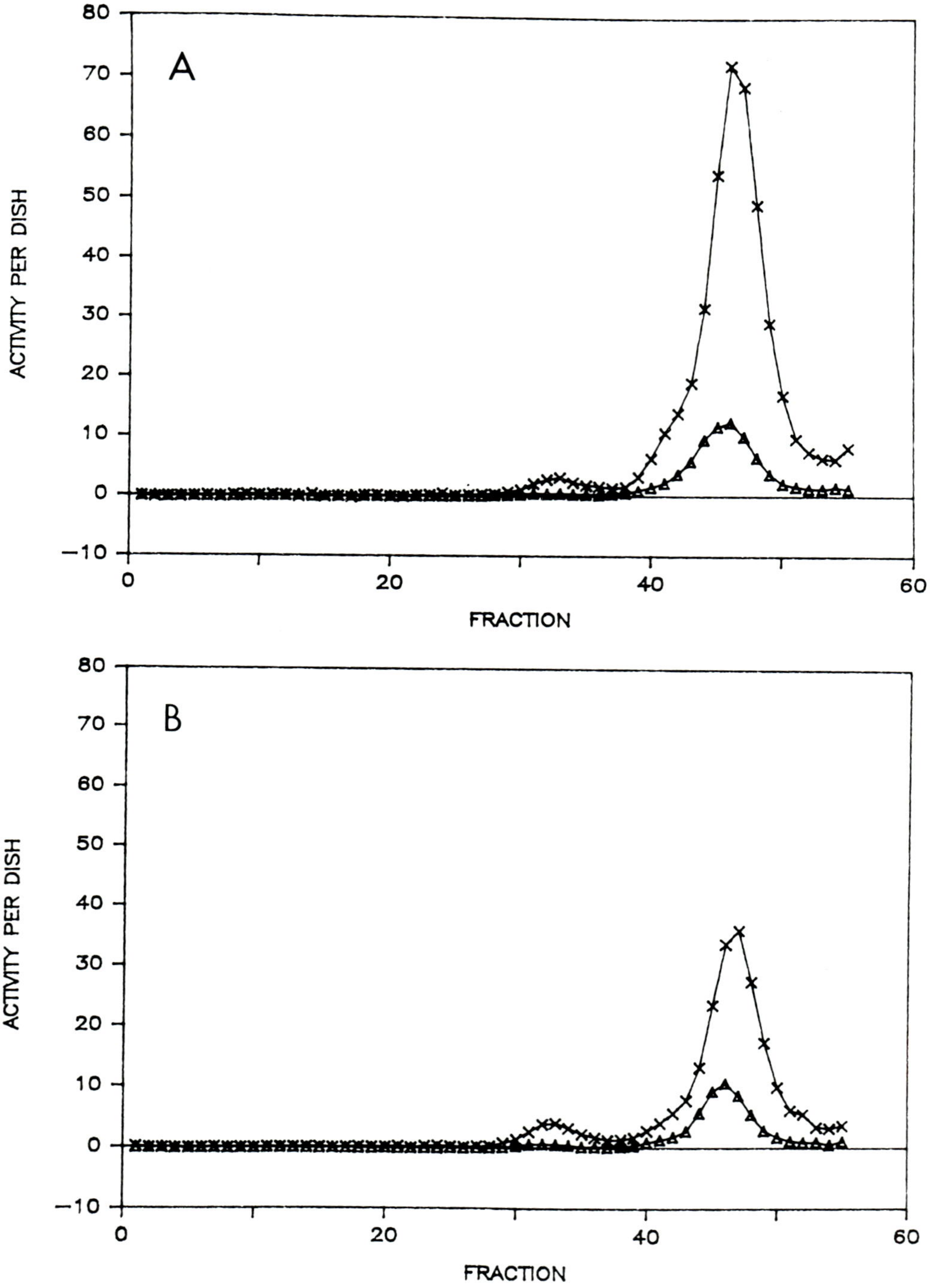

FIGURE 1. Measurement of labeled globular AChE forms. A pulse-chase experiment was performed as described in the text. Shown are globular forms present in the S1 fraction from the sequential extraction procedure before (△) and after (x) 2-PAM reactivation of the samples. (A) Forms present immediately after labeling with PAX; (B) Forms present 120 min after labeling. The forms shown in each panel have sedimentation coefficients of 10S and 4S (from left to right) and are G_4 and G_1 forms, respectively. In each panel the difference between the gradients represents labeled AChE (these differences are shown in Figure 2). The gradients shown represent the average of the gradients obtained in four separate experiments. The total AChE activity (picomoles of [^{3}H]-ACh hydrolyzed per minute) shown on each gradient is the activity present in the S1 fraction of a single dish of myotubes.

in Figure 2. The data in Figures 1 and 2 establish that the reproducibility of the pulse-chase paradigm and the resolution of G_4 AChE were sufficient to demonstrate that labeled G_4 enzyme increases during the chase period as the amount of labeled G_1 AChE declines.

In a second series in which individual asymmetric forms were additionally examined, five pulse-chase experiments were performed. The changes observed were similar to those seen in the initial series: labeled G_1 AChE decreased 55% during the 130-min chase, whereas labeled G_4 enzyme increased 58%, labeled asymmetric AChE increased 80%, and labeled nonextractable AChE increased 43%. Overall, in the course of these two series, we observed a marked decrease in labeled G_1 AChE and an increase in labeled G_4 and asymmetric/nonextractable AChE during the chase in each of nine consecutive experiments. These data establish that active monomeric AChE is assembled into larger multimeric forms.

The changes that occurred in individual asymmetric forms are shown in Figures 3 and 4. Figure 3 shows the asymmetric forms present immediately after labeling (Figure 3A) and following a 130-min chase (Figure 3B). Each plot shows the activity both before (△) and after (x) 2-PAM reactivation. The gradients shown are the average of the gradients obtained in five separate experiments. Labeled asymmetric AChE, calculated from Figure 3 as the difference in AChE activity before and after 2-PAM reactivation, is shown in Figure 4A. Shown in Figure 4B for the purpose of identifying individual forms is the gradient profile of asymmetric forms obtained from control cells. Quantitatively the amount of labeled 16S AChE (gradient fractions 14 to 23) increased from 1.1 to 4.2 (pmol $[^3H]$-ACh hydrolyzed per minute per culture dish), an increase of 380% ($p < .05$). In addition, there was a large increase in labeled AChE sedimenting at about 12.5S (gradient fractions 24 to 27) and 10S (gradient fractions 28 to 40). There was essentially no change in the amount of labeled AChE sedimenting at about 4S (gradient fractions 41 to 51).

The A12 form of AChE is of particular interest because it is a functionally important enzyme localized in high concentration at the mammalian neuromuscular junction. Our data show that this form is assembled from enzymatically active globular precursors (Figure 4), but they do not permit us to determine whether assembly occurs directly from monomers or from tetrameric and/or dimeric forms. In this study, the labeled monomer that appeared in more complex forms during the chase period represented less than 15% of the labeled monomer lost. It is apparent, therefore, that most G_1 AChE is not destined for assembly into more complex cell-associated forms. In separate experiments on embryonic rat myotubes cultured for 7 days we observed that 28% of the AChE that turns over in 2 hr appears in the medium.[59] Thus, most of the active enzyme produced in rat myotubes is either degraded or inactivated, and appreciable fraction is secreted, and only a small fraction is assembled into G_4 and asymmetric forms. Comparison of the various studies that have been performed, each using a different method, shows that AChE metabolism in rat myotubes, chick myotubes, and T28 cells is similar in many respects. In each system there is rapid loss of newly synthesized enzyme. In the case of chick myotubes, inactive molecules (which could not be detected in the present study or in the study of T28 cells) are rapidly degraded. In the case of T28 cells and rat myotubes, most of the enzyme synthesized is G_1 enzyme that is rapidly lost through degradation or inactivation. Thus, rapid loss of newly synthesized AChE is the rule, although degradation may occur at different levels in the different cell types. In each system, there is significant secretion of AChE, and only a small fraction of the enzyme produced is cell associated multimeric forms. In the case of rat myotubes, it is clear that multimeric forms are assembled from active monomeric AChE.

Our results established that multimeric forms of AChE are assembled from G_1 precursor, but they do not precisely define the pathway for assembly of G_4 and asymmetric forms and are compatible with a number of models. The multiple forms of AChE might be produced through a process in which monomers derived from a single mRNA are assembled into tetramers that are then assembled into asymmetric forms with modification occurring at

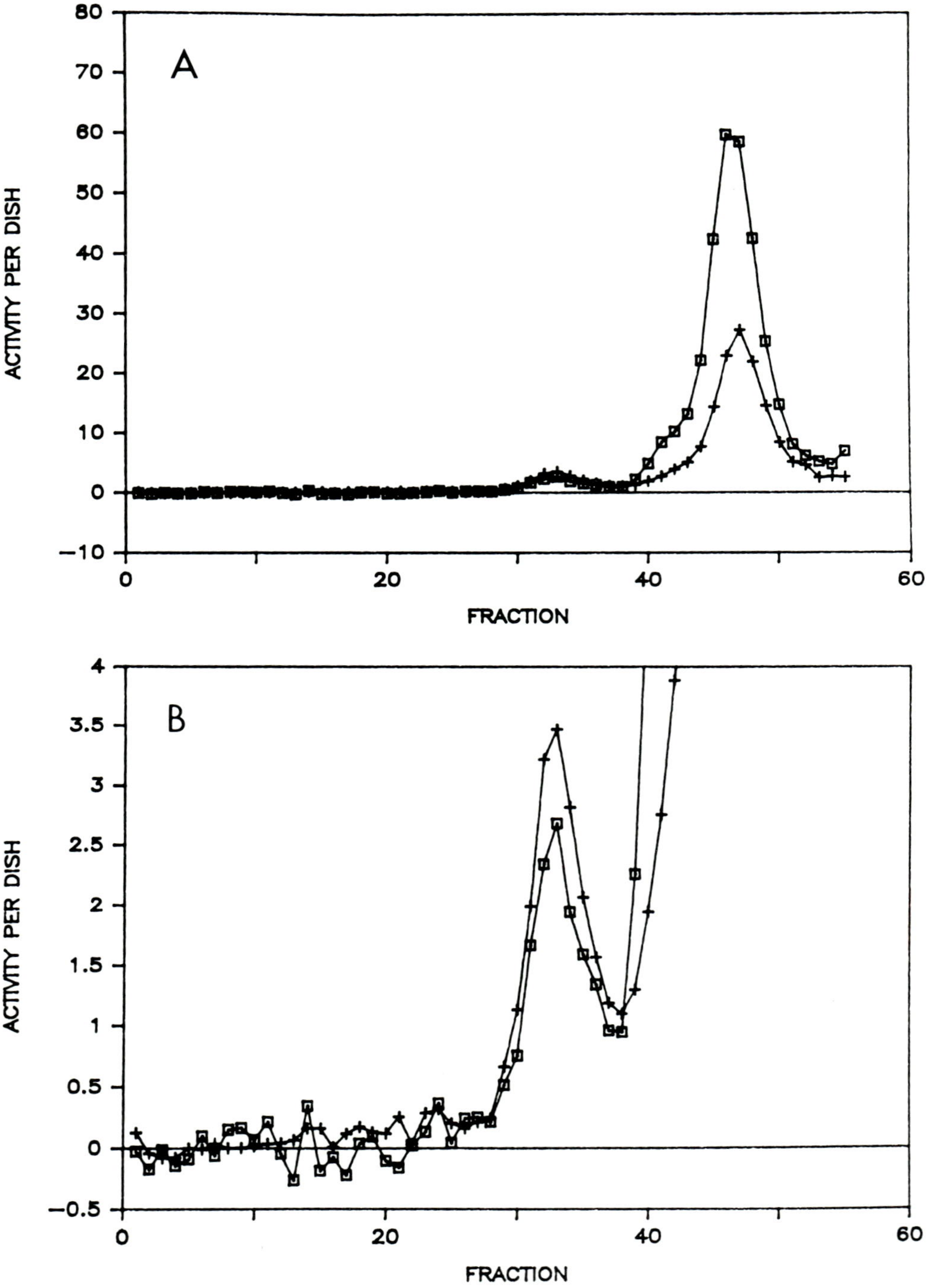

FIGURE 2. Labeled globular AChE. (A) The differences between the gradients of unlabeled and total (2-PAM-reactivated) AChE shown in Figure 1 are plotted. These differences represent labeled AChE present immediately after the labeling period (+) and 120 min later (□). The forms have sedimentation coefficients of 10S and 4S (from left to right) and are G_4 and G_1 forms, respectively. (B) The same data are plotted on an expanded scale to show the G_4 form better. The total AChE activity (picomoles of [^{3}H]-ACh hydrolyzed per minute) shown on each gradient is the labeled activity in the S1 fraction of a single dish of myotubes.

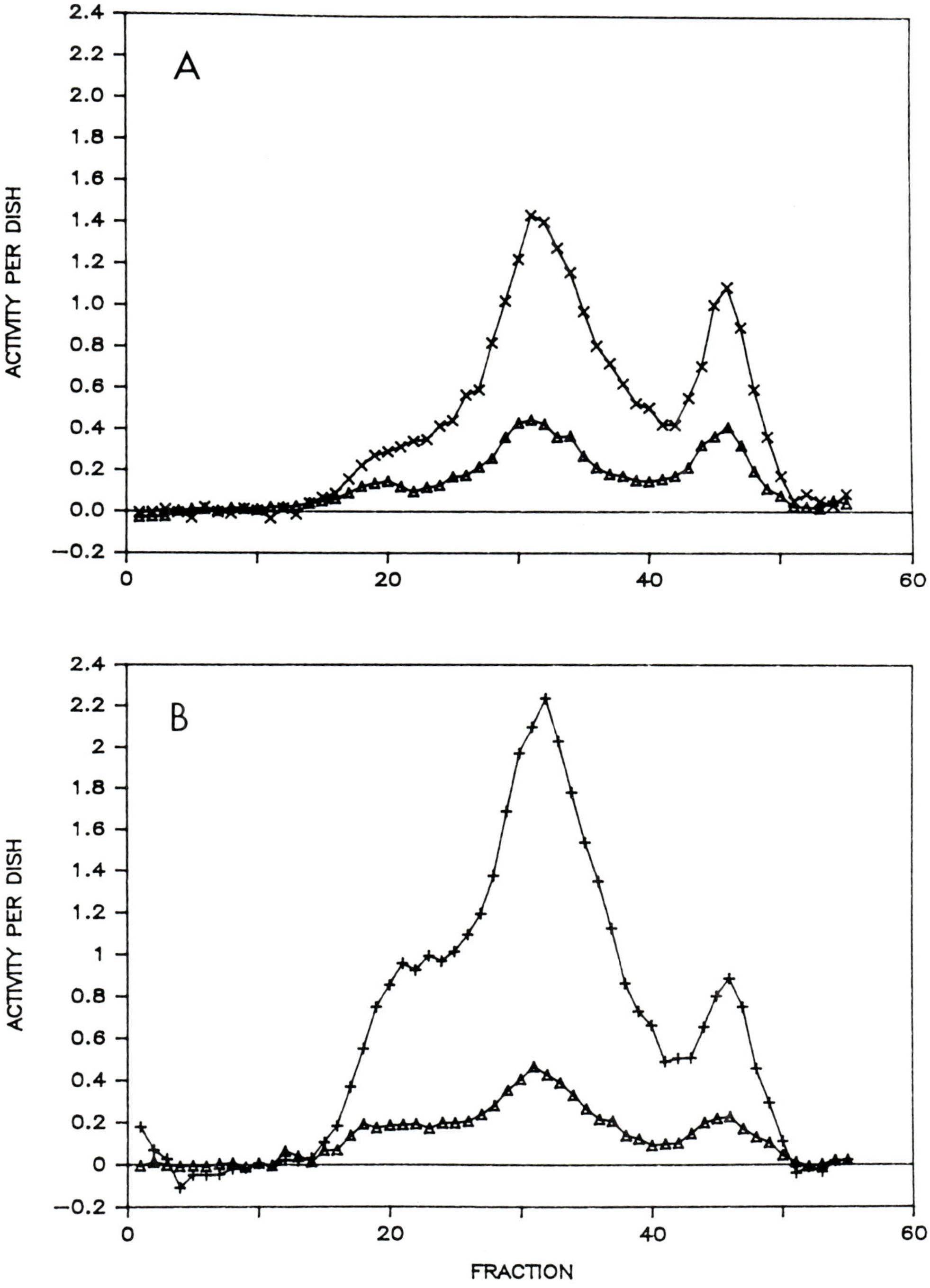

FIGURE 3. Measurement of labeled asymmetric AChE forms. A pulse-chase experiment was performed as described in the text. Shown are the forms present in the S4 fraction from the sequential extraction procedure before (△) and after (x) 2-PAM-reactivation of the samples. (A) Forms present immediately after labeling with PAX; (B) forms present 130 min after labeling. The major forms of AChE present have sedimentation coefficients of about 16S, 10S, and 4S (from left to right). In each panel the difference between the gradients represents labeled AChE (these differences are shown in Figure 4A). The gradients shown represent the average of five experiments. The total AChE activity (picomoles of [^{3}H]-ACh hydrolyzed per minute) shown on each gradient is the activity in the S4 fraction of a single dish of myotubes.

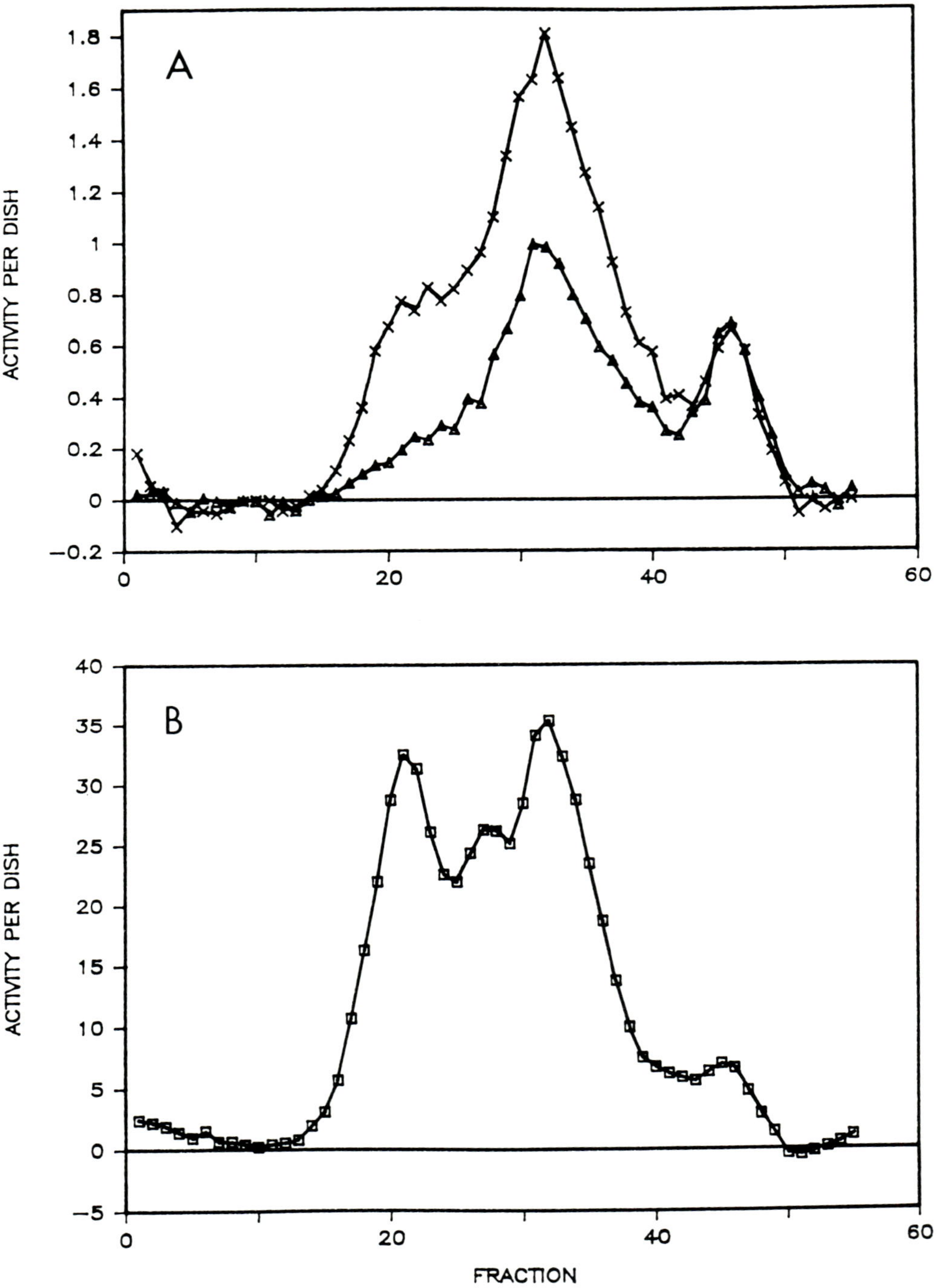

FIGURE 4. Labeled asymmetric AChE. (A) Differences between the gradients of unlabeled and total (2-PAM-reactivated) AChE shown in Figure 3. These differences represent labeled AChE present immediately after the labeling period (△) and 130 min later (x). The major forms have sedimentation coefficients of about 16S, 10S, and 4S (from left to right). (B) Average gradient profile of asymmetric forms obtained from control cells in five separate experiments. The total AChE activity (picomoles of [^{3}H]-ACh hydrolyzed per minute) shown on each gradient is the labeled activity in the S4 fraction of a single dish of myotubes.

selected points during posttranslational processing to give the various soluble, secreted, and membrane-associated globular forms. At the other extreme, assembly could be a process directed at the mRNA level in which many different mRNAs are produced, each generating a distinct monomer targeted for assembly into a particular form of AChE. Several recent reports suggest that multiple mRNAs may be involved in the generation of the various forms of AChE. Sikorav et al.[60] translated poly(A)-rich mRNA from *Torpedo marmorata* in a reticulocyte lysate system and produced at least two distinct precursors, suggesting that several mRNAs code for AChE in this species. Schumacher et al.[61] isolated a cDNA clone coding for AChE from a *Torpedo californía* electric organ λgt10 library and have shown that this clone hybridizes with multiple sizes of *Torpedo* RNA by northern blot analysis. Soreq et al.[62] studied mRNA fractions from primary gliomas, meningiomas, and embryonic human brain on sucrose gradients and found three broad bands of mRNA that produced AChE when injected into *Xenopus* oocytes. Velocity sedimentation of the AChE produced by mRNA-injected oocytes suggested that the different size classes of mRNA may produce different forms of AChE. There was however, no clear relationship between the forms of AChE produced in mRNA-injected oocytes and the forms produced in the original tissue.

B. Influence of Electromechanical Activity on Acetylcholinesterase Metabolism

Rieger et al.[18] examined the effect of spontaneous fibrillatory activity on AChE in cultured rat myotubes derived from 18-day-old embryos. They found that the 16S (A12) form of the enzyme appears shortly after fibrillation begins and that blockade of fibrillation with TTX reversibly inhibits the appearance of this form. Based on our analysis of AChE in innervated and denervated rat diaphragm, we developed the working hypothesis that AChE synthesis is increased by the activity set up in muscle by nerve. Cultured embryonic myotubes appeared to be an excellent model system in which to test this hypothesis.

We therefore compared spontaneously fibrillating myotubes (derived from 20-day rat embryos) with myotubes in which fibrillation was blocked on the fourth day of culture with 1 μM TTX.[63] Both globular and asymmetric forms of AChE were present on the fifth day of culture when myotubes were just beginning to fibrillate. Between days 5 and 8, the G_1, G_4, and A12 forms increased dramatically in fibrillating myotubes, and appreciable A8 AChE appeared. When fibrillation was blocked, the increase in the A12 and A8 forms was prevented, and the increases in the G_1 and G_4 forms were significantly impaired. At 8 days, fibrillating myotubes had 19 times more A12 AChE and over 4 times more G_1 and G_4 enzyme than did the nonfibrillating myotubes.

Careful measurement of the surface area occupied by myotubes on days 5 to 8 showed that the area occupied by fibrillating the TTX-treated myotubes was the same; thus, the effect of TTX on AChE was not due to gross impairment of morphological development. When TTX was removed from 7-day myotubes exposed on day 4, all forms of AChE recovered; thus, the effect of TTX was reversible. Titration of AChE-active sites in 8-day fibrillating and TTX-treated myotubes with O-ethyl-S^2-diisopropylaminoethyl methyl-phosphonothionate (MTP) showed that the reduction in AChE activity in TTX-treated myotubes was primarily, if not exclusively, due to a reduction in the number of AChE molecules.

To evaluate AChE metabolism in fibrillating and nonfibrillating myotubes, we labeled the enzyme in 7-day cultures with PAX as described above and then simultaneously evaluated synthesis (by following the newly synthesized active AChE) and turnover (by following the 2-PAM-reactivatable AChE) for 2 hr. Both globular and asymmetric forms of AChE were synthesized more rapidly in fibrillating than in nonfibrillating myotubes. There was significant turnover of globular forms in the 2-hr period during which measurements were made, but the rate of turnover was essentially identical in fibrillating and nonfibrillating cultures. These findings are consistent with our earlier results on adult muscle and strongly suggest that the synthesis of globular and asymmetric forms of AChE in innervated muscle is increased by muscle electromechanical activity.

Table 1
COMPARISON OF AChE PRODUCED IN OOCYTES BY mRNA PREPARED FROM MATCHED FIBRILLATING AND NONFIBRILLATING CULTURES

	AChE (pm ACh/min-dish)		mRNA (μg/dish)		AChE produced by mRNA (pm ACh/min-oocyte)	
Culture	−TTX	+TTX	−TTX	+TTX	−TTX	+TTX
1	3550	1031	34.8	23.5	0.47	0.05
2	2965	1192	25.2	29.8	0.70	0.08
3	4923	2295	41.6	40.0	0.83	0.39
4	6116	2113	20.8	18.9	0.27	0.00
Mean	4389	1658[a]	30.6	28.1	0.57	0.13[a]
SD	1225	553	8.1	7.9	0.21	0.15

[a] $p < 0.01$ by paired t test.

Soreq et al.[64] have shown that microinjected *Xenopus* oocytes can be used to evaluate AChE mRNA. In recent experiments we have employed this system to compare the level of AChE mRNA in fibrillating and nonfibrillating myotubes. In these experiments, the method of Chirgwin et al.[65] was employed to purify mRNA from 8- to 9-day fibrillating cultures and from matched cultures in which fibrillation was blocked on day 4 with TTX. To evaluate AChE mRNA, each of 20 *Xenopus* oocytes was injected with 50 nℓ of 2.0 mg/mℓ mRNA in 10 m*M*, pH 7.6, Hepes buffer, and each of 20 control oocytes was injected with buffer alone. After 22 hr of incubation, the 20 oocytes were homogenized in 0.3 mℓ of 0.1 *M* phosphate buffer containing 1% triton X-100, and 0.1 mℓ aliquots were assayed in triplicate for AChE in the presence of 0.91 m*M* *p*-chloromercuriphenyl sulfonic acid (*p*CMPSA). The *p*CMPSA in the assay buffer eliminated virtually all of the endogenous cholinesterase activity in the oocytes and made it possible to measure the AChE produced by myotube mRNA reliably. Table 1 shows the results of four separate experiments in which we compared the AChE produced in oocytes by mRNA prepared from matched fibrillating and nonfibrillating cultures. Fibrillating and nonfibrillating cultures had essentially identical levels of total mRNA, but mRNA from the fibrillating cultures produced significantly more AChE when injected into *Xenopus* oocytes. This result correlates well with our analysis of AChE metabolism by PAX labeling and indicates that electromechanical activity has an effect on the amount and/or the translational efficiency of AChE mRNA.

V. SUMMARY AND CONCLUDING REMARKS

Innervation has a profound influence on AChE of mammalian skeletal muscle. When mammalian skeletal muscle is denervated both non-endplate AChE and the AChE specifically associated with endplates decrease dramatically within several days. Available data indicate that the trophic influence of motor nerve on mammalian skeletal muscle AChE is mediated in part by the electromechanical activity evoked in muscle by nerve and in part by a mechanism which is activity-independent (see Chapter 5). The influence of innervation on non-endplate mammalian AChE appears to be mediated primarily, if not exclusively, through electromechanical activity. The effect of innervation on endplate AChE appears to be mediated in part by activity and in part by trophic substance(s) delivered to muscle by nerve (see Chapter 11).

In the rat diaphragm, non-endplate AChE is primarily sarcolemmal G_4 and intracellular G_1 forms. Following denervation the intracellular G_1 form decreased rapidly, reaching a minimum of 38% of normal 2 days after denervation, whereas the G_4 form decreased more slowly, reaching a minimum of 33% of normal 7 days after denervation. This observation, coupled with our finding that the turnover of ECHO-labeled non-endplate AChE is unaffected by denervation, led us to propose that AChE synthesis is increased by the electromechanical activity set up in muscle by nerve. In subsequent experiments, we tested this hypothesis in cultured embryonic rat myotubes by comparing AChE metabolism in spontaneously active (fibrillating) and TTX-treated (nonfibrillating) cultures. This experimentation, which was based on PAX-labeled AChE, revealed that both asymmetric and globular forms of AChE are synthesized more rapidly in fibrillating than in nonfibrillating myotubes. AChE turnover was not affected by spontaneous electromechanical activity in the 2-hr interval over which it was evaluated. Fibrillating and nonfibrillating myotubes had similar levels of total mRNA, but the mRNA from fibrillating myotubes produced significantly more AChE when injected into *Xenopus* oocytes. Taken together these results strongly support the concept that the electromechanical activity set up in muscle by nerve increases non-endplate AChE by increasing the rate of AChE synthesis.

The influence of innervation on the metabolism of endplate-specific AChE is less certain. Endplate-specific AChE is composed primarily of asymmetric forms that appear to be associated with the junctional basal lamina. After denervation, the A12 form in the junctional region decreases very rapidly, the A8 form decreases more slowly, and the A4 form increases transiently as the larger forms decrease. This pattern suggested to us that denervation might accelerate the rate at which asymmetric junctional AChE is degraded. Subsequent experiments comparing innervated and 18 to 42-hr denervated rat diaphragms showed that ECHO-labeled endplate AChE is, in fact, degraded more rapidly in denervated than in innervated rat diaphragm muscle. These findings suggest that innervation may decrease the rate at which asymmetric forms are degraded at the endplate, but additional experimentation is needed to establish this point definitively. The observation that spontaneous electromechanical activity increases the rate of synthesis of both AChE asymmetric and globular forms in cultured embryonic rat myotubes suggests that muscle electromechanical activity may increase endplate-specific as well as non-endplate AChE by increasing AChE synthesis. Thus, the influence of innervation on endplate AChE may involve both an increased rate of synthesis and a decreased rate of degradation of the asymmetric forms of the enzyme.

REFERENCES

1. **Mumenthaler, E. and Engel, W. K.,** Cytological localization of cholinesterase in developing chick embryo skeletal muscle, *Acta Anat.*, 47, 274, 1961.
2. **Kovács, T., Kovér, A., and Balogh, G.,** Studies on the lcoalization of cholinesterase in various types of muscle, *J. Cell. Comp. Physiol.*, 57, 63, 1961.
3. **Filogamo, G. and Gabella, C.,** Cholinesterase behaviour in the denervated and reinnervated muscles, *Acta Anat.*, 63, 199, 1966.
4. **Filogamo, G. and Gabella, C.,** The development of neuromuscular correlations, in vertebrates, *Arch. Biol. (Liege)*, 78, 9, 1967.
5. **Vigny, M., Koenig, J., and Rieger, F.,** The motor end-plate specific form of acetylcholinesterase: appearance during embryogenesis and reinnervation of rat muscle, *J. Neurochem.*, 27, 1347, 1976.
6. **Guth, L., Albers, R. W., and Brown, W. C.,** Quantitative changes in cholinesterase activity of denervated muscle fibers and sole plates, *Exp. Neurol.*, 10, 236, 1964.
7. **Guth, L., Brown, W. C., and Watson, P. K.,** Studies on the role of nerve impulses and acetylcholine release in the regulation of the cholinesterase activity of muscle, *Exp. Neurol.*, 18, 443, 1967.

8. **Drachman, D. B.** Neurotrophic regulation of muscle cholinesterase: effects of botulinum toxin and denervation, *J. Physiol. (London),* 226, 619, 1972.
9. **Hall, Z. W.,** Multiple forms of acetylcholinesterase and their distribution in endplate and non-endplate regions of rat diaphragm muscle, *J. Neurobiol.,* 4, 343, 1973.
10. **Davey, B. and Younkin, S. G.** Effect of nerve-stump length on cholinesterase in denervated rat diaphragm, *Exp. Neurol.,* 59, 168, 1978.
11. **Guth, L. and Zalewski, A. A.,** Disposition of cholinesterase following implantation of nerve into innervated and denervated muscle, *Exp. Neurol.,* 7, 316, 1963.
12. **Weinberg, C. B. and Hall, Z. W.,** Junctional form of acetylcholinesterase restored at nerve-free endplates, *Dev. Biol.,* 68, 631, 1979.
13. **Fischbach, G. D. and Robbins, N.,** Changes in contractile properties of disused soleus muscles, *J. Physiol. (London),* 201, 305, 1969.
14. **Guth, L.,** Effect of immobilization on sole-plate and background cholinesterase of rat skeletal muscle, *Exp. Neurol.,* 24, 508, 1969.
15. **Lømo, T., Massoulié, J., and Vigny, M.,** Stimulation of denervated rat soleus muscle with fast and slow activity patterns induces different expression of acetylcholinesterase molecular forms, *J. Neurosci.,* 5, 1180, 1985.
16. **Rubin, L. L., Schuetze, S. M., Weill, C. L., and Fischbach, G. D.,** Regulation of acetylcholinesterase appearance at neuromuscular junctions *in vitro, Nature (London),* 283, 264, 1980.
17. **Lømo, T. and Slater, C. R.,** Control of junctional acetylcholinesterase by neural and muscular influence in the rat, *J. Physiol. (London),* 303, 191, 1980.
18. **Rieger, F., Koenig, J., and Vigny, M.,** Spontaneous contractile activity and the presence of the 16S form of acetylcholinesterase in rat muscle cells in culture: reversible suppressive action of tetrodotoxin, *Dev. Biol.,* 76, 358, 1980.
19. **Fernandez, H. L., Duell, M. J., and Festoff, B. W.,** Neurotrophic control of 16S acetylcholinesterase at the vertebrate neuromuscular junction, *J. Neurobiol.,* 10, 441, 1979.
20. **Ranish, N. A., Dettbarn, W. D., and Wecker, L.,** Nerve stump length-dependent loss of acetylcholinesterase activity in endplate regions of rat diaphragm, *Brain Res.,* 191, 379, 1980.
21. **Guth, L,** "Trophic" influences of nerve on muscle, *Physiol. Rev.,* 48, 645, 1968.
22. **Lentz, T.,** Neurotrophic regulation at the neuromuscular junction, *Ann. N.Y. Acad. Sci.,* 228, 323, 1974.
23. **Oh, T. H.,** Neurotrophic effects: characterization of the nerve extract that stimulates muscle development in culture, *Exp. Neurol.,* 46, 432, 1975.
24. **Max, S. R., Deshpande, S., and Albuquerque, E. X.,** Neural regulation of muscle acetylcholinesterase: effects of batrachotoxin and 6-aminonicotinamide, *Brain Res.,* 130, 101, 1977.
25. **Fernandez, H. L. and Inestrosa, N. C.,** Role of axoplasmic transport in neurotrophic regulation of muscle end-plate acetylcholinesterase, *Nature (London),* 262, 55, 1976.
26. **Lentz, T.,** Nerve trophic function: *in vitro* assay of effects of nerve tissue on muscle cholinesterase activity, *Science,* 171, 187, 1971.
27. **Lentz, T.,** Development of the neuromuscular junction. III. Degeneration of motor endplates after denervation and maintenance *in vitro* by nerve explants, *J. Cell Biol.,* 55, 93, 1972.
28. **Lentz, T.,** Effect of brain extracts on cholinesterase activity of cultured skeletal muscle, *Exp. Neurol.,* 45, 520, 1974.
29. **Oh, T. H., Johnson, D. D., and Kim, S. U.,** Neurotrophic effect of isolated chick embryo muscle in culture, *Science,* 178, 1298, 1972.
30. **Oh, T. H.,** Neurotrophic effects: characterization of the nerve extract that stimulates muscle development in culture, *Exp. Neurol.,* 46, 432, 1975.
31. **Younkin, S. G., Brett, R. S., Davey, B., and Younkin, L. H.,** Substances moved by axonal transport and released by neural stimulation have an innervation-like effect on muscle, *Science,* 200, 1292, 1978.
32. **Davey, B., Younkin, L. H., and Younkin, S. G.** Neural control of skeletal muscle cholinesterase: a study using organ cultured muscle, *J. Physiol. (London),* 288, 501, 1979.
33. **Fernandez, H. L., Patterson, M. R., and Duell, M. J.,** Neurotrophic control of the 16S acetylcholinesterase from mammalian skeletal muscle in organ culture, *J. Neurobiol.,* 11, 557, 1980.
34. **Younkin, S. G., Younkin, L. H., Zoller, M., and Rosenstein, C.,** Partial purification by vacuum distillation of a factor that increases the junctional acetylcholinesterase of organ cultured rat diaphragm, *Brain Res.,* 249, 192, 1982.
35. **Massoulié, J.,** The polymorphism of cholinesterases and its physiological significance, *Trends Biochem. Sci.,* 5, 160, 1980.
36. **Bon, S., Vigny, M., and Massoulié, J.,** Asymmetric and globular forms of acetylcholinesterase in mammals and birds, *Proc. Natl. Acad. Sci. U.S.A.* 76, 2546, 1979.
37. **Vigny, M., Bon, S., Massoulié, J., and Gisiger, V.,** The subunit structure of mammalian acetylcholinesterase: catalytic subunits, dissociating effect of proteolysis and disulphide reduction on the polymeric forms, *J. Neurochem.,* 33, 559, 1979.

38. **Younkin, S. G., Rosenstein, C., Collins, P. L., and Rosenberry, T. L.,** Cellular localization of the molecular forms of AChE in rat diaphragm, *J. Biol. Chem.*, 257, 13630, 1982.
39. **Collins, P. L. and Younkin, S. G.** Effect of denervation on the molecular forms of acetylcholinesterase in rat diaphragm, *J. Biol. Chem.*, 257, 13638, 1982.
40. **Carter, J. L. and Brimijoin, S.,** Effects of acute and chronic denervation on release of acetylcholinesterase and its molecular forms in rat diaphragms, *J. Neurochem.*, 36, 1018, 1981.
41. **Inestrosa, N. C., Ramirez, B. U., and Fernandez, H. L.,** Effect of denervation and of axoplasmic transport blockage on the *in vitro* release of muscle endplate acetylcholinesterase, *J. Neurochem.*, 28, 941, 1977.
42. **Newman, J. R., Virgin, J. B., Younkin, L. H., and Younkin, S. G.,** Turnover of acetylcholinesterase in innervated and denervated rat diaphragm, *J. Physiol. (London)*, 352, 305, 1984.
43. **Hall, Z. W. and Kelly, R. B.,** Enzymatic detachment of endplate acetylcholinesterase from muscle, *Nature (London) New Biol.*, 232, 62, 1971.
44. **Betz, W. and Sakmann, B.,** Effects of proteolytic enzymes on function and structure of frog neuromuscular junctions, *J. Physiol. (London)*, 230, 673, 1973.
45. **McMahan, U. J., Sanes, J. R., and Marshall, L. M.,** Cholinesterase is associated with the basal laminal at the neuromuscular junction, *Nature (London)*, 271, 172, 1978.
46. **Rotundo, R. L. and Fambrough, D. M.,** Synthesis, transport, and fate of acetylcholinesterase in cultured chick embryo muscle cells, *Cell*, 22, 583, 1980.
47. **Fambrough, D. M.,** Control of acetylcholine receptors in skeletal muscle, *Physiol. Rev.*, 59, 165, 1979.
48. **Rotundo, R. L. and Fambrough, D. M.,** Secretion of acetylcholinesterase: relation to acetylcholine receptor metabolism, *Cell*, 22, 595, 1980.
49. **Rotundo, R. L.** Synthesis, assembly, processing, and secretion of acetylcholinesterase by chick muscle cells in culture, *Fed. Proc.*, 1986, in press.
50. **Rotundo, R. L.,** Asymmetric acetylcholinesterase is assembled in the Golgi apparatus, *Proc. Natl. Acad. Sci. U.S.A.*, 81, 479, 1984.
51. **Lazar, M., Salmeron, E., Vigny, M., and Massoulié, J.,** Heavy isotope-labeling study of the metabolism of monomeric and tetrameric acetylcholinesterase forms in the murine neuronal-like T28 hybrid cell line, *J. Biol. Chem.*, 259, 3703, 1984.
52. **Rieger, F. and Vigny, M.,** Solubilization and physicochemical characterization of rat brain acetylcholinesterase: development and maturation of its molecular forms, *J. Neurochem.*, 27, 121, 1976.
53. **Villafruela, M. J., Barat, A., Manrique, E., Villa, S., and Ramirez, G.,** Molecular forms of acetylcholinesterase in the developing chick visual system, *Dev. Neurosci.*, 4, 25, 1981.
54. **Cisson, C. M., McQuarrie, C. H., Sketelj, J., McNamee, M. G., and Wilson, B. W.,** Molecular forms of acetylcholinesterase in chick embryonic fast muscle: developmental changes and effects of DFP treatment, *Dev. Neurosci.*, 4, 157, 1981.
55. **Koenig, J. and Vigny, M.,** Neural induction of the 16S acetylcholinesterase in muscle cell cultures, *Nature (London)*, 271, 75, 1978.
56. **Brimijoin, S. and Carter, J.,** Turnover of the molecular forms of acetylcholinesterase in the rat diaphragm, *J. Neurochem.*, 38, 588, 1982.
57. **Brockman, S. K., Usiak, M. F., and Younkin, S. G.,** Assembly of monomeric acetylcholinesterase into tetrameric and asymmetric forms, *J. Biol. Chem.*, 261, 1201, 1986.
58. **Brockman, S. K., Przybylski, R. J., and Younkin, S. G.,** Cellular localization of the molecular forms of acetylcholinesterase in cultured embryonic rat myotubes, *J. Neurosci.*, 2, 1775, 1982.
59. **Brockman, S. K. and Younkin, S. G.,** The effect of spontaneous electromechanical activity on the secretion of acetylcholinesterase from cultured embryonic rat myotubes, *Brain Res.*, 376, 407, 1986.
60. **Sikorav, J., Grassi, J., and Bon, S.,** Synthesis *in vitro* of precursors of the catalytic subunits of acetylcholinesterase from *Torpedo marmorata* and *Electrophorus electricus*, *Eur. J. Biochem.*, 145, 519, 1984.
61. **Schumacher, M., Maulet, Y., Newton, M., Camp, S., MacPhee-Quigley, K., Taylor, S., Friedmann, T., and Taylor, P.,** Isolation of a cDNA clone coding for acetylcholinesterase, *Nature (London)*, 1986, in press.
62. **Soreq, H., Zevin-Sonkin, D., and Razon, N.,** Expression of cholinesterase gene(s) in human brain tissues: translational evidence for multiple mRNA species, *EMBO J.*, 3, 1371, 1984.
63. **Brockman, S. K., Younkin, L. H., and Younkin, S. G.,** The effect of spontaneous electromechanical activity on the metabolism of acetylcholinesterase in cultured embryonic rat myotubes, *J. Neurosci.*, 4, 131, 1984.
64. **Soreq, H., Parvari, R., and Silman, I.,** Biosynthesis and secretion of catalytically active acetylcholinesterase in *Xenopus* oocytes microinjected with mRNA from rat brain and from Torpedo electric organ, *Proc. Natl. Acad. Sci. U.S.A.*, 79, 830, 1982.
65. **Chirgwin, J. M., Przbyla, A. E., MacDonald, R. J., and Rutter, W. J.,** Isolation of biologically active ribonucleic acid from sources enriched in ribonuclease, *Biochemistry*, 18, 5294, 1979.

Chapter 5

TROPHIC CONTROL OF SKELETAL MUSCLE MEMBRANE PROPERTIES

Terje Lømo and Kristian Gundersen

TABLE OF CONTENTS

I. INTRODUCTION

Normal neuromuscular function depends on several kinds of specializations among motoneurons and muscle fibers. Some motoneurons fire tonically at low frequency, innervate slowly contracting fatigue-resistant muscle fibers, and thus form slow motor units (type S). Other motoneurons fire at high frequency and innervate rapidly contracting fibers to form fast motor units which may be fatigue resistant (type FR) or fatigue sensitive (type FF).[1,2] Some muscles, like the soleus (SOL), primarily contain type S motor units; others, like the extensor digitorum longus (EDL), mainly contain type FR and FF motor units.[2-4] Such differences in muscle contractile properties accompany differences in the properties of neuromuscular junctions,[5-9] nonjunctional membranes,[10-14] and excitation-contraction coupling.[15] Presumably, such muscle specializations, together with a remarkable potential for plastic changes (see Chapter 10),[16-23] enable the organism to effectively cope with the different and ever-changing motor demands of daily life.

The most important long-term influences on muscle are exerted by the nerve, as shown by the devastating effects of denervation and the restoration of normal membrane and contractile properties after reinnervation.[24,25] The transformation of contractile properties by cross-reinnervation[16,26] clearly illustrates the importance of neural inputs and the plasticity of muscle. However, muscles are also affected by other factors, such as hormones[18,27,28] and stretch.[29-31]

Muscle properties may be affected by different types of neural signals: nerve terminals may release trophic substances,[16,32] generate electrical fields,[33,34] or interact directly by contact[35] or indirectly via the basal lamina,[36-38] with the muscle fiber to induce changes at the neuromuscular junction (see Chapters 1, 2, and 11). Whether such signals, particularly trophic substances, also regulate the properties along the rest of the fiber is not clear. While some workers favor[39-42] the existence of such control, others question it[43-45] (see Chapters 3 and 4).

This chapter deals with the mechanisms underlying the control of junctional and nonjunctional membrane specializations in adult muscle fibers, while the control of contractile properties is described elsewhere.[46,47] The purpose is to review here some of the efforts made to identify these mechanisms and to discuss why similar experimental results are often given very different interpretations. In addition, we present our own view based on earlier and recent published material. Part of the material to be presented here has also been reviewed and discussed by others.[18,24,45,48-52]

II. COMPARISON OF JUNCTIONAL, PERIJUNCTIONAL, AND EXTRAJUNCTIONAL MEMBRANE PROPERTIES

Many muscle membrane constituents are unevenly distributed along the myofiber. Acetylcholine receptors (AChR) are densely packed underneath motor nerve terminals and almost absent along the rest of most normal muscle fibers.[48] Sodium channels are more concentrated near the endplate, while chloride channels show the opposite distribution.[53,54] The perijunctional region also differs in other respects from junctional and extrajunctional regions. For example, during reinnervation regenerating axons terminate precisely on sites of previous nerve contact, while avoiding the perijunctional regions.[55] Following transplantation of the fibular nerve onto SOL muscle of adult rats the transplanted nerve will not form ectopic endplates closer than 0.5 to 1.0 mm to the original denervated endplates, but will either reinnervate the old endplates or form new ectopic endplates father away.[56,57] A relative refractoriness to innervation in perijunctional regions, extending for 1 to 2 mm from the endplate in SOL muscles, may account for the tendency of neuromuscular junctions to be distributed at intervals of about 1 mm or more.[50,58-60]

Most mammalian skeletal muscle fibers are sensitive to ACh only at the endplates. Some fibers of the slow SOL muscle, however, show a background ACh sensitivity which is lowest near the endplate and rises toward the tendon.[10] ACh sensitivity increases in both slow and fast fibers after inactivation by cordotomy or blockage of nerve impulse conduction by tetrodotoxin (TTX).[14] The increase is smallest in the perijunctional region, suggesting that this region is subject to an activity-independent suppressive influence from nerve. Evoked muscle activity provides an additional suppressive influence which affects the entire peri- and extrajunctional membrane in an activity-dependent manner.[43] Low-frequency activity is less effective than high-frequency activity, a fact that may partly account for the background sensitivity of some normally innervated slow fibers.

A neural influence restricted to the perijunctional region may have important functional consequences. For example, it may be responsible for the high density of sodium channels and the high safety factor involved in the generation of muscle action potentials at the endplate. It may also cause depletion of recognition/adhesion molecules in perijunctional regions[61] and hence contribute to control of the number and distribution of synapses on single muscle fibers.

The mechanism by which nerves affect the properties of junctional and perijunctional membranes is unclear. In a recent work on cultured *Xenopus* myotubes,[62] nerve contact caused a loss of preexisting AChR clusters and many hours later, apparently by a separate mechanism, clustering of AChR underneath nerve terminals. Impulse-blocking drugs did not prevent either of these effects. Similar results were also obtained after contact of myotubes by polyornithin-coated latex beads, suggesting that simple nerve contact with the muscle fiber may be important.[35,63]

Thus, motoneurons appear to control some junctional and perijunctional properties by mechanisms independent of impulse activity perhaps involving contact, electrical fields, or trophic substances provided by nerve terminals. For complete control, however, evoked muscle activity is essential.

III. CONTROL OF EXTRAJUNCTIONAL MEMBRANE PROPERTIES

A. Role of Evoked Muscle Activity

Mammalian skeletal muscle fibers respond to denervation by a striking increase in ACh sensitivity outside the endplate, a fall in resting membrane potential (RMP),[64-67] and a decay of acetylcholinesterase (AChE) activity.[68-70] Also, denervated fibers undergo atrophy despite the appearance of spontaneous activity (fibrillations). Direct electrical stimulation of the denervated fibers reverses these changes and usually restores normal fiber properties.[43,65,66,70,71] The effects of stimulation are strongly stimulus pattern dependent. In rat SOL muscles patterns similar to those of normal fast muscles induce fast muscle properties, while slow patterns have opposite effects.[46,70,72,73] It is sometimes stated that direct stimulation may restore normal extrajunctional properties in slow, but not in fast, muscles.[74,75] We have reexamined this question and found that stimulation (one train of 60 stimuli at 100 Hz every 1 min for 5, 14, or 21 days) completely prevents the appearance of extrajunctional ACh sensitivity and the fall in RMP in denervated fibers of rat EDL muscles (fast). In two muscles all fibers ($n = 35$) examined at random were insensitive to ACh 4 mm from the tendon, compared to a mean sensitivity of 407 mV/nC for fibers ($n = 21$) in denervated control muscles. Three other muscles contained restricted regions, usually along one edge, where every fiber was supersensitive, while all the fibers ($n = 44$) examined across the rest of the muscle were insensitive. It was highly likely that the sensitive regions resulted from poor positioning of the electrodes and ineffective stimulation of some of the muscle fibers. Thus, we conclude that direct stimulation can restore normal extrajunctional membrane properties in both slow and fast denervated muscles.

It has recently become possible to record continuously for 24 hr the normal activity of single motor units in SOL and EDL muscles of freely moving rats.[2] In some EDL units (probably FF units) the total amount of activity is very low (average firing frequency, 0.03 to 0.2 Hz), whereas in denervated fibers, at least in organ culture, the fibrillatory activity may be quite high (about 1 Hz averaged over several days).[76] If activity maintains normal extrajunctional membrane properties, why do denervated fibrillating fibers become atrophic and supersensitive to ACh, while less active fibers in the EDL do not? The explanation is probably the particular pattern of activity in denervated fibers. In organ-cultured rat diaphragms the denervated fibers fibrillate for about 22 hr and, thereafter, remain inactive for 2 to 3 days, when a new cycle is repeated.[76] In contrast, even the least active EDL units are active hundreds or thousands of times every day. Since each discharge usually shows only one or a few impulses, however, the average frequency becomes low.[2]

That the stimulation pattern rather than the amount of activity is the critical factor can be demonstrated by direct stimulation of denervated muscles. For example, a given number of brief stimulus trains per day has a much greater effect on denervation supersensitivity when the trains are applied regularly than when they are applied during only a few hours every day.[66] Delivered at intervals of about 2 hr, brief 100-Hz stimulus trains prevent denervation supersensitivity and most of the atrophy even though the average impulse frequency (0.01 Hz) is lower than in the least active EDL units (0.03 Hz).[2] These and other stimulation experiments suggest that inactivity triggers the production of ACh supersensitivity (and other membrane changes) whenever a critical duration of inactivity is exceeded, as occurs after denervation.

In conclusion, direct electrical stimulation with patterns comparable to the normal firing of fast and slow motoneurons can both maintain and restore normal extrajunctional membrane properties in denervated muscles. Accordingly, motoneurons would appear to control muscle properties through evoked muscle activity. To be effective, however, the activity must be evoked at intervals that are much shorter than the inactive periods in spontaneously fibrillating fibers.

B. Effects of Nerve Degeneration

Results supporting a trophic mechanism(s) have been obtained by comparing the effects of cutting the nerve at different distances from the muscle. ACh supersensitivity and other effects of denervation appear earlier when the nerve stump is short,[77-81] possibly because the nerve terminals degenerate more rapidly, thereby causing putative trophic substances to disappear earlier.[82] Muscles also respond earlier and more strongly to cutting the nerve than to inactivity produced by a TTX nerve block,[83-85] possibly because the release of putative trophic substances continues during a TTX block, but not after a cut.

In trying to interpret these results, it is important to consider the possibility that some of the effects of denervation are due to substances released or produced when the intramuscular nerves degenerate. Evidence supporting this possibility has been obtained by Cangiano and collaborators. They first reported a weak but clear increase in TTX-resistant action potentials (an effect of denervation) in innervated fibers of partially denervated muscle.[86] This finding was criticized by Tiedt et al.,[87] but satisfactorily answered by Cangiano and Lutzemberger,[88,89] who went on to show that if the remaining intact axons were blocked by TTX, then the innervated muscle fibers responded much more strongly than before.[88] Indeed, under such conditions the innervated fibers became as sensitive to ACh and as resistant to TTX as the denervated fibers. Furthermore, in both denervated and inactive innervated fibers the response increased in magnitude with the degree of partial denervation and in latency with the distance from the muscle to the site of partial denervation. These results are inconsistent with the view that trophic substances carried by axoplasmic transport and released independently of activity from motor nerve terminals are responsible for maintaining normal membrane properties in extrajunctional regions. In that case the muscle fibers innervated by the TTX-blocked axons would have been expected to have a lower ACh sensitivity and resistance to TTX than the denervated fibers, because TTX does not appear to block axoplasmic transport.[42,90] On the other hand, the results are consistent with the view that nerve degeneration causes the appearance of substances that affect all adjacent muscle fibers whether they are innervated or not.

Further support for this view was obtained by transplanting the fibular nerve to the proximal endplate-free region of rat SOL muscles.[91] In those experiments the fibular nerve did not form ectopic endplates because the SOL nerve was intact. Some months after the transplantation the fibular nerve was cut in the thigh and the tibial nerve which innervates the SOL was blocked by TTX. After a further 2 to 3 days the muscle fibers were found to be markedly sensitive to ACh and resistant to TTX underneath the degenerating nerve terminals. These local changes were not caused by the TTX block alone, because in the absence of a transplanted degenerating nerve the effects of the TTX block which appeared later were widespread and less intense. Nor were they caused by the degenerating nerve alone because in the absence of the TTX block the properties of the (now active) muscle fibers remained normal underneath the degenerating nerve. Evidently, such changes resulted from the combined effects of inactivity and nerve degeneration. Since in those experiments the SOL was always innervated by its normal nerve and not by the transplanted foreign nerve, it seems unlikely for the muscle to become trophically dependent on the foreign nerve. Hence, the effects of cutting the transplanted nerve probably resulted from the appearance of substances as the nerve degenerated rather than the disappearance of trophic substances. Brown and colleagues[44,92,93] reached a similar conclusion when they destroyed the sensory nerve fibers to the muscle and found a small increase in ACh sensitivity. They also compared denervation with paralysis by botulinum toxin (BT) and found that denervation only initially produced a greater ACh sensitivity which they attributed to products of nerve degeneration.

The effects of nerve degeneration have also been studied in muscle fibers activated by another nerve or by electrical stimulation.[43] The results were as follows. First, when rat SOL muscles were denervated and stimulated at low frequency (slow pattern), a transient supersensitivity lasting 8 to 10 days developed. No supersensitivity developed in muscles stimulated at high frequency (fast pattern). Second, when the stimulation was stopped after 2 weeks, supersensitivity appeared earlier in muscles stimulated at low frequency. The most rapid response, however, was observed in acutely denervated muscles where the effects of nerve degeneration were presumably most intense. Third, when SOL muscle fibers reinnervated at separate sites by a fast and a slow nerve (dual innervation) were partially denervated by cutting the fast nerve (retaining slow nerve activity), then a marked transient supersensitivity developed around the degenerating fast endplates. No supersensitivity developed, however, when the slow nerve was cut and fast activity was maintained by an intact fibular nerve.

These results show that evoked muscle activity can suppress the development of extrajunctional ACh sensitivity around degenerating endplates in agreement with the suppressive effects of activity on the synthesis of extrajunctional AChR.[48] The degree of suppression, however, depends on the pattern of activity. Fast activity is more effective than slow activity.[43] This explains why fast activity (but no slow activity) may prevent supersensitivity during the acute stages of nerve degeneration. The supersensitivity around degenerating endplates in fibers with slow activity was both local and transient, suggesting that it was due to products of nerve degeneration. If it were due to absence of trophic substances, which normally maintain a low extrajunctional ACh sensitivity, then a generalized supersensitivity would have been expected in the electrically stimulated muscles. In turn, no supersensitivity would have been expected in the innervated muscles because the remaining intact nerve would supply both activity and trophic substances.

C. Possible Mechanisms behind the Effects of Nerve Degeneration

Denervation has drastic effects on muscle tissue apart from paralyzing the muscle fibers and possibly interrupting the delivery of putative trophic substances. As nerves degenerate, potent mitogens appear partly from disintegrating myelin.[94,95] These mitogens can diffuse over considerable distances and induce proliferation of Schwann cells, connective tissue, and vascular cells.[95-97] It is significant, therefore, that acutely denervated muscles show a marked transient increase in the number of nonmuscle cells. Cell proliferation starts at the endplate, spreads quickly along the muscle, and is maximal after 4 days, when the number of proliferating nonmuscle cells is increased more than 20 times.[98] It is probably caused by a breakdown of the nerve, because cell proliferation is not elicited by blocking nerve impulse conduction with TTX, which leaves the nerve intact.[98] Murray and Robbins[98] argue that the mitogenic stimulus derives for the most part from the denervated muscle fibers. To us, however, it seems more likely that it derives, at least initially, from the disintegrating nerves, although its effects may become potentiated later by mitogens released by the denervated muscle fibers.

It is possible that mitogens and other inflammatory substances are partly responsible for the effects of denervation on extrajunctional membrane properties. Thus, supersensitivity to ACh may develop transiently after injury[66,99] or after placing a small piece of nerve or thread on the surface of the muscle, which then leads to a local accumulation of inflammatory cells.[100-102] On the other hand, Cangiano[45,91] has suggested that the effects of nerve degeneration are due to axonal substances released by degenerating nerves, rather than inflammatory substances. The axonal substances could be the same as those recently extracted from nerves and shown to promote AChR synthesis and clustering.[103-105] Cangiano bases his suggestion partly on the greater effects of cutting motor nerves rather than sensory nerves[92] and partly on the inability of cortisone to suppress denervation supersensitivity. In addition to an antiinflammatory effect, however, cortisone may have a direct effect on denervated muscle fibers, since cortisone receptors increase after denervation.[106] Given that denervation leads to such striking cell proliferation and inflammation together with a marked ACh supersensitivity, we favor the idea that degenerating nerve terminals release substances that affect the muscle partly directly and partly indirectly via inflammatory or other nonmuscle cells.

It appears from these results that two types of opposing influences affect ACh sensitivity and presumably other properties of the extrajunctional membrane. Under normal conditions evoked muscle activity suppresses the appearance of extrajunctional AChR and other abnormal membrane properties. Under certain conditions substances may arise as a result of nerve degeneration, inflammation, or other insults, which induce the appearance of extrajunctional AChR, TTX-resistant sodium channels, and other abnormal properties. When such substances appear in active muscles, they counteract the suppressive effects of activity in a graded, competitive manner. This scheme explains why nerve degeneration and inflammation have much more drastic effects on inactive than on active muscles. It also explains why supersensitivity and other changes in the extrajunctional membrane appear earlier and at greater intensity after cutting the nerve than after just blocking nerve impulse conduction. Earlier suggestions that intracellular free calcium levels may mediate the effects of activity have recently been substantiated, both for AChR[107-110] and for sodium channels.[111] Cyclic nucleotides[111-113] may play an additional role, possibly in association with intracellular calcium-binding proteins such as calmodulin.[111] Hence, several possibilities exist with respect to sites where muscle activity and substances on the outside of the muscle fiber might interact.

D. Attempts to Separate the Effects of Nerve Degeneration and Trophic Substances

One approach to separate the effects of nerve degeneration and putative trophic substances rests on the premise that nerve degeneration is a transient process. Consequently, the effects if the hypothesis of products of nerve degeneration is correct, but permanently greater if the hypothesis of trophic control is correct. A transient difference has been observed with respect to muscle RMP in rats[42,85] and ACh sensitivity in mice.[44] In similar experiments by Bray et al.,[39] however, the difference appears to be permanent. Gilliat et al.,[114] who studied baboon lumbrical muscles, found that the extrajunctional ACh sensitivity was higher (even after 70 days) in denervated fibers than in those paralyzed by acute compression of the nerve, which caused local demyelination and nerve impulse conduction block. This important result implies a trophic influence on extrajunctional ACh sensitivity and needs to be confirmed, particularly in other species. The possibility that nerve degeneration may have prolonged effects on the muscle should also be considered, particularly since substances affecting muscle properties can still be detected in the peripheral nerve stump 3 weeks after a nerve cut.[115]

A second approach is to allow denervated muscles to become reinnervated by axons blocked with TTX. If the reinnervating axons release trophic substances, an increase in RMP and a decrease in extrajunctional ACh sensitivity would be expected. Unfortunately, this well-designed experiment has so far given conflicting results. Bray et al.[39] did find an increase in RMP and a decrease in ACh sensitivity, whereas Cangiano and Lutzemberger[116] did not except in a few muscles, where the nerve block was presumed to be incomplete. Such discrepancies raise questions about the effectiveness of the TTX block, since very small amounts of activity may suppress the supersensitivity once it is established.[66,114]

E. Is ACh a Trophic Substance?

Johns and Thesleff[117] have shown that after cutting the spinal cord of rats extrajunctional ACh sensitivity is increased only slightly despite a marked reduction in muscle activity. Thus, they proposed that the low ACh sensitivity might be due to spontaneously released ACh. Mathers and Thesleff[118] provided further evidence for this idea by showing that a neurotoxin which blocks postjunctional AChR potentiates the denervation-like effects of BT, which paralyzes the muscles without completely preventing spontaneous ACh release. Drachman et al.[42] perfused α-bungarotoxin (α-BTX) over rat SOL muscles and found that it produced exactly the same increase in AChR density as did denervation, while the increase after a TTX block of the nerve was considerably smaller.[119] They concluded that the effects of denervation may be entirely due to the combined effects of inactivity and a trophic influence of spontaneously released ACh.

The possibility must be considered, however, that application of toxins has direct effects on paralyzed muscles similar to those suggested for the products of nerve degeneration. For example, implantation of cannulas for delivery of α-BTX may cause inflammation and ACh supersensitivity, particularly in paralyzed muscles. To control for this possibility Drachman et al.[42] perfused saline through similarly implanted cannulas onto normally innervated SOL muscles and observed no increase in extrajunctional AChR density. However, during saline perfusion, the muscles remained active, whereas during α-BTX perfusion they were inactive and therefore likely to be much more responsive to the effects of inflammation. A more appropriate control, therefore, would have been to perfuse the saline onto TTX-paralyzed muscles and then to look for a potentiating, possibly saturating, effect of inflammation. Surgical trauma and intramuscular injection of saline can in some cases increase AChR density or extrajunctional ACh sensitivity even in innervated muscles.[66,120]

The case for a trophic function of spontaneously released ACh is further complicated by several conflicting results in the literature. For example, the RMP is reported by some to be reduced 3 to 6 hr after denervation.[121,122] Spontaneous ACh release disappears later and cannot be responsible for such a fall. Other workers, however, have reported a fall in RMP only after 15 to 18 hr, i.e., after the disappearance of quantal and presumably nonquantal release of ACh.[67,79,81,123] In experiments by Robbins and colleagues[81,123,124] the effects of denervation occurred later when the nerve was transected far from the muscle (long nerve stump) even in the presence of α-BTX and other nicotinic blockers, i.e., a result which would not be expected if spontaneously released ACh were responsible for the nerve stump effect. Thus, at present there appears to be no unequivocal evidence that spontaneously released ACh plays a role in regulating the properties of the extrajunctional membrane (but see Chapter 3).

F. Attempts to Selectively Block the Delivery of Putative Trophic Substances

Colchicine, which blocks axoplasmic transport,[125] induces ACh supersensitivity and other denervation-like changes in muscles after local treatment of the nerve.[126-128] It is difficult, however, to be sure that this effect is due to arrested transport of putative trophic substances. When larger doses are injected into the nerve,[126] colchicine may act directly on the muscle through a systemic effect.[129,130] The systemic effects may be avoided by applying colchicine from cuffs around the nerve,[127] but this technique easily leads to axonal damage owing to compression or toxic effects.[130] When cotton pieces soaked in colchicine are placed on the nerve, compression is avoided, but even then some muscles have to be discarded (up to 25%) because of partial denervation. Furthermore, even in the muscles that are accepted, there are signs of depressed synaptic transmission.[131] In similar experiments on postganglionic sympathetic nerves a considerable fraction of neurons cannot be antidromically invaded after application of colchicine, and some ganglion cells eventually degenerate although signs of systemic effects appear to be absent. Moreover, in the electron microscope some axons appear grossly swollen.[132] The possibility must therefore be considered that when colchicine is given in doses just sufficient to have effects on the muscle, some axons may fail to conduct or transmit impulses for a shorter or longer period of time. This would be difficult to detect and yet could affect the muscle fibers. ACh supersensitivity lasting several days can develop after periods of inactivity as short as 1 to 2 days,[133] and denervation of only a few motor units may affect adjacent innervated fibers due to products of nerve degeneration.[86,88] These objections do not, of course, exclude that colchicine may act partly by blocking transport of trophic substances. If that is the case, however, it seems important to consider whether the effect is on junctional properties such as AChE activity,[134,135] where a trophic control mechanism seems well substantiated, or on extrajunctional properties. Given existing uncertainties about the mechanisms and sites of action of colchicine and the methods of application, we are led to conclude that the use of colchicine provides as yet no unequivocal evidence for trophic control of extrajunctional membrane properties.

G. The Search for Trophic Substances

Attempts to isolate trophic substances that maintain normal extrajunctional membrane properties have so far been unsuccessful. Substances have been extracted from nerves and injected into denervated muscles in vivo[120,136] or added to the medium of organ-cultured muscles.[137] These substances, however, lead to an increased number of extrajunctional AChR and not to the decrease that would be expected if they were responsible for the low extrajunctional ACh sensitivity of normally innervated fibers. These substances also lead to an increased AChE activity at denervated endplates, but do not appear to affect nonjunctional AChE activity.[137,138] Therefore, there is good evidence that these substances may regulate junctional properties, but little evidence so far that they regulate extrajunctional properties. In agreement with this view junctional, but not extrajunctional, AChE is sensitive to cutting the nerve at different distances from the muscle.[139]

One of the proteins extracted from sciatic nerves, named sciatin, not only induces synthesis of AChR and AChE but also promotes maturation and long-term survival of muscle cells in vitro (see Chapter 6).[140] Sciatin, however, has recently been shown to be similar or identical to transferrin,[51,141] a normal constituent of blood, which is taken up but not synthesized by some adult muscles.[142] It is therefore possible that both nerves and muscles take up transferrin and other substances from the blood to serve their own trophic needs and that a trophic effect of such substances is unrelated to their presence in the nerve. It is also possible that these substances normally act only on junctional receptors which appear in extrajunctional regions after denervation. In that case both application or intramuscular injection of these substances might have a generalized effect on noninnervated fibers which is unrelated to their normal action on innervated fibers. Finally, it is worth noting that trophic proteins are present in nerves after nerve section and degeneration of the axons.[115] For these reasons the presence of trophic proteins in motor nerves does not necessarily mean that these proteins are normally used by the motoneurons to regulate the properties of the muscle fiber outside the neuromuscular junction.

So far, it has been impossible to provide unequivocal evidence that trophic substances mediate the neural control of extrajunctional membrane properties. Some results, often taken as evidence of such control, may be interpreted as being due to products of nerve degeneration rather than the absence of trophic substances. The issue, however, is far from resolved. Further progress will partly require reliable techniques for selective and long-term blockade of nerve impulse conduction to firmly establish whether reinnervation by impulse-blocked nerves will restore normal properties in extrajunctional regions and whether a nerve impulse conduction block will have the same effects as denervation when all the processes associated with nerve degeneration have subsided. It will also be important to continue the search for trophic substances in motor nerves (see Chapter 7), but as discussed above the results may be difficult to interpret (see Chapters 3, 4, and 6).

We propose that a substance should be considered to play an essential neurotrophic role only when it can be demonstrated that (1) it is released from motor nerve endings; (2) it cannot be supplied by the blood or nonneural cells; and (3) skeletal muscles develop abnormal properties in its absence despite electrical stimulation (see Chapter 1). Since electrical stimulation appears to be able to substitute for the nerve in maintaining the integrity of several important extrajunctional properties, we may conclude that trophic substances are not essential in their normal regulation. It is still possible, of course, that neural substances affecting nonjunctional properties exist and that they may become important under certain conditions. Nevertheless, at present it appears that extrajunctional membrane properties are regulated primarily by evoked muscle activity with muscle stretch and substances carried in the blood probably playing modulatory roles.

IV. CONTROL OF JUNCTIONAL PROPERTIES

A. Formation of Neuromuscular Junctions

Formation of neuromuscular junctions can be studied in adult rat SOL muscles by first transplanting a foreign nerve (the fast fibular nerve) to the proximal part of the SOL, then cutting the SOL nerve 3 weeks later to induce the appearance of extrajunctional AChR and receptivity to innervation, and finally by examining the ectopic junctions which begin to form 2 to 3 days after the SOL nerve is cut.[143,144] The first signs of synapse formation are clusters of AChR appearing underneath the often profusely growing fibular nerve terminals.[57,144] The clusters vary greatly in size, are irregularly distributed, and may occur in large numbers on a single fiber.[57] Initially the new junctions lack postjunctional folds[145] and AChE activity.[144,146] The AChR turn over rapidly and have long open-channel times,[147] and the fibers show multiple innervation.[57,143,148] These immature properties make transmission more effective, and explain why action potentials and contractions can be evoked within a few hours by the fibular nerve despite low quantal content of ACh.[133] During the next 1 to 2 weeks, the multiple clusters on each fiber condense into one or two separate clusters or collections of closely adjacent clusters which gradually take on the appearance of mature neuromuscular junctions with intense AChE activity,[57,144,146] stable AChR channel complexes with short open times, and extensive postjunctional folds.[145,147,149] The number of axonal inputs per fiber is considerably reduced,[57,143,148] Thus, the essential steps of normal ontogenetic synapse formation in the rat are recapitulated.[150,151] In the embryo, however, only one junction usually forms on each muscle fiber, whereas in the adult more than one synapse may form at separate sites along the fiber. The reason for this may be that in the embryo the muscle fiber is only a few 100 μm long when the synapses form,[152] whereas in the adult the nerve sprouts may cover the fiber over a length of several millimeters, where they usually form synapses at 1-mm intervals.[43,57,148]

The earliest events in synaptogenesis do not depend on impulse activity.[153,154] Remarkably quickly, however, they lead to effective transmission and evoked muscle activity, [133] which is essential for further maturation of the synapse. The effect of activity per se can be studied in the adult fibular nerve-SOL muscle preparation by cutting the transplanted fibular nerve at an early stage and then examining the subsequent fate of the denervated immature postsynaptic sites in the presence or absence of direct stimulation of the muscle via implanted electrodes.

In the absence of stimulation the fibers undergo atrophy, AchR clusters remain irregularly distributed, and AChE fails to appear.[57,146] Upon stimulation, however, processes very similar to those seen with an intact nerve occur. Within 1 to 2 weeks the clusters condense into one or two separate clusters or collections of closely adjacent clusters which acquire the same size as the innervated clusters as well as high ACh sensitivity, AChE activity, and short open-channel time.[57,143,146] These results show that postjunctional impulse activity induces several important postjunctional specializations, provided that an early priming influence of the nerve is present. On the other hand, the continued presence of the nerve is not needed.

From these observations the following picture of synaptogenesis emerges (see Chapter 10). When nerve terminals first contact receptive muscle fibers, they induce local changes (imprints) in the fibers. Such imprints form independently of impulse activity, possibly by trophic substances and/or contact interactions with the nerve, and persist after destruction of the nerve at an early stage. The imprints do not themselves lead to mature synapses. This requires muscle activity, which is normally evoked soon after nerve contact. Thus, muscle activity, by interacting in some unknown way with the imprints laid down early by the nerve, becomes responsible for several important events in synaptogenesis. First, activity leads to elimination of redundant AChR clusters and growth of surviving clusters to a specified size. Second, it causes the appearance of AChE activity, AChR channels with short open times, and postjunctional folds. Third, it suppresses synthesis of AChR and promotes other changes in the extrajunctional region which result in refractoriness to subsequent additional innervation. Finally, impulse activity appears to be involved in eliminating polyneuronal innervation during postnatal development[155-158] and in regulating the competition between motoneurons for muscle cells and, hence, in regulating the size of their terminal fields (see Chapters 9 and 10).[159] Through these effects impulse activity becomes instrumental in stabilizing the individual synapse and in regulating the number and distribution of synapses on single muscle fibers. Similar processes may take place on neurons in the central nervous system, where impulse activity also appears to play an important role in synaptogenesis and synapse elimination.[160-162]

The nature of the imprint is unknown. It may reside in the basal lamina, which acquires several special properties after contact with the nerve, including the ability to induce clusters of AChR in the underlying muscle fiber membrane and sites for transmitter release in the overlying nerve terminal.[36-38,163,164] The synaptic basal lamina may also be responsible for the precise reinnervation of denervated original synaptic sites.[55,163] Recently, monoclonal

Table 1
CHARACTERISTICS OF MOTOR ENDPLATES IN FAST AND SLOW MUSCLE FIBERS[7-9,11,145]

	Slow fibers	Fast fibers
Axon terminals	Relatively thick, elongated, and irregular	Many boutons connected by thin branches
Presynaptic vesicles	Less closely packed	More closely packed
MEPP[a]		
Frequency (Hz)	2.4	8.8
Amplitude (mV)	1.0	0.7
Duration (msec)	2.9	1.3
AChE	Mainly six molecular forms	Mainly three molecular forms
No. of functional ACHR/μm^2	343	837
Postjunctional folds	Sparse, broad shallow, and irregular	Closely spaced, deep, branched, and regular

[a] Miniature endplate potentials.

antibodies have been raised which distinguish between junctional and extrajunctional basal lamina components.[36] Perhaps the imprint resides in such nerve-induced molecular specializations (see Chapter 8).

The experiments on rat muscles discussed so far indicate that impulse activity plays an important role in synaptogenesis. In *Xenopus laevis,* however, many of the maturational processes at the neuromuscular junction, such as appearance of the AChE and conversion of AChR open-channel time, do not appear to depend on activity.[165] Accordingly, we emphasize that the relative importance of activity, trophic influences, and intrinsic developmental programs may vary considerably depending on the system and species studied.

B. Determination of Junctional Properties in Fast and Slow Muscles

The properties of neuromuscular junctions differ in fast and slow muscles (Table 1), possibly because they transmit different patterns of impulses[2] or the motoneurons and muscle fibers have different inherent properties. Indications that inherent properties are important have been obtained by comparing the capacity of the fibular, the soleus, and the sural nerves to form ectopic junctions with SOL and EDL muscles. In contrast to the fibular nerve, which grows profusely and usually forms numerous ectopic junctions, the soleus nerve, when transplanted to the same region of the SOL muscle, extends very few regenerating axons and forms ectopic junctions only very rarely.[9] The sural nerve behaves as the fibular nerve in that it forms many ectopic junctions with the SOL muscle. On the other hand, when it is transplanted onto the EDL, it almost never forms ectopic junctions, regardless of where it is placed on the muscle, but reinnervates the original denervated junctions.[166] Thus, motoneurons appear to be intrinsically different in their capacity to form ectopic endplates, while muscles appear to be inherently different in their ability to accept ectopic reinnervation by competent axons. Such differences could arise through the irreversible differentiation of motoneurons and muscles into different types early in development[167-171] and through the expression of different molecular labels, depending on segmental origin and position of motoneurons and muscle fibers.[172]

Single SOL muscle fibers may become reinnervated first ectopically by the fibular nerve and then at the original endplates by the soleus nerve (dual innervation). This offers an opportunity to study the interactions between different types of synapses on the same fiber. Six months after reinnervation by the fibular nerve alone, the junctions appear fast and the fibers contract rapidly. After reinnervation by the soleus nerve alone, the junctions appear slow and the fibers contract slowly.[9] Dually innervated fibers, on the other hand, show an intermediate contraction speed, while the fibular nerve junctions retain their fast appearance and the soleus nerve junctions their slow appearance in the light microscope.[9] These results suggest that the properties of a neuromuscular junction are determined locally by the motoneuron regardless of contraction speed or presence of junctions elsewhere on the fiber. In turn, the contractile properties appear to be determined by the integrated influence of all the synaptic inputs to the fiber.

Fast and slow neuromuscular junctions are probably different because effective transmission of fast and slow impulse patterns requires different functional properties. Therefore, it is possible that neuromuscular junctions respond to changes in activity by changing their properties, just as muscle fibers respond to altered activity by changing their contraction speed or fatigue resistance.[19,23,72] This would require modifiable neuromuscular junctions, and recent evidence for such plasticity has accumulated. First, axon terminals at normal adult neuromuscular junctions may extend and retract sprouts at rates which appear to be influenced by activity, suggesting that continuous remodeling of the endplate is possible (see Chapter 10).[173,174] Second, in denervated adult rat SOL muscles intermittent high-frequency (fast-pattern) and tonic low-frequency (slow-pattern) stimulation promotes the appearance of AChE molecular forms typical of fast and slow muscles, respectively, possibly reflecting a mechanism for adjusting junctional AChE function to different impulse patterns.[70] Third, impulse patterns appear to influence the formation of neuromuscular junctions in the normal immature crayfish[175] or in the adult rat SOL,[9] since long-term slow-pattern stimulation leads to some slow characteristics at junctions which normally develop into fast junctions. The transformation is not complete, however, perhaps because some fast impulse patterns persist in the fast nerves or because inherent properties of the fast motoneurons prevent it.[9] Finally, the amount of transmitter released per unit length of nerve terminal may be increased by removing other terminals belonging to the same motoneuron,[176] or changes in the size or release characteristics of a junction may be altered by denervating muscles in the contralateral leg.[177] These and other experimental results show that neuromuscular junctions are dynamic structures capable of changing their physiological and morphological characteristics in response to a variety of influences. Patterns of impulse activity probably play a central role in adjusting the properties of the neuromuscular junction, as they do in adjusting the contractile properties of the muscle itself. These adjustments seem to be of a tuning kind, occurring within adaptive ranges set by the inherent properties of the different types of motoneurons and muscle fibers.

V. SUMMARY AND CONCLUDING REMARKS

It is important to distinguish between junctional, perijunctional, and extrajunctional regions of muscle fibers because the properties of these regions appear to be under different types of control.

Junctional membrane properties are affected at very early stages of synapse formation by activity-independent neural influences. These may include trophic substances, nerve-muscle contact, or electrical fields generated by axon terminals. As a result, muscle activity is evoked early and is essential for further maturation of the junction. Such activity promotes the appearance of junctional AChR with short open-channel times, AChE activity, and postjunctional folds. It also influences the size, number, and distribution of endplates; mediates competitive interactions between motoneurons for footholds on the muscle fibers; and prevents further innervation. Many properties of the neuromuscular junction appear to be determined by the motoneuron, to be functionally matched to the impulse patterns that it transmits, and to have the capacity to change when the pattern of impulses changes.

Some of the muscle's perijunctional regional properties differ from those of junctional and extrajunctional regions. Little is known about their regulation, but suggestions exist that both activity and activity-independent neural influences may be involved.

The properties of extrajunctional membranes are primarily controlled by evoked muscle activity, as shown by the ability of direct stimulation to restore normal properties in denervated fast and slow rat muscles. Denervation has some effects, however, which cannot be due to inactivity, but may possibly result from the absence of a trophic substance(s) or the action of products of nerve degeneration. So far, it has not been possible to attribute all activity-independent effects of denervation unequivocally to either of these two mechanisms (see Chapter 11). The search for trophic substances in nerve extracts has so far been unsuccessful (but see Chapters 4, 6, and 7), and the results of attempting to block axoplasmic transport of such substances by colchicine seem to be difficult to interpret. Thus, we favor the view that trophic substances have few, if any, effects on the extrajunctional membrane, whose properties appear to be controlled primarily by nerve-evoked activity.

REFERENCES

1. **Burke, R.,** Motor units: anatomy, physiology, and functional organization, in *Handbook of Physiology*, Section 1, Vol. 2, Part 1, Brookhart, J. M. and Mountcastle, V. B., Eds., Brooks, V. B., Ed., American Physiological Society, Bethesda, Md., 1981, 345.
2. **Hennig, R. and Lømo, T.,** Firing patterns of motor units in normal rats, *Nature (London)*, 314, 164, 1985.
3. **Kugelberg, E.,** Adaptive transformation of rat soleus motor units during growth, *J. Neurol. Sci.*, 27, 269, 1976.
4. **Niederle, B. and Mayr, R.,** Course of denervation atrophy in type I and type II fibers of rat extensor digitorum longus muscle, *Anat. Embryol. (Berlin)*, 153, 9, 1978.
5. **Padykula, H. A. and Gauthier, G. F.,** The ultrastructure of the neuromuscular junctions of mammalian red, white and intermediate skeletal muscle fibers, *J. Cell Biol.*, 46, 27, 1970.
6. **Robbins, N.,** Plasticity at the mature neuromuscular junction, *Trends Neurosci.*, 3, 120, 1980.
7. **Sterz, R., Pagala, M., and Peper, K.,** Postjunctional characteristics of the endplates in mammalian fast and slow muscles, *Pfluegers Archiv*, 398, 48, 1983.
8. **Florendo, J. A., Reger, J. F., and Law, P. K.,** Electrophysiological differences between mammalian fast and slow myofibers, *Soc. Neurosci.*, 9, 13, 1983.
9. **Lømo, T. and Waerhaug, O.,** Motor endplates in fast and slow muscles of the rat: what determines their difference?, *J. Physiol. (Paris)*, 180, 290, 1985.
10. **Miledi, R. and Zelena, J.,** Sensitivity to acetycholine in rat slow muscle, *Nature (London)*, 210, 855, 1966.
11. **Ellisman, M. H., Rash, J. E., Staehelin, L. A., and Porter, K. R.,** Studies of excitable membranes. II. A comparison of specializations at neuromuscular junctions and non-junctional sarcolemmas of mammalian fast and slow twitch muscle fibers, *J. Cell. Biol.*, 68, 752, 1976.
12. **Smith, P. B. and Appel, S. H.,** Isolation and characterization of the surface membranes of fast and slow mammalian skeletal muscle, *Biochim. Biophys. Acta.* 466, 109, 1977.
13. **McArdle, J. J., Michelson, L., and D'Alonzo, A. J.,** Action potentials in fast and slow twitch mammalian muscles during reinnervation and development, *J. Gen. Physiol.*, 75, 655, 1980.
14. **Yoshioka, K. and Miyata, Y.,** Changes in the distribution of the extrajunctional acetylcholine sensitivity along muscle fibers during development and following cordotomy in the rat, *Neuroscience*, 9, 437, 1983.
15. **Dulhunty, A. F. and Gage, P. W.,** Asymmetrical charge movement in slow- and fast-twitch mammalian muscle fibers in normal and paraplegic rats, *J. Physiol. (London)*, 341, 213, 1983.
16. **Buller, A. J., Eccles, J. C., and Eccles, R. M.,** Interactions between motoneurons and muscles in respect of the characteristic speeds of their responses, *J. Physiol. (London)*, 150, 417, 1960.
17. **Lømo, T.,** The role of activity in the control of membrane and contractile properties in skeletal muscle, in *Motor Innervation of Muscle*, Thesleff, S., Ed., Academic Press, London, 1977, 289.
18. **Gutmann, E.,** Neurotrophic relations, *Annu. Rev. Physiol.*, 38, 177, 1976.
19. **Salmons, S. and Henriksson, J.,** Adaptive response of skeletal muscle to increased use, *Muscle Nerve*, 4, 94, 1981.

20. **Jolesz, F. and Sreter, F. A.,** Development, innervation, and activity-pattern induced changes in skeletal muscle, *Annu. Rev. Physiol.*, 43, 531, 1981.
21. **Saltin, B. and Gollnick, P. D.,** Skeletal muscle adaptability: significance for metabolism and performance, in *Handbook of Physiology,* Section 10, Pearchey, L. D., Ed., American Physiological Society, Bethesda, Md., 1983, 555.
22. **Eerbeek, O., Kernell, D., and Verhey, B. A.,** Effects of fast and slow patterns of tonic long-term stimulation on contractile properties of fast muscle in the rat, *J. Physiol. (London),* 352, 73, 1984.
23. **Pette, D.,** Activity-induced fast to slow transitions in mammalian muscle, *Med. Sci. Sports Exercise,* 16, 517, 1984.
24. **Purves, D.,** Long-term regulation in the vertebrate peripheral nervous system, *Int. Rev. Physiol.*, 10, 125, 1976.
25. **Gordon, T. and Stein, R. B.,** Time course and extent of recovery in reinnervated motor units of cat triceps surae muscles, *J. Physiol. (London),* 307, 323, 1982.
26. **Close, R. I.,** Dynamic properties of mammalian skeletal muscles, *Physiol. Rev.*, 52, 129, 1972.
27. **Dhoot, G. K. and Perry, S. V.,** Effects of thyroidectomy on the distribution of the fast and slow forms of troponin I in rat soleus muscle, *FEBS Lett.*, 133, 225, 1981.
28. **Hall-Craggs, E. C. B., Wines, M. M., and Max, S. R.,** Fiber type changes in denervated soleus muscles of the hyperthyroid rat, *Exp. Neurol.*, 80, 252, 1983.
29. **Stewart, D. M.,** The role of tension in muscle growth, in *Regulation of Organ and Tissue Growth,* Goss R. J., Ed., Academic Press, New York, 1972, 77.
30. **Vandenburg, H. H. and Kaufman, S.,** *In vitro* skeletal muscle hypertrophy and Na pump activity, in *Plasticity of Muscle,* Pette, D., Ed., Walter de Gruyter, Berlin, 1979, 494.
31. **Frankeny, J., Holly, R., and Ashmore, C.,** Effects of graded duration of stretch on normal and dystrophic skeletal muscle, *Muscle Nerve,* 6, 269, 1983.
32. **Miledi, R.,** An influence of nerve not mediated by impulses, in *The Effect of Use and Disuse of Neuromuscular Functions,* Gutmann, E. and Hník, R., Eds., Publishing House of the Czechoslvak Academy of Sciences, Prague, 1963, 35.
33. **Freeman, J. A., Mayes, B., Snipes, G. J., and Wikswo, J. P.,** Real-time measurement of minute neuronal current with a circularly vibrating microprobe, *Soc. Neurosci.*, 8, 302, 1982.
34. **Luther, P. W. and Peng, H. B.,** Membrane-related specializations associated with acetylcholine receptor aggregates induced by electrical fields, *J. Cell Biol.*, 100, 235, 1985.
35. **Peng, H. B. and Cheng, P.-C.,** Formation of postsynaptic specializations induced by latex beads in cultured muscle cells, *J. Neurosci.*, 2, 1760, 1982.
36. **Sanes, J. R. and Chiu, A. Y.,** The basal lamina of the neuromuscular junction, *Cold Spring Harbor Symp. Quant. Biol.*, 48, 667, 1983.
37. **Fallon, J. R., Nitkin, R. M., Reist, N. E., Wallace, B. G., and McMahan, U. J.,** Acetylcholine receptor-aggregating factor is similar to molecules concentrated at neuromuscular junctions, *Nature (London),* 315, 571, 1985.
38. **Wallace, B. G., Nitkin, R. M., Reist, N. E., Fallon, J. R., Moayeri, N. N., and McMahan, U. J.,** Aggregates of acetylcholinesterase induced by acetylcholine receptor-aggregating factor, *Nature (London),* 315, 574, 1985.
39. **Bray, J. J., Hubbard, J. F., and Mills, R. G.,** The trophic influence of tetrodotoxin-inactive nerves on normal and reinnervated rat skeletal muscles, *J. Physiol. (London),* 297, 479, 1979.
40. **Thesleff, S. and Sellin, L. C.,** Denervation supersensitivity, *Trends Neurosci.*, 3, 122, 1980.
41. **Guth, L., Kemerer, V. F., Samras, T. A., Warnick, J. E., and Albuquerque, E. X.,** The roles of disuse and loss of neurotrophic function in denervation atrophy of skeletal muscle, *Exp. Neurol.*, 73, 20, 1981.
42. **Drachman, D. B., Stanley, E. F., Pestronk, A., Griffin, J. W., and Price, D. L.,** Neurotrophic regulation of two properties of skeletal muscle by impulse-dependent and spontaneous acetylcholine transmission, *J. Neurosci.*, 2, 232, 1982.
43. **Lømo, T. and Westgaard, R. H.,** Control of ACh sensitivity in rat muscle fibers, *Cold Spring Harbor Symp. Quant. Biol.*, 40, 263, 1975.
44. **Brown, M. C., Hopkins, W. G., and Keynes, R. J.,** Comparison of effects of denervation and botulinum toxin paralysis on muscle properties in mice, *J. Physiol. (London),* 327, 29, 1982.
45. **Cangiano, A.,** Denervation supersensitivity as a model for the neural control of muscle, *Neuroscience,* 14, 963, 1985.
46. **Lømo, T., Westgaard, R. H., Henning, R., and Gundersen, K.,** The response of denervated muscle to long-term electrical stimulation, in *Cell Biology and Clinical Management in Functional Electron Stimulation of Neurones and Muscles,* Carraro, U. and Angelini, C., Eds., Cleup Editore, Padova, 1985, 81.
47. **Lømo, T.,** Neural regulation of membrane and contractile properties of rat skeletal muscles, in *International Series on Sport Sciences,* Vol. 16, Saltin, B., Ed., Human Kinetics Publ., Champaign, Ill., 1986, 27.

48. **Fambrough, D. M.,** Control of acetylcholine receptors in skeletal muscle, *Physiol. Rev.*, 59, 165, 1979.
49. **Harris, W. A.,** Neural activity and development, *Ann. Rev. Physiol.*, 43, 689, 1981.
50. **Grinnell, A. D. and Herrera, A. A.,** Specificity and plasticity of neuromuscular connections: long-term regulation of motoneuron function, *Prog. Neurobiol.*, 17, 203, 1981.
51. **Max, S. R. and Markelonis, G. J.,** Neural control of muscle, *Neurochem. Int.*, 5, 675, 1983.
52. **Fernandez, H. L. and Donoso, J. A.,** The relevance of axoplasmic transport in neurotrophic functions, in *Axoplasmic Transport,* Iqbal, Z., Ed., CRC Press, Boca Raton, Fla., 1986, 209.
53. **Beam, K. G., Caldwell, J. H., and Campbell, D. T.,** Na-channels in skeletal muscle concentrated near the neuromuscular junction, *Nature (London),* 313, 588, 1985.
54. **Betz, W. J., Caldwell, J. H., and Kinnamon, S. C.,** Physiological basis of steady endogenous current in rat lumbrical muscle, *J. Gen. Physiol.*, 83, 175, 1984.
55. **Letinsky, M. S., Fishbeck, K. H., and McMahan, U. J.,** Precision of reinnervation of original postsynaptic sites in frog muscle after a nerve crush, *J. Neurocytol.*, 5, 691, 1976.
56. **Brown, M. C., Jansen, J. K. S., and VanEssen, D.,** Polyneuronal innervation of skeletal muscle in new-born rats and its elimination during maturation, *J. Physiol. (London),* 261, 387, 1976.
57. **Lømo, T., Mirsky, R., and Pockett, S.,** Formation of neuromuscular junctions in adult rats: role of postsynaptic impulse activity, in *Neuromuscular Diseases,* Serratrice, G. et al., Eds., Raven Press, New York, 1984, 393.
58. **Haimann, C., Mallart, A., Ferre, J. T., and Zilber-Gaechelin, N. F.,** Patterns of motor innervation in the pectoral muscle of adult *Xenopus laevis:* evidence for possible synaptic remodelling, *J. Physiol. (London),* 310, 241, 1981.
59. **Nudell, B. M. and Grinnell, A. D.,** Regulation of synaptic position, size, and strength in anuran skeletal muscle, *J. Neurosci.*, 3, 161, 1983.
60. **Bennett, M. R.,** Development of neuromuscular synapses, *Physiol. Rev.*, 63, 915, 1983.
61. **Poo, M.-M.,** Mobility and localization of proteins in excitable membranes, *Annu. Rev. Neurosci.*, 8, 369, 1985.
62. **Kuromi, H. and Kidokoro, Y.,** Nerve disperses pre-existing acetylcholine receptor clusters prior to induction of receptor accumulation in *Xenopus* muscle cultures, *Dev. Biol.*, 103, 53, 1984.
63. **Peng, H. B.,** On the time course of the formation and the elimination of ACh receptor clusters, *Soc. Neurosci.*, 9, 1180, 1983.
64. **Axelsson, J. and Thesleff, S.,** A study of supersensitivity in denervated mammalian skeletal muscle, *J. Physiol. (London),* 147, 178, 1959.
65. **Lømo, T. and Rosenthal, J.,** Control of ACh sensitivity by muscle activity in the rat, *J. Physiol. (London),* 221, 493, 1972.
66. **Lømo, T. and Westgaard, R. H.,** Further studies on the control of ACh sensitivity by muscle activity in the rat, *J. Physiol. (London),* 252, 603, 1975.
67. **Stanley, E. F. and Drachman, D. B.,** Denervation and the time course of resting membrane potential changes in skeletal muscle *in vivo, Exp. Neurol.*, 69, 253, 1980.
68. **Hall, Z. W.,** Multiple forms of acetylcholinesterase and their distribution in endplate and non-endplate regions of rat diaphragm muscle, *J. Neurobiol.*, 4, 343, 1973.
69. **Vigny, M., Koenig, J., and Rieger, F.,** The motor endplate specific form of acetylcholinesterase: appearance during embryogenesis and reinnervation of rat muscle, *J. Neurochem.*, 27, 1347, 1976.
70. **Lømo, T., Massoulié, J., and Vigny, M.,** Stimulation of denervated rat soleus muscle with fast and slow activity patterns induces different expression of acetylcholinesterase molecular forms, *J. Neurosci.*, 5, 1180, 1985.
71. **Westgaard, R. H.,** Influence of activity on the passive electrical properties of denervated soleus muscle fibers in the rat, *J. Physiol. (London),* 251, 683, 1975.
72. **Lømo, T., Westgaard, R. H., and Dahl, H.,** Contractile properties of muscle: control by pattern of muscle activity in the rat, *Proc. R. Soc. London Ser. B,* 187, 99, 1974.
73. **Lømo, T., Westgaard, R. H., and Engebretsen, L.,** Different stimulation patterns affect contractile properties of denervated rat soleus muscles, in *Plasticity of Muscle,* Pette, D., Ed., Watter de Gruyter, New York, 1980, 297.
74. **Guth, L. and Albuquerque, E. X.,** The neurotrophic regulation of resting membrane potential and extrajunctional acetylcholine sensitivity in mammalian skeletal muscle, in *Muscle Regeneration,* Mauro, A., Ed., Raven Press, New York, 1979, 405.
75. **Davis, H. L.,** Is electrostimulation beneficial to denervated muscles? A review of results from basic research, *J. Physiol. Can.*, 35, 306, 1983.
76. **Purves, D. and Sakmann, B.,** The effect of contractile activity on fibrillation and extrajunctional acetylcholine-sensitivity in rat muscle maintained in organ culture, *J. Physiol. (London),* 237, 157, 1974.
77. **Luco, J. V. and Eyzaguirre, C.,** Fibrillation and hypersensitivity to ACh in denervated muscle: effect of length of degenerating nerve fibers, *J. Neurophysiol.*, 18, 65, 1955.

78. **Harris, J. B. and Thesleff, S.,** Nerve stump length and membrane changes in denervated skeletal muscle, *Nature (London) New Biol.,* 236, 60, 1972.
79. **Card, D. J.,** Denervation: sequence of neuromuscular degenerative changes in rats and the effect of stimulation, *Exp. Neurol.,* 54, 251, 1977.
80. **Fernandez, H. L., Duell, M. J., and Festoff, B. W.,** Neurotrophic control of 16S acetylcholinesterase at the vertebrate neuromuscular junction, *J. Neurobiol.,* 10, 441, 1979.
81. **Olek, A. J. and Robbins, N.,** A nerve stump dependent appearance of junctional and perijunctional acetylcholine receptors in organ culture, *Neuroscience,* 9, 1771, 1981.
82. **Miledi, R. and Slater, C. R.,** Electrophysiology and electronmicroscopy of rat neuromuscular junctions after nerve degeneration, *Proc. R. Soc. London Ser. B,* 169, 289, 1968.
83. **Pestronk, A., Drachman, D. B., and Griffin, J. W.,** Effect of muscle disuse on acetylcholine receptors, *Nature (London),* 260, 352, 1976.
84. **Cangiano, A., Lutzemberger, L., and Nicotra, L.,** Non-equivalence of impulse blockade and denervation in the production of membrane changes in rat skeletal muscles, *J. Physiol. (London),* 273, 691, 1977.
85. **Stanley, E. F. and Drachman, D. B.,** Effect of disuse on the resting membrane potential of skeletal muscle, *Exp. Neurol.,* 64, 231, 1979.
86. **Cangiano, A. and Lutzemberger, L.,** Partial denervation affects both denervated and innervated fibers in the mammalian skeletal system, *Science,* 196, 542, 1977.
87. **Tiedt, T. N., Albuquerque, E. X., and Guth, L.,** Degenerating nerve fiber products do not alter physiological properties of adjacent innervated skeletal muscle fibers, *Science,* 198, 839, 1978.
88. **Cangiano, A. and Lutzemberger, L.,** Partial denervation in inactive muscle affects innervated and denervated fibers equally, *Nature (London),* 285, 233, 1980.
89. **Cangiano, A. and Lutzemberger, L.,** Normal EDL and diaphragm muscles differ in their sensitivity to tetrodotoxin, *Acta Physiol. Scand.,* 108, 205, 1980.
90. **Catterall, W. A.,** Neurotoxins that act on voltage-sensitive sodium channels in excitable membranes, *Annu. Rev. Pharmacol. Toxicol.,* 20, 15, 1980.
91. **Cangiano, A., Magherini, P. C., Pasino, E., Pellegrino, M., and Risaliti, R.,** Interaction of inactivity and nerve breakdown products in the origin of acute denervation changes in rat skeletal muscles, *J. Physiol. (London),* 335, 345, 1984.
92. **Brown, M. C., Holland, R. L., and Ironton, R.,** Degenerating nerve products affect innervated muscle fibers, *Nature (London),* 275, 652, 1978.
93. **Brown, M. C. and Holland, R. L.,** A central role for denervated tissues in causing nerve sprouting, *Nature (London),* 282, 724, 1979.
94. **Abercombie, M.,** Localized formation of new tissue in an adult mammal, *Symp. Soc. Exp. Biol.,* 11, 235, 1957.
95. **Mescher, A.,** Growth factors from nerves and their roles during limb regeneration, in *Limb Development and Regeneration,* Fallon, J. F. and Caplan, A. I., Eds., Alan R. Liss, New York, 1982, 501.
96. **Yachnis, A. T. and Mescher, A. L.,** Stimulation of DNA synthesis in Balb/c3T3 cells by peripheral nerve degenerating *in vitro, Exp. Neurol.,* 76, 139, 1982.
97. **Yoshino, J. E., Dinneen, M. P., Lewis, B. L., Meador-Woodruff, J. H., and Devries, G. H.,** Differential proliferative responses of cultured Schwann cells to axolemma- and myelin-enriched fractions. I. Biochemical studies, *J. Cell Biol.,* 99, 2309, 1984.
98. **Murray, M. A. and Robbins, N.,** Cell proliferation in denervated muscle: identity and origin of dividing cells, *Neuroscience,* 7, 1823, 1982.
99. **Katz, B. and Miledi, R.,** The development of acetylcholine sensitivity in nerve-free segments of skeletal muscle, *J. Physiol. (London),* 170, 389, 1964.
100. **Jones, R. and Vrbová, G.,** Two factors responsible for the development of denervation hypersensitivity, *J. Physiol. (London),* 236, 517, 1974.
101. **Jones, R. and Vyskocil, F.,** An electrophysiological examination of the changes in skeletal muscle fibers in response to degenerating nerve tissue, *Brain Res.,* 88, 309, 1975.
102. **Guth, L., Richman, E., Barrett, C. P., Warnick, J. E., and Albuquerque, E. X.,** The mechanism by which degenerating peripheral nerve produces extrajunctional acetylcholine sensitivity in mammalian skeletal muscle, *Exp. Neurol.,* 68, 465, 1980.
103. **Podleski, T. R., Axelrod, D., Ravdin, P., Greenberg, I., Johnson, M. M., and Salpeter, M. M.,** Nerve extract induces increase and redistribution of acetylcholine receptors on cloned muscle cells, *Proc. Natl. Acad. Sci. U.S. A.,* 75, 2035, 1978.
104. **Christian, C. N., Daniels, M. P., Sugiyama, H., Vogel, Z., Jacques, L., and Nelson, P. G.,** A factor from neurons increases the number of acetylcholine receptor aggregates on cultured muscle cells, *Proc. Natl. Acad. Sci. U.S.A.,* 75, 4011, 1978.
105. **Jessel, T. M., Siegel, R. E., and Fischbach, G. D.,** Induction of acetylcholine receptors on cultured skeletal muscle by a factor extracted from brain and spinal cord, *Proc. Natl. Acad. Sci. U.S.A.,* 76, 5397, 1979.

106. **Karpati, G.,** Denervation and disuse atrophy of skeletal muscles-involvement of endogenous glucocorticoid hormones?, *Trends Neurosci.*, 7, 61, 1984.
107. **Shainberg, A. and Burstein, M.,** Decrease of acetylcholine receptor synthesis in muscle cultures by electrical stimulation, *Nature (London)*, 264, 368, 1976.
108. **Forrest, J. W., Mills, R. G., Bray, J. J., and Hubbard, J. I.,** Calcium-dependent regulation of membrane potential and extrajunctional acetylcholine receptors of rat skeletal muscle, *Neuroscience*, 6, 741, 1981.
109. **McManaman, J. L., Blosser, J. C., and Appel, S. H.,** Inhibitors of membrane depolarization regulate acetylcholine receptor synthesis by a calcium-dependent, cyclic nucleotide-independent mechanism, *Biochim. Biophys. Acta.*, 720, 28, 1982.
110. **Kallo, J. R. and Steinhardt, R. A.,** The regulation of extrajunctional acetylcholine receptors in the denervated rat diaphragm muscle in culture, *J. Physiol. (London)*, 344, 433, 1983.
111. **Sherman, S. J., Chrivia, J., and Catterall, W. A.,** Cyclic adenosine 3′:5′-monophosphate and cytosolic calcium exert opposing effects on biosynthesis of tetrodotoxin-sensitive sodium channels in rat muscle cells, *J. Neurosci.*, 5, 1570, 1985.
112. **Betz, H. and Changeux, J. -P.,** Regulation of muscle acetylcholine receptor synthesis *in vitro* by cyclic nucleotide derivatives, *Nature (London)*, 278, 749, 1979.
113. **Rubin, L. L., Schuetze, S. M., Weill, C. L. and Fischbach, G. D.,** Regulation of acetylcholinesterase appearance at neuromuscular junctions *in vitro, Nature (London)*, 283, 264, 1980.
114. **Gilliatt, R. W., Westgaard, R. H., and Williams, I. R.,** Extrajunctional acetylcholine sensitivity of inactive muscle fibers in the baboon during prolonged nerve pressure block, *J. Physiol. (London)*, 280, 499, 1978.
115. **Oh, T. H., Markelonis, G. J., Reier, P. J., and Zalewski, A. A.,** Persistence in degenerating sciatic nerve of substances having a trophic influence upon cultured muscle, *Exp. Neurol.*, 67, 646, 1980.
116. **Cangiano, A., Lutzemberger, L., and Magherini, P. C.,** Extrajunctional membrane properties of denervated skeletal muscles reinnervated by intact and impulse-blocked nerves, *Soc. Neurosci.*, 6, 384, 1980.
117. **Johns, T. R. and Thesleff, S.,** Effects of motor inactivation on the chemical sensitivity of skeletal muscle, *Acta Physiol. Scand.*, 51, 136, 1961.
118. **Mathers, D. A. and Thesleff, S.,** Studies on neurotrophic regulation of murine skeletal muscle, *J. Physiol. (London)*, 282, 105, 1978.
119. **Berg, D. K. and Hall, Z. W.,** Increased extrajunctional acetylcholine sensitivity produced by chronic post-synaptic neuromuscular blockade, *J. Physiol. (London)*, 244, 659, 1975.
120. **Davis, H. L. and Kiernan, J. A.,** Effect of nerve extract on number of acetylcholine receptors in denervated muscles of rats, *Exp. Neurol.*, 83, 108, 1984.
121. **Albuquerque, E. X., Schuh, F. T., and Kauffman, F. C.,** Early membrane depolarization of the fast mammalian muscle after denervation, *Pfluegers Arch.*, 328, 36, 1971.
122. **Bray, J. J., Forrest, J. W., and Hubbard, J. I.,** Evidence for the role of non-quantal acetylcholine in the maintenance of the membrane potential of rat skeletal muscle, *J. Physiol. (London)*, 326, 285, 1982.
123. **Rochel-Landau, S. and Robbins, N.,** Denervation induced degeneration of neuromuscular junction: the role of nicotinic interactions, *Soc. Neurosci.*, 9, 934, 1983.
124. **Robbins, N.,** Neurotrophic regulation of rat muscle glucose-6-phosphate dehydrogenase *in vitro, Brain Res.*, 225, 387, 1981.
125. **Ochs, S.,** *Axoplasmic Transport and Its Relation to Other Nerve Functions*, John Wiley & Sons, New York, 1982, 1.
126. **Hofmann, W. W. and Thesleff, S.,** Studies of the trophic influence of nerve on skeletal muscle, *Eur. J. Pharmacol.*, 20, 256, 1972.
127. **Albuquerque, E. X., Warnick, J. E., Tasse, J. R., and Sansone, F. M.,** Effects of vinblastine and colchicine on neural regulation of the fast and slow skeletal muscles of the rat, *Exp. Neurol.*, 37, 607, 1972.
128. **Cangiano, A.,** Acetylcholine supersensitivity: the role of neurotrophic factors, *Brain Res.*, 58, 255, 1973.
129. **Lømo, T.,** Neurotrophic control of colchicine effects on muscle?, *Nature (London)*, 249, 473, 1974.
130. **Cangiano, A. and Fried, J. A.,** The production of denervation-like changes in rat muscle by colchicine, without interference with axonal transport or muscle activity, *J. Physiol. (London)*, 265, 63, 1977.
131. **Tiedt, T. N., Wisler, P. L., and Younkin, S. G.,** Neurotrophic regulation of resting membrane potential and acetylcholine sensitivity in rat extensor digitorum longus muscle, *Exp. Neurol.*, 57, 766, 1977.
132. **Purves, D.,** Functional and structural changes in mammalian sympathetic neurons following colchicine application to post-ganglionic nerves, *J. Physiol. (London)*, 259, 159, 1976.
133. **Lømo, T. and Slater, C. R.,** Control of acetylcholine sensitivity and synapse formation by muscle activity, *J. Physiol. (London)*, 275, 391, 1978.
134. **Fernandez, H. L. and Inestrosa, N. C.,** Role of axoplasmic transport in neurotrophic regulation of muscle endplate acetylcholinesterase, *Nature (London)*, 262, 55, 1976.
135. **Fernandez, H. L., Patterson, M. R., and Duell, M. J.,** Neurotrophic control of 16S acetylcholinesterase from mammalian skeletal muscle in organ culture, *J. Neurobiol.*, 11, 557, 1980.

136. **Davis, H. L.,** Trophic action of nerve extract on denervated skeletal muscle *in vivo:* dose dependency, species specificity, and timing of treatment, *Exp. Neurol.,* 80, 383, 1983.
137. **Younkin, S. G., Breet, R. S., Davey, B., and Younkin, L. H.,** Substances moved by axonal transport and released by nerve stimulation have an innervation-like effect on muscle, *Science,* 200, 1292, 1978.
138. **Davey, B., Younkin, L. H., and Younkin, S. G.,** Neural control of skeletal muscle cholinesterase: a study using organ-cultured rat muscle, *J. Physiol. (London),* 289, 501, 1979.
139. **Ranish, N. A., Dettbarn, W. -D., and Wecker, L.,** Nerve stump length-dependent loss of acetylcholinesterase activity in endplate regions of rat diaphragm, *Brain Res.,* 191, 379, 1980.
140. **Markelonis, G. J. and Oh, T. H.,** A sciatic nerve protein has a trophic effect on development and maintenance of skeletal muscle cells in culture, *Proc. Natl. Acad. Sci. U.S.A.,* 76, 2470, 1979.
141. **Markelonis, G. J., Oh, T. H., Eldefrawi, M. E., and Guth, L.,** Sciatin: a myotrophic protein increases the number of acetycholine receptors and receptor clusters in cultured skeletal muscle, *Dev. Biol.,* 89, 353, 1982.
142. **Matsuda, R., Spector, D., and Skohman, R. C.,** There is selective accumulation of a growth factor in chicken skeletal muscle. I. Transferrin accumulation in adult anterior latissimus dorsi, *Dev. Biol.,* 103, 267, 1984.
143. **Lømo, T. and Slater, C. R.,** Acetylcholine sensitivity of developing ectopic nerve-muscle junctions in adult rat soleus muscles, *J. Physiol. (London),* 303, 173, 1980.
144. **Weinberg, C. B., Sanes, J. R., and Hall, Z. W.,** Formation of neuromuscular junctions in adult rats: accumulation of acetylcholine receptors, acetylcholinesterase and components of synaptic basal lamina, *Dev. Biol.,* 84, 255, 1981.
145. **Waerhaug, O., Korneliussen, H., and Sommerschild, H.,** Morphology of motor nerve terminals on rat soleus muscle fibers reinnervated by the original and by a "foreign" nerve, *Anat. Embryol. (Berlin),* 151, 1, 1977.
146. **Lømo, T. and Slater, C. R.,** Control of junctional acetylcholinesterase by neural and muscular influences in the rat, *J. Physiol. (London),* 303, 191, 1980.
147. **Brenner, H. R. and Sakmann, B.,** Neurotrophic control of channel properties at neuromuscular synapses of rat muscle, *J. Physiol. (London),* 337, 159, 1983.
148. **Kuffler, D. P., Thompson, W. J., and Jansen, J. K. S.,** The fate of foreign endplates in cross-innervated rat soleus muscle, *Proc. R. Soc. London Ser. B,* 208, 189, 1980.
149. **Brenner, H. R., Meier, T., and Widmer, B.,** Functional and structural differentiation of subsynaptic membrane in rat muscle denervated at early stages of endplate development, *Nature (London),* 305, 536, 1983.
150. **Dennis, M. J., Ziskind-Conhaim, L., and Harris, A. J.,** Development of neuromuscular junctions in rat embryos, *Dev. Biol.,* 81, 266, 1981.
151. **Dennis, M. J.** , Development of the neuromuscular junction: inductive interactions between cells, *Annu. Rev. Neurosci.,* 4, 43, 1981.
152. **Bennett, M. R. and Pettigrew, A. G.,** The formation of neuromuscular synapses, *Cold Spring Harbor Symp. Quant. Biol.,* 40, 409, 1975.
153. **Jansen, J. K. S., Lømo, T., Nicolaysen, K., and Westgaard, R. H.,** Hyperinnervation of skeletal muscle fibers: dependence on muscle activity, *Science,* 181, 559, 1973.
154. **Cangiano, A., Lømo, T., Lutzemberger, L., and Sveen, O.,** Effects of chronic nerve conduction block on formation of neuromuscular junctions and junctional AChE in the rat, *Acta Physiol. Scand.,* 109, 283, 1980.
155. **Benoit, P. and Changeux, J. -P.** Consequences of tenotomy on the evolution of multi-innervation in developing rat soleus muscle, *Brain Res.,* 99, 354, 1975.
156. **Thompson, W. J. and Jansen, J. K. S.,** The extent of sprouting of remaining motor units in partly denervated immature and adult rat soleus muscle, *Neuroscience,* 2, 523, 1977.
157. **O'Brien, R. A. D., Ostberg, A. J. C., and Verbová, G.,** Observations on the elimination of polyneural innervation in developing mammalian skeletal muscles, *J. Physiol. (London),* 282, 571, 1978.
158. **Thompson, W. J.,** Synapse elimination in neonatal rat muscle is sensitive to pattern of muscle use, *Nature (London),* 302, 614, 1983.
159. **Ribchester, R. R. and Taxt, R.,** Motor unit size and synaptic competition in rat lumbrical muscles reinnervated by active and inactive motor axons, *J. Physiol. (London),* 344, 89, 1983.
160. **Archer, S. M., Dubin, M. W., and Stark, L. A.,** Abnormal development of kitten retino-geniculate connectivity in the absence of action potentials, *Science,* 217, 743, 1982.
161. **Stryker, M. P., Frank, E., and Schmidt, J. T.,** Developmental strategies for selective synapse formation, *Soc. Neurosci.,* 9, 452, 1983.
162. **Reh, T. A. and Constantine-Paton, M.,** Eye-specific segregation requires neural activity in three-eyed *Rana pipiens, J. Neurosci.,* 5, 1132, 1985.

163. **Sanes, J. R., Marshall, L. M., and McMahan, U. J.,** Reinnervation of muscle fiber basal lamina after removal of myofibers. Differentiation of regenerating axons at original synaptic sites, *J. Cell Biol.*, 78, 176, 1978.
164. **Burden, S. J., Sargent, P. B., and McMahan, U. J.,** Acetylcholine receptors in regenerating muscle accumulate at original synaptic sites in the absence of the nerve, *J. Cell Biol.*, 82, 412, 1979.
165. **Kullberg, R., Owens, J. L., and Vickers, J.,** Development of synaptic currents in immobilized muscle of *Xenopus laevis*, *J. Physiol. (London)*, 364, 57, 1985.
166. **Taxt, T.,** Cross-innervation of fast and slow twitch muscles by motor axons of the sural nerve in the mouse, *Acta Physiol. Scand.*, 117, 331, 1983.
167. **Frémont, P. H., LeRay, C. F., and Le Douarin, G. H.,** *In vitro* differentiation in the absence of nerve of avian myoblasts derived from slow and fast muscle rudiments, *Cell Differ.*, 13, 325, 1983.
168. **McLennan, I. S.,** The development of the pattern of reinnervation in chicken hindlimb muscles: evidence for specification of nerve-muscle connections, *Dev. Biol.*, 97, 229, 1983.
169. **Thompson, W. J., Sutton, L. A., and Riley, D. A.,** Fiber type composition of single motor units during synapse elimination in neonatal rat soleus muscle, *Nature (London)*, 309, 709, 1984.
170. **Landmesser, L.,** The development of specific motor pathways in the chicken embryo, *Trends Neurosci.* 7, 336, 1984.
171. **Phillips, W. D. and Bennett, M. R.,** Differentiation of fiber types in wing muscles during embryonic development: effect of neural tube removal, *Dev. Biol.*, 106, 457, 1984.
172. **Wigston, D. J. and Sanes, J. R.,** Selective reinnervation of intercostal muscles transplanted from different segmental levels to a common site, *J. Neurosci.*, 5, 1208, 1985.
173. **Werning, A., Carmody, J. J., Anzil, A. P., Hansert, E., Marciniak, M., and Zucker, H.,** Persistence of nerve sprouting with features of synapse remodelling in soleus muscles of adult mice, *Neuroscience*, 11, 241, 1984.
174. **Cardasis, C. A. and Padykula, H. A.,** Ultrastructural evidence indicating reorganization at the neuromuscular junction in the normal rat soleus muscle, *Anat. Rec.*, 200, 41, 1981.
175. **Lnenicka, G. A. and Atwood, H. L.,** Age-dependent conversion of synaptic properties at the crayfish neuromuscular junction through altered motoneuron activity, *Soc. Neurosci.*, 9, 53, 1983.
176. **Herrera, A. A. and Grinnell, A. D.,** Transmitter release from frog motor nerve terminals depends on motor unit size, *Nature (London)*, 287, 649, 1980.
177. **Trussell, L. O. and Grinnell, A. D.,** The regulation of synaptic strength within motor units of the frog cutaneous pectoris muscle, *J. Neurosci.*, 5, 243, 1985.

Chapter 6

TROPHIC INFLUENCES OF NEUROGENIC SUBSTANCES ON SKELETAL MUSCLE DIFFERENTIATION AND GROWTH IN VITRO

Tae H. Oh, George J. Markelonis, and Sang H. Shim

TABLE OF CONTENTS

I. INTRODUCTION

Innervation exerts a trophic influence on the morphological, physiological, and metabolic properties of skeletal muscle.[1-3] For example, innervation regulates the rate of morphological maturation and the degree of fetal muscle differentiation;[4-9] resting membrane potentials (RMP);[10] junctional acetylcholine (Ach) sensitivity;[11-13] endplate acetylcholinesterase (AChE) activity;[14] and the structural integrity of adult muscle.[15] To explain the mechanism(s) by which motor nerves regulate muscle properties, three hypotheses have been proposed: (1) muscle activity,[16-18] (2) ACh transmission,[19-20] and (3) neurally derived trophic substances.[1-3] Although considerable data support each of these hypotheses their relative contributions to the neural control of muscle metabolic processes has yet to be determined.

Since it has proved to be extraordinarily difficult to assess the trophic influences of nerve on muscle in vivo, we, along with many others, have employed an in vitro bioassay system in order to test trophic substances from neural tissues. Using muscle cultures, a number of laboratories have identified putative trophic neurogenic substances which affect muscle (Table 1).[21-31] A variety of neurally derived trophic substances which are unrelated to impulse activity have been shown to regulate many properties of cultured muscle cells, including morphological maturation, macromolecular synthesis, AChE activity, and ACh receptors (AChR). With the exception of transferrin,[23-25] however, the biochemical characterization of such substances has not been fully accomplished. Although no unique trophic substances synthesized by the motor neuron have yet been identified, the possibility of the existence of one or more such substances remains. Theoretically, one would anticipate that such substances would (1) be present in the motor nerve; (2) be synthesized in the cell body of the motor neuron; (3) be carried to the motor nerve terminal by axoplasmic transport; (4) be released from the nerve terminal; (5) transverse the synaptic cleft and interact with the target cells; and (6) initiate or maintain molecular modifications in skeletal muscle or mimic the effects of innervation (see Chapter 1).

The objectives of this chapter are to review the data on neurally derived trophic substances which have been characterized and partially purified and to discuss the biological activities of these substances in vitro. In addition, we summarize our recent data regarding transferrin (sciatin).

II. MORPHOLOGICAL DEVELOPMENT

The development of skeletal muscle *in vivo* requires innervation.[1] The initial differentiation of myoblasts and the development of myotubes take place even in denervated fetal muscle.[4] However, the process of maturation and further differentiation is slowed down as compared to innervated muscles, and in some fibers differentiation is arrested at the myotube stage. Ultimately, denervation atrophy and degeneration of muscle fibers occur.[4-9]

Further differentiation, maturation, and long-term maintenance of cultured muscle fibers are dependent on innervation by spinal neurons.[32-34] Long-term suppression of neuromuscular transmission by *d*-tubocurarine does not interfere with the trophic support of muscle development by spinal cord neurons in vitro.[34] This demonstrates that the trophic effect of the innervation on muscle maturation is not dependent on muscle activity (but see Chapter 4). Furthermore, dorsal root ganglia physically separated from adult newt forelimb by Millipore® filters simulate blastemal cell mitosis.[35] Thus, these results indicate that the nerve effect is probably mediated by a diffusible substance produced by sensory ganglia.

Over the past several years, a number of trophic substances isolated from neural tissues have been shown to promote morphological maturation and maintenance of cultured muscle (Table 2). Soluble extracts of brain and spinal cord tissues enhance the morphological maturation of muscle cells and maintain nature myotubes for several months in the absence

Table 1
MUSCLE TROPHIC SUBSTANCES

Source	Molecular weight	Target muscle culture	Trophic activity	Ref.
Rat brain	34,000	Adult newt	AChE activity	21
Chicken brain and liver	≳5,000	Chick embryo	DNA synthesis	22
Chicken sciatic nerve	84,000	Chick embryo	Myogenesis	23
Chicken ischiatic-peroneal nerve	80,000	Chick embryo	DNA synthesis	24
Chicken serum	80,000	Chick embryo	Myogenesis	25
Chicken brain	<20,000	Adult newt	AChE activity	26
Mouse spinal cord	<10,000	Adult mouse	TTX sensitivity	27
Chicken brain and spinal cord	1,000	Chick embryo	AChR synthesis and aggregation	28
Rat embryo brain and spinal cord	– 100,000	Rat L_6 clone	AChR synthesis and aggregation	29
Medium conditioned by neuroblastoma-glioma cells	>150,000	Rat embryo	AChR aggregation	30
Medium conditioned by rat embryo spinal cord neurons	≥50,000	Rat embryo	AChR aggregation	31

From Oh, T. H. and Markelonis, G. J., in *Growth and Maturation Factors,* Vol. 2, Guroff, G., Ed., John Wiley & Sons, New York, 1984, 55. With permission.

Table 2
TROPHIC EFFECTS OF NERVE TISSUES AND NERVE EXTRACTS ON MUSCLE MORPHOLOGY IN VITRO

Source of trophic activity	Target muscle culture	Trophic effect	Ref.
Innervation	Adult and fetal rat	Myofiber maturation and maintenance	32, 33
Spinal cord tissue	Chick embryo	Myotube maturation	34
Dorsal root ganglia	Adult newt	Blastema cell proliferation	35
Brain and spinal cord extract	Chick embryo	Myotube maturation and maintenance	36
Brain extract	Fetal calf	Myoblast proliferation	37
Pituitary fibroblast growth factor	Adult rat	Satellite cell proliferation	38
Sciatic nerve extract	Chick embryo	Myoblast proliferation, myotube maturation, and maintenance	41—44
Degenerated sciatic nerve extract	Chick embryo	Myotube maturation and maintenance	46
Innervated and denervated muscle extract	Chick embryo	Myotube maturation	44, 47

of innervation.[36] Bovine brain and pituitary extracts stimulate the proliferation of myoblasts seeded at low densities.[37] Bovine pituitary fibroblast growth factor (FGF) also has a mitogenic effect on satellite cells from adult muscles in culture.[38] Since trophic influences appear to be transmitted from the cell body of a motor neuron to its peripheral axon by axoplasmic transport,[39,40] it seems appropriate to analyze the peripheral nerve for possible trophic properties (see Chapter 3). In fact, several investigators have shown that soluble extracts of adult sciatic nerves have a profound effect on proliferation, maturation, and maintenance of muscle cells in culture.[41-44] The active principle in these extracts is a protein of high molecular weight.[36,37,41-43] Interestingly, our laboratory has succeeded purifying such a trophic protein of high molecular weight from adult chicken sciatic nerves.[23,45]

Evidence which appears to contradict the trophic substance hypothesis has also been obtained from tissue culture experiments. Soluble proteins from normal and degenerated sciatic nerves are equally effective in promoting the morphological maturation and long-term maintenance of cultured muscle cells.[46] These results appear to rule out the possibilities that the trophic protein is progressively depleted in transected nerves or is produced by motor neurons. Extracts obtained from either adult innervated or denervated muscle also enhance the maturation of muscle cells in culture.[44,47] However, when extracts are prepared from innervated and/or denervated muscles exsanguinated in vivo prior to tissue dissection, they show only marginal trophic activity, suggesting that serum components may contribute significantly to the trophic effect of tissue extracts.[47] We have shown that a protein (sciatin) isolated from chicken sciatic nerves is practically identical to serum transferrin.[48,49]

III. MACROMOLECULAR SYNTHESIS

Innervation has an important role in promoting the homeostatic maintenance of macromolecular synthesis in muscle.[50,51] Limb regeneration in the adult newt has been one of the best model systems for the study of neural influences on macromolecular synthesis.[52] Regeneration of the newt's limb is dependent upon the presence of nerves at the amputation surface.[53] When the regenerating limb is denervated, the rates of DNA, RNA, and protein synthesis ultimately decrease.[52,54-56] Moreover, the nerve affects the absolute rate of protein synthesis without altering the rate of protein degradation or the size of the amino acid pool.[55,57] Because infusion of nerve extracts into denervated newt limbs restores to normal the rate of DNA and protein synthesis in blastemas *in vivo*,[55,58,59] it has been suggested that neural regulation of macromolecular synthesis in regenerating limbs is mediated by a diffusible substance produced by nerves.[58]

Tissue culture of newt limb regenerates has proven useful for assaying trophic substances regulating rates of macromolecular synthesis. For example, addition of chick embryonic brain extracts to the culture medium of newt limb regenerates prevents the decrease in protein synthesis that otherwise results from denervation.[60] Further studies show that the active substance is likely to be a basic protein with a molecular weight of 13.5 kdaltons.[61] Similar extracts also stimulate DNA synthesis in chick muscle cells in culture.[22] The biologically active substances in these extracts consist of two proteins, one of high molecular weight (>5 kdaltons) and one of low molecular weight (<5 kdaltons), which must be present together in order to produce full activity.[22] Crude extracts of adult chicken sciatic nerves enhance DNA and protein syntheses in cultured chick muscle cells.[41,44] A trophic protein (sciatin)[23,45] purified from chicken sciatic nerves in our laboratory selectively stimulates the rate of protein synthesis without affecting the rate of protein degradation.[62] Furthermore, the presence of tetrodotoxin (TTX), an agent which abolished spontaneous muscle contraction, has no effect on sciatin-stimulated protein synthesis, suggesting that this action of sciatin is not dependent upon muscle activity.[62] A similar protein has been identified and partially purified from chicken ischiatic-peroneal nerves; it has a stimulatory effect on DNA synthesis in chick muscle cultures.[24] However, recent evidence indicates that both this protein and sciatin are identical to chicken serum transferrin.[48,49,63]

It has been known for some time that the motor nerve exerts a regulatory influence on the synthesis of muscle-specific contractile proteins.[64-68] Although the motor nerve is necessary for the regulation of tropomyosin and troponin expression,[71] recent evidence indicates that innervation is not required for the normal sequential transition in the myosin heavy chain (MHC) isoforms from the embryonic to neonatal and to adult fast myosin type during muscle development.[69-71]

Several in vitro studies have revealed that only embryonic MHC is found in aneural muscle cultures, although such cultures show fully mature, cross-striated myotubes.[72-74] In long-term nerve-muscle cultures, however, the adult fast MHC appears in the innervated, regenerated muscle fibers.[75] Since innervation does not appear to be necessary for adult fast MHC expression in vivo, the role of the nerve in the induction of adult fast myosin in the nerve-muscle cultures remains to be determined. A soluble extract of adult chicken sciatic nerve also stimulates the synthesis of the adult slow MHC in chick muscle cells in culture,[76] suggesting that a trophic substance in the extract can activate slow MHC gene expression.

IV. ACETYLCHOLINESTERASE

In vertebrate muscles AChE is especially concentrated at the motor endplates.[77-80] At normal adult motor endplates, AChE is partly associated with the basal lamina of the synaptic cleft.[81-83] Furthermore, sedimentation analysis of the enzyme has revealed the existence of several molecular forms, and the high-molecular-weight form (asymmetric A_{12} AChE) is concentrated at the endplate.[84] During synaptogenesis, AChE becomes localized at the developing motor endplate shortly after the nerve has established contact with the region of the muscle fiber.[85-92] When adult muscle is denervated, endplate AChE decreases, and, conversely, reinnervation of the denervated muscle restores AChE to normal levels.[14,93,94] Similarly, the endplate form of AChE decreases after denervation and reappears following reinnervation at either the original endplate or at ectopic sites.[95,96] Furthermore, following the irreversible inhibition of AChE by diisopropylfluorophosphate (DFP), *de novo* synthesis of the enzyme occurs in normally innervated muscles, but not in denervated muscles.[97,98] Thus, it is apparent that the motor nerve regulates muscle endplate AChE. Although the mechanism by which the nerve regulates endplate AChE is not known, trophic substances,[14,99] ACh transmission,[100] muscle activity,[91,100-105] and nerve-muscle contact[106,107] have all been implicated in this phenomenon (see Chapter 4).

Numerous studies have demonstrated that AChE becomes localized at sites of nerve-muscle contact in co-cultures of spinal cord neurons and muscle cells.[108-110] AChE accumulates at synaptic contacts in co-cultures of *Xenopus* spinal cord and muscle cells in the absence of muscle contraction.[111] On the other hand, several studies have indicated that muscle activity is important for the localization of synaptic AChE in vitro (see Chapter 4). For example, agents which block neuromuscular transmission prevent the accumulation of synaptic AChE in rat and chick muscle cultures.[112,113] Therefore, it appears that the role of muscle activity in regulating the accumulation of synaptic AChE may be different in cultured muscle cells from different species. Furthermore, when *Xenopus* muscle cells are cultured without nerves, AChE patches are seen at sites where the sarcolemma contains basal lamina-like material.[114] Thus, the role of nerve in regulating the accumulation of synaptic AChE remains to be determined.

A number of studies have indicated that neurally derived substances regulate AChE activity or molecular forms of the enzyme in cultured muscle cells (Table 3). For example, AChE activity in chick muscle cultures is increased by the presence of spinal cord explants regardless of whether they innervate or do not innervate the muscle.[115] This indicates that the regulation of muscle AChE activity may be mediated by a diffusible substance produced by spinal cord tissue. Spinal cord neurons also induce the high-molecular-weight A_{12} AChE form in muscle cells in culture.[116] When muscle cells from 13- to 14-day-old rat embryos are co-cultured with spinal cord neurons, the A_{12} AChE form appears in the mixed cultures, whereas muscle cells alone do not synthesize the heavy form. By contrast, muscle cells from 18-day-old embryos, which have already made contact with neurons in vivo before being plated, synthesize the endplate form of AChE in the absence of neurons.[116] The synthesis of A_{12} AChE by these cells in the absence of nerve seems inconsistent with the fact that this enzymatic

Table 3
TROPHIC EFFECTS OF NERVE TISSUES AND NERVE EXTRACTS ON MUSCLE AChE IN VITRO

Source of trophic activity	Target muscle culture	Trophic effect on AChE	Ref.
Innervation and spinal cord tissue	Chick embryo	Total	115
Spinal cord neuron	Rat embryo	A_{12} (16S)	116
Sensory ganglion	Adult newt	Total	117
Brain extract	Chick embryo and adult newt	Total	36, 118
Brain proteins			
Mol wt <20,000	Adult newt	A_8 (12S) and A_{12} (16S)	26
Mol wt 34,000	Adult newt	A_{12} (16S)	21
Sciatic nerve extract	Chick embryo and adult chicken	Total	41, 119, 120
Medium conditioned by neurally stimulated muscle	Adult rat	Total	121
Pituitary fibroblast growth factor	Adult newt	G_1 (4S), G_2 (5S), G_4 (10S), and A_8 (12S)	122
β-Endorphin	Rat embryo	G_1 (4S) and G_2 (6S)	123

form disappears in the denervated adult muscles. In organ-cultured adult newt muscles, sensory ganglia and ganglia separated by a Millipore® filter prevent the denervation-induced loss of muscle AChE activity,[117] suggesting that this trophic effect is mediated by a substance released by nerves. Brain extracts also increase AChE activity of cultured chick embryo muscle cells[36] and maintain the enzyme's activity in denervated adult newt muscles in organ culture.[118] Further studies indicate that the active substance in the extracts is a protein of 34 kdaltons and maintains the A_{12} form of AChE in cultured denervated muscle.[21] Similarly, a protein with a molecular weight less than 20 kdaltons isolated from brains is effective in maintaining A_g and A_{12} AChE of denervated newt muscle in organ culture.[26]

Soluble extracts of peripheral nerves have an effect similar to that of brain extracts in regulating muscle AChE activity. Addition of sciatic nerve extracts produces an increase in AChE activity in normal,[41] but not in dystrophic, chick muscle in culture.[119] Therefore, the failure of dystrophic muscle to respond to the trophic substance may be due to a lack of a receptor mechanism for internalizing the mediator of the trophic effect.[119] A protein from sciatic nerves (transferrin) also maintains the level of AChE in cultures of aneural embryonic or denervated adult chick muscle.[120] Furthermore, the maintenance of AChE levels by the serum protein is regulated by a posttranscriptional mechanism.[120] Further evidence suggests that a trophic substance extracted from rat sciatic nerves that regulates muscle AChE is moved by axoplasmic transport and released upon neural stimulation.[121]

Recently, several well-characterized neural peptides have been shown to regulate muscle AChE in vitro. Addition of bovine pituitary FGF to the culture medium of denervated adult newt muscles maintains low-molecular-weight forms (globular) of AChE.[122] However, it does not prevent the postdenervation loss of A_{12} AChE in cultured muscle. When rat embryonic muscle cells are cultured in the presence of synthetic β-endorphin or conditioned medium by spinal neurons, A_{12} AChE decreases, and the globular AChE forms increase in activity in a manner identical to that seen in muscles in vivo during early postnatal development.[123] Furthermore, analysis of the neuron-conditioned medium reveals the release of β-endorphin by neurons, suggesting that the neuropeptide is a regulator of AChE molecular forms during early postnatal muscle development.[123] However, proof of this β-endorphin hypothesis requires further investigation.

V. ACETYLCHOLINE RECEPTORS

The motor nerve exerts a regulatory effect on ACh sensitivity (i.e., AChR density) in skeletal muscle fibers.[1,2] AChRs in innervated adult muscle are localized at the postsynaptic membrane of neuromuscular junctions.[11,12,124] Following denervation, however, extrajunctional AChRs appear on the entire surface of the muscle membrane.[11-13] Furthermore, denervation causes an increase in the turnover rate of junctional AChR which is a typical characteristic of extrajunctional receptors.[125,126] Conversely, reinnervation of the denervated muscle suppresses the number of extrajunctional AChR and restores the junctional receptors.[127] Fetal muscle fibers, like denervated adult muscle, exhibit AChR over their entire surface.[128] After functional neuromuscular junctions are established, the extrajunctional AChRs decrease rapidly, while high-density AChR clusters appear in the region of nerve-muscle contact.[129,130] Although trophic substances,[1,2,39,40] muscle activity,[17,18,131,132] ACh,[19,20] and degenerating nerve products[133] have all been proposed to account for the neural control of muscle AChR, it is apparent that muscle activity plays a key role in the regulation of extrajunctional AChR.[134]

AChRs are uniformly distributed on the surface of noninnervated muscle cells grown in culture, whereas innervated muscle cultures show a high receptor density at the nerve-muscle contact which exhibits transmitter function.[135-137] When cloned muscle cells are grown in co-culture with neuroblastoma cells, a high AChR density occurs at the site of nerve-muscle contact even though no synaptic functions are apparent.[138-139] This suggests that the release of ACh is not required for the localization of AChR in vitro (but see Chapter 3). In the absence of nerve cells, clusters of high AChR densities also appear on the surface of muscle cells in culture.[137,140-143] However, innervation causes a loss of pre-existing AChR clusters and prevents the formation of new receptor clusters elsewhere on the muscle cells in vitro.[143-146] In addition, the nerve-induced localization of AChR occurs in the absence of muscle activity,[139,144,145] and the nerve is required for the maintenance of receptor localization at the sites of nerve-muscle contact.[147]

Tissue culture studies have recently demonstrated that a number of neurally derived substances can mimic the effects of innervation on the AChR density and distribution (Table 4). In spinal cord explant-muscle co-cultures, there is an increase in the number of AChR and AChR clusters on the myotube.[29,148] Furthermore, the increase in the AChR density on myotubes adjacent to the explant is greater than on myotubes located farther away in the same culture.[29,148] This graded distribution suggests the diffusion of a substance from the explant. A 100-kdalton protein isolated from rat embryo brain and spinal cord increases AChR numbers on L_6 myotubes and also promotes the formation of receptor clusters.[29] Soluble extracts from chick embryo brains also stimulate the synthesis and aggregation of AChR in cultured chick myotubes.[149] The active substance in the extracts appears to be a small, 1-kdalton peptide.[28] Brain extracts also cause the aggregation of AChR on the surface of cultured rat myotubes by redistributing the preexisting diffuse AChR. This aggregation mimics an early event at developing neuromuscular junctions.[150,151] Either brain extracts or ascorbic acid induce both the aggregation of AChR and collagen synthesis in cultured rat muscle cells.[152] Thus, it seems likely that collagen synthesis induced by a substance (or ascorbic acid) present in neural extracts may be involved in the induction of AChR aggregation on myotubes in vitro. In fact, ascorbic acid has been identified as the agent in fetal calf brain extracts responsible for causing the increase in AChR in L_6 muscle cultures.[153] Media conditioned by nerve cells in culture also induce the aggregation of AChR, but not the synthesis of receptors on rat myotubes[30,31,154] Transferrin isolated from chicken sciatic nerves stimulates the synthesis of AChR in cultured chick muscle cells.[155] A number of recent studies have shown that an insoluble 80-kdalton substance isolated from *Torpedo* electric organ extracellular matrix causes the aggregation of AChR on cultured chick muscle

Table 4
TROPHIC EFFECTS OF NERVE EXTRACTS ON MUSCLE AChRs IN VITRO

Source of trophic activity	Mol wt	Target muscle culture	Trophic effect on AChR	Ref.
Rat embryo brain and spinal cord	~100,000	Rat L_6 clone	Synthesis and aggregation	29
Chicken brain	1,000	Chick embryo	Synthesis and aggregation	28
Rat and pig embryo brain	—	Rat embryo	Aggregation	150—152
Medium conditioned by neuroblastoma-glioma cells	>150,000	Rat embryo	Aggregation	30
Medium conditioned by rat spinal neurons	≥50,000	Rat embryo	Aggregation	31
Chicken sciatic nerve transferrin	84,000	Chick embryo	Synthesis	155
Fetal calf brain ascorbic acid	—	Rat embryo and L_6 clone	Synthesis	153
Torpedo electric organ extracellular matrix component	80,000	Chick embryo	Aggregation	156
Torpedo electric organ	<5,000	Chick embryo	Synthesis and aggregation	159

cells.[156-158] Also, a soluble substance (<5 kdaltons) from the electric organ shows similar effects on cultured chick myotube AChR[159] (see Chapter 8).

VI. OTHER MEMBRANE PROPERTIES

The motor nerve exerts a regulatory effect on some membrane properties of skeletal muscle.[1,2] For example, denervation causes a decrease in RMP[10] and the development of TTX-resistant action potentials. In innervated muscles, TTX blocks generation of the action potential. Following denervation, however, the action potential becomes resistant to TTX, suggesting that denervation reduces the density of TTX-sensitive sodium channels in the muscle membrane.[160] Such denervation-induced changes in muscle can be attributed to the interaction of the effects of loss of muscle activity, ACh transmission, quantal release of ACh, and putative trophic substances.[161] However, there is evidence that the changes produced in the properties of muscle membranes may all result from the loss of trophic substances released by the nerve. For example, the decrease in the RMP and the appearance of TTX-resistant action potentials develop more rapidly if the nerve is cut close to (rather than far from) the muscle.[10,162-164] These results indicate that muscle activity alone cannot be responsible for the regulation of RMP and TTX-sensitive action potentials, since muscle activity is abolished instantaneously regardless of the level at which the nerve is cut. Also, the quantal release of ACh has no effect on this regulation, since the change in RMP precedes the disappearance of miniature endplate potentials (but see Chapter 3).

Probably the best evidence for the trophic influence of neurally derived substances in the regulation of membrane properties is obtained from tissue culture studies. Extracts of neural tissues partially reverse the decrease in RMP and prevent the development of TTX-resistant action potentials in denervated adult muscle in organ culture.[165-167] Although extracts of many nonnervous tissues are ineffective,[166] these effects are not completely specific to nervous tissue, since extracts of muscle or liver produce an effect similar to the nerve extracts.[165] A peptide (<10 kdaltons) which is responsible for regulating TTX-sensitive action potentials has been partially purified from spinal cord tissue.[27] This peptide reduces the TTX-resistant action potentials in rat myotube cultures, but it has no stimulatory effects on morphological development, protein synthesis, or AChR density in the same cultures.[168] A number of laboratories have shown that TTX-sensitive sodium channels can develop and be maintained in cultured embryonic or clonal myotubes in the absence of innervation.[169-174] Thus, it seems that, unlike the situation in adult muscles, the innervation alone cannot account for the complete regulation of TTX-sensitive sodium channels in cultured myotubes. However, nerve extracts increase the density of TTX-sensitive sodium channels and the RMP of chick myotubes in culture.[175] Addition of spinal cord explants to cultured chick myotubes also changes the passive electrical properties (i.e., distribution of transverse membrane resistance) of the myotubes from those characteristic of slow-fiber types to those of fast muscle fibers in vivo (but see Chapter 5).[176]

VII. TRANSFERRIN

We have identified and purified an 86.4-kdalton glycoprotein from chicken sciatic nerve which exerts trophic effects on cultured chick muscle cells.[23,177] This trophic protein, referred to as sciatin,[45] enhances the rate of morphological maturation of cultured muscle cells, maintains noninnervated cross-striated myotubes, and increases protein and AChR synthesis in culture muscle cells.[23,62,155] Furthermore, sciatin is essential for the initial differentiation of chick myogenic cells in vitro, since it can support normal myogenesis even when chick embryo extract (EE) is omitted from the culture medium.[45] Immunocytochemical studies have revealed that sciatin is localized in the perikarya of spinal cord neurons and the axoplasm of sciatic nerves of adult chickens.[178] Preliminary studies indicate that sciatin is transported by anterograde axoplasmic transport in chicken sciatic nerves and is released from chicken nerve-muscle preparations by electrical stimulation.[179]

Further studies have indicated that sciatin is structurally very similar to chicken serum transferrin, as evidenced by their amino acid composition, HN^2-terminal sequence immunological cross-reactivity, and iron-binding capacity.[48] Serum transferrin also duplicates the trophic influences of sciatin on chick muscle cells in culture.[49] Both sciatin and transferrin increase muscle protein and AChR synthesis, increase AChE activity, and promote myogenesis in the absence of EE.[49] Thus, sciatin is structurally and biologically very similar or identical to transferrin, and it is likely that sciatin exerts its trophic influence by transporting iron to muscle cells.

Recent studies have revealed that certain trophic substances isolated from chicken serum or neural tissues are identical to serum transferrin. For example, a trophic protein purified from chicken serum has been identified as transferrin.[25] Also, transferrin is found to be an essential component of EE.[180] A serum component has been shown to be responsible for the trophic effect of muscle extracts on cultured muscle cells, since extracts of serum-free muscles have little or no trophic activity.[47]Indeed, transferrin has been identified as the component of the extracts of chicken slow muscle which stimulates avian myogenesis in vitro.[181] Transferrin also has a trophic effect similar to that of nerve extracts on DNA synthesis in cultured newt limb blastemas.[182] Furthermore, another trophic protein isolated from chicken ischiatic-peroneal nerves is identical to chicken serum transferrin in terms of its molecular weight and biological activity.[63] On the other hand, transferrin is not a trophic substance responsible for preventing muscle atrophy in vivo, since its removal from rat sciatic nerve extracts does not abolish the effects of the extract on denervated rat muscles (see Chapter 7). Also, the injection of iron-bound rat transferrin fails to prevent the muscle atrophy observed.[183]

In view of the trophic effects of transferrin on cultured muscle and its localization in the cell body and axon of the motor neuron, it is anticipated that the protein might represent a trophic substance synthesized by motor neurons. In fact, several reports indicate that transferrin is synthesized by embryonic and adult neural tissues.[184,185] However, our recent studies

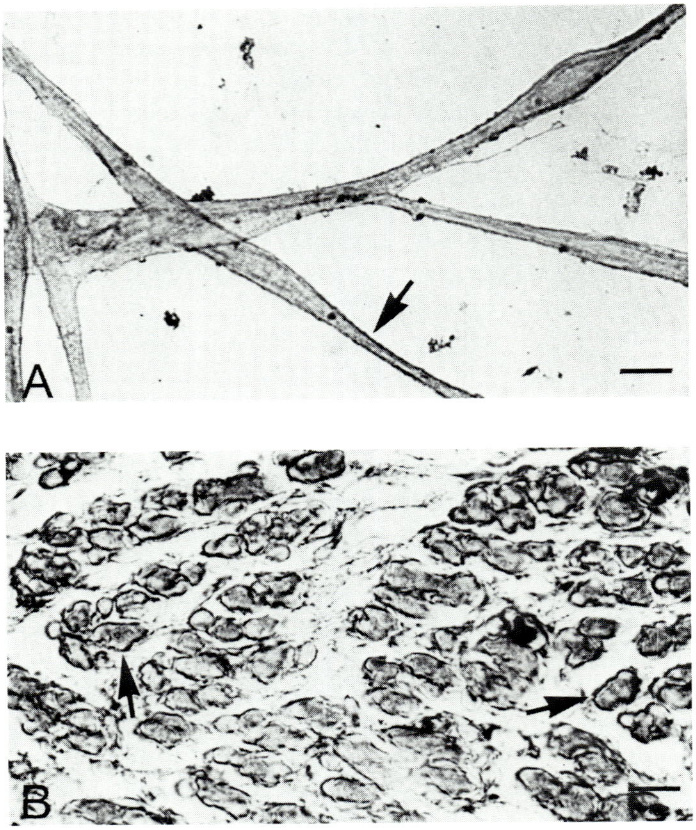

FIGURE 1. Immunocytochemical localization of transferrin receptor in cultured and embryonic chick muscle. Cultured and embryonic chicken muscles were fixed in cold 4% paraformaldehyde-0.1% glutaraldehyde and processed for immunocytochemical staining (peroxidase-antiperoxidase, or PAP) of transferrin receptors using rabbit antitransferrin receptor serum. (A) Muscle cultures (3 days old). Note that the staining reaction is more intense on the surface of myotubes (arrow). (B) Transverse section of breast muscle from 10-day-old chicken embryos. Arrows indicate myotubes.(Bar = 20 μm.)

demonstrate that transferrin is not synthesized by cultures of spinal cord neurons.[186] Instead, these cultures synthesize the neuronal receptor for transferrin. Our preliminary studies also demonstrate that transferrin receptors are present on the surface of developing chick muscle in vitro as well as in vivo (Figure 1), and ^{125}I-labeled transferrin is specifically taken up by cultured muscle via receptor-mediated endocytosis (Figure 2). Furthermore, transferrin induces a significant increase in the muscle-specific form of aldolase in cultured muscle cells (Figure 3).[187] These observations suggest that transferrin plays a critical role in muscle development.

VIII. SUMMARY AND CONCLUDING REMARKS

Numerous in vivo experiments have shown that spinal motor nerves exert long-term, trophic, or regulatory influences on many properties of skeletal muscle. While the precise mechanism for the neural regulation of muscle properties is unknown, evidence suggests that this regulation is accomplished in part by a trophic substance(s) released by nerves.

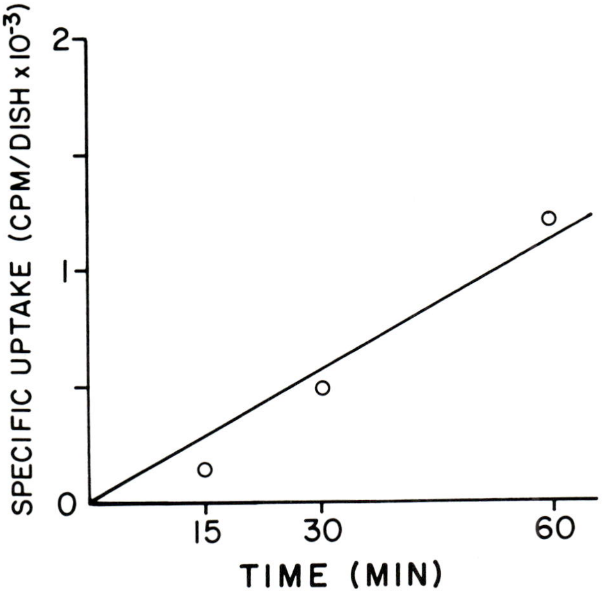

FIGURE 2. Specific uptake of ^{125}I-transferrin by myoblasts in culture. Myoblast cultures were washed twice in Dulbecco's modified Eagle's medium containing 0.5% bovine serum albumin (DMEM/BSA) and preincubated for 1 hr at 37°C in DMEM/BSA. The DMEM/BSA was then replaced with fresh DMEM/BSA, and 20 μℓ of ^{125}I-transferrin (300 ng; 1.32 μCi; 3.75 n*M*) ± 100 μg unlabeled transferrin were added. Cultures were washed at timed intervals; the cells were dissolved in 10 m*M* potassium phosphate (pH, 7.5), 150 m*M* NaCl, and 1.0% triton-X100; and the extracts were counted in a gamma counter. Specific uptake of labeled transferrin was defined as the uptake in the absence of unlabeled transferrin minus the uptake in the presence of unlabeled transferrin. Each point represents the mean of triplicate determinations; the line shown was fitted by regression analysis.

Using muscle tissue culture as an in vitro bioassay system, many investigators have attempted to identify a trophic substance(s) from neural tissues. Over the past decade, investigators in a number of laboratories have reported neurally derived trophic substances which mimic the effects of innervation in vivo. These substance(s) have been shown to regulate the morphological development, macromolecular synthesis, AChE activity, AChR, and other membrane properties of muscle cells in culture. With the exception of transferrin, however, the biochemical purification and characterization of the putative trophic substances have not been fully accomplished. These neurally derived substances have been shown to fulfill some of the criteria for a trophic substance(s), but the existence of a unique trophic agent has not yet been demonstrated. Although the possibility remains that a unique trophic agent(s) is synthesized and released by neural tissue, the current failure to identify such a substance, despite numerous efforts, has led investigators to consider alternative mechanisms. Based upon recent findings regarding transferrin as a growth substance, it is likely that nerve exerts its trophic effect in part by modulating the ability of myogenic tissue to respond to various substances (e.g., transferrin and hormones). For example, mitogen-induced T lymphocyte proliferation is dependent on the presence of both interleukin (IL-2, or T cell growth substance) and transferrin, even though resting lymphocytes do not have receptors for either.[188] Exposure to mitogen (phytohemagglutinin) alone induces the transient generation of IL-2 receptors, and their presence leads to the appearance of transferrin receptors on T cells.[188]

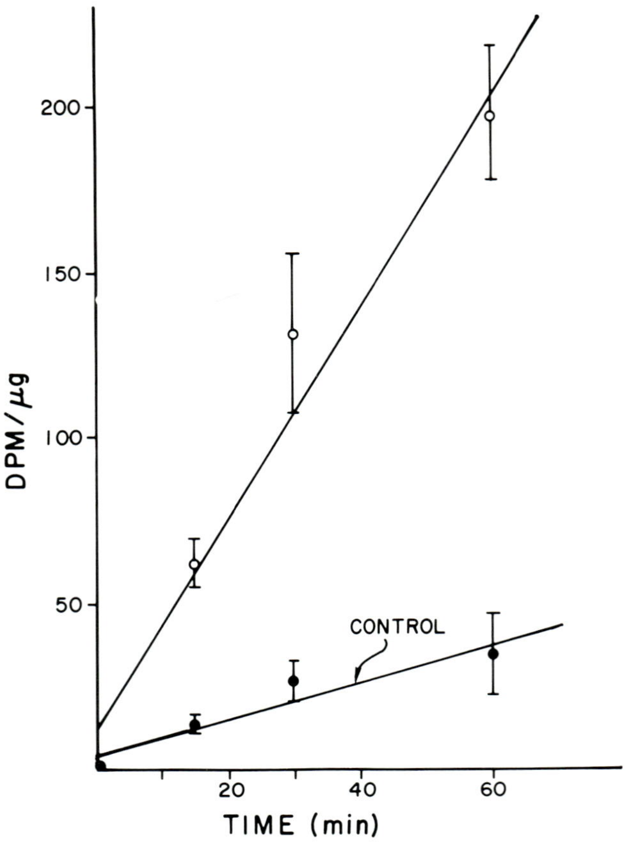

FIGURE 3. Effect of transferrin on aldolase synthesis in cultured myotubes. Myogenic cells were plated and grown continuously in the standard culture medium. On days 1 and 4 after plating, some cultures were treated with transferrin (30 µg per dish), while the untreated cultures served as controls. After 5 days in culture, cells were labeled with 10 µCi ^{3}H-L amino acids. Incorporation of labeled amino acids into aldolase was assayed by immunoprecipitating ^{3}H-aldolase from the soluble cellular fraction with goat antialdolase serum. Each value is the mean plus or minus the SE of triplicate determinations. (●, Control; ○, transferrin, 30 µg per dish.) (From Oh, T. H., Markelonis, G. J., Dion Guidera, T., Hobbs, S. L., and Park, L. P., in *Gene Expression in Muscle,* Strohman, R. C. and Wolf, S., Eds., Plenum Press, New York, 1985, 129. With permission.)

This indicates that IL-2 stimulates T cell proliferation solely by induction of transferrin receptors on these cells. Similarly, transferrin is required for kidney differentiation.[189] However, only nephrogenic mesenchyme exposed to an inductive tissue (e.g., spinal cord) becomes responsive to transferrin, whereas unexposed mesenchyme remains totally unresponsive.[189] Thus, the interaction between nephrogenic mesenchyme and spinal cord acts by stimulating the expression of transferrin receptors in the developing kidney. In this regard, transferrin may serve as a useful model substance for further elucidation of neural influences on myogenesis and maintenance of adult muscle.

ACKNOWLEDGMENTS

The authors wish to thank Ms. T. Guidera and Ms. S. Hobbs for their technical assistance and to Mrs. Eloise DeLong for the excellent editorial help. This work was supported by grants (NS15013 and 20490) from the National Institutes of Health.

REFERENCES

1. **Guth, L.,** Trophic influences of nerve on muscle, *Physiol. Rev.*, 48, 645, 1968.
2. **Gutmann, E.,** Neurotrophic relations, *Annu. Rev. Physiol.*, 38, 177, 1976.
3. **Singer, M.,** On the nature of the neurotrophic phenomenon in Urodele limb regeneration, *Am. Zool.*, 18, 829, 1978.
4. **Zelená, J.,** The effect of denervation on muscle development, in *The Denervated Muscle*, Gutmann, E., Ed., Publishing House of the Czechoslovak Academy of Sciences, Prague, 1962, 103.
5. **Engel, W. K. and Karpati, G.,** Impaired skeletal muscle maturation following neonatal neurectomy, *Dev. Biol.*, 17, 713, 1968.
6. **Shafiq, S. A., Siedu, S. A., and Milhorat, A. T.,** Effect of neonatal neurectomy on differentiation of fiber types in rat skeletal muscle, *Exp. Neurol.*, 35, 529, 1972.
7. **Hanzliková, V. and Shiaffino, S.,** Studies on the effect of denervation in developing muscle. III. Diversification of myofibrillar structure and origin of the heterogeneity of muscle fiber types, *Z. Zelforsch.*, 147, 75, 1973.
8. **Sohal, G. S. and Holt, R. K.,** Role of innervation on the embryonic development of skeletal muscle, *Cell Tissue Res.*, 210, 383, 1980.
9. **Popiela, H.,** *In vivo* limb muscle development in the absence of nerves: quantitative study, *Exp. Neurol.*, 53, 214, 1976.
10. **Albuquerque, E. X., Schuh, F. T., and Kauffman, F. C.,** Early membrane depolarization of the fast mammalian muscle after denervation, *Pfluegers Arch.*, 328, 36, 1971.
11. **Axelsson, J. and Thesleff, S.,** A study of supersensitivity in denervated mammalian skeletal muscle, *J. Physiol. (London)*, 147, 178, 1959.
12. **Miledi, R.,** Junctional and extrajunctional acetylcholine receptors in skeletal muscle fibers, *J. Physiol. (London)*, 151, 24, 1960.
13. **Hartzell, H. L. and Fambrough, D. M.,** Acetylcholine receptors: distribution and extrajunctional density in rat diaphragm after denervation correlated with acetylcholine sensitivity, *J. Gen. Physiol.*, 60, 248, 1972.
14. **Guth, L., Albers, R. W., and Brown, W. C.,** Quantitative changes in cholinesterase activity of denervated muscle fibers and sole plates, *Exp. Neurol.*, 10, 236, 1964.
15. **Gutman, E.,** *The Denervated Muscle*, Publishing House of the Czechoslovak Academy of Sciences, Prague, 1962, 1.
16. **Fischbach, G. D. and Robbins, N.,** Effects of chronic disuse of rat soleus neuromuscular junctions on post-synaptic membrane, *J. Neurophysiol.*, 34, 562, 1971.
17. **Lømo, T. and Rosenthal, J.,** Control of ACh sensitivity by muscle activity in the rat, *J. Physiol. (London)*, 221, 439, 1972.
18. **Drachman, D. B. and Witzke, F.,** Trophic regulation of acetylcholine sensitivity of muscle: effect of electrical stimulation, *Science*, 176, 514, 1972.
19. **Thesleff, S.,** Supersensitivity of skeletal muscle produced by botulinum toxin, *J. Physiol. (London)*, 151, 598, 1960.
20. **Drachman, D. B.,** The role of acetylcholine as a neurotrophic transmitter, *Ann. N.Y. Acad. Sci.*, 228, 160, 1974.
21. **Lentz, T. L., Addis, J. S., and Chester, J.,** Partial purification and characterization of a nerve trophic factor regulating muscle acetylcholinesterase activity, *Exp. Neurol.*, 73, 542, 1981.
22. **Jabaily, J. and Singer, M.,** Neurotrophic and hepatotrophic stimulation of proliferation of embryonic chick muscle cells in vitro: assay and partial characterization of mitogenic activity in chick embryonic organ and tissue extracts, *Dev. Biol.*, 64, 189, 1978.
23. **Markelonis, G. J. and Oh, T. H.,** A sciatic nerve protein has a trophic effect on development and maintenance of skeletal muscle cells in culture, *Proc. Natl. Acad. Sci. U.S.A.*, 76, 2470, 1979.
24. **Popiela, H. and Ellis, S.,** Neurotrophic factor: characterization and partial purification, *Dev. Biol.*, 83, 266, 1981.
25. **Kimura, I., Hasegawa, T., Miura, T., and Ozawa, E.,** Muscle trophic factor is identical to transferrin, *Proc. Jpn. Acad.*, 57, 200, 1981.
26. **Rathbone, M. P., Vickers, J. D., and Logan, D. M.,** Neural regulation of cholinesterase in newt skeletal muscle. III. Regulation *in vitro* by proteins from chick embryo brains, *J. Exp. Zool.*, 210, 463, 1979.
27. **Kuromi, H., Gonoi, T., and Hasegawa, S.,** Partial purification and characterization of neurotrophic substance affecting tetrodotoxin sensitivity of organ-cultured mouse muscle, *Brain Res.*, 175, 109, 1979.
28. **Buc-Caron, M.-H., Nystrom, P., and Fischbach, G. D.,** Induction of acetylcholine receptor synthesis and aggregation: partial purification of low-molecular-weight activity, *Dev. Biol.*, 95, 378, 1983.
29. **Podleski, T. R., Axelrod, D., Ravdin, P., Greenberg, I., Johnson, M. M., and Salpeter, M. M.,** Nerve extract induces increase and redistribution of acetylcholine receptors on cloned muscle cells, *Proc. Natl. Acad. Sci. U.S.A.*, 75, 2035, 1978.

30. **Bauer, H. C., Daniels, M. P., Pudimat, P. A., Jacques, L., Sugiyama, H., and Christian, C. N.,** Characterization and partial purification of a neuronal factor which increases acetylcholine receptor aggregation on cultured muscle cells, *Brain Res.*, 209, 395, 1981.
31. **Schaffner, A. E. and Daniels, M. P.,** Conditioned medium from cultures of embryonic neurons contains a high molecular weight factor which induces acetylcholine receptor aggregation on cultured myotubes, *J. Neurosci.*, 2, 623, 1982.
32. **Peterson, E. R. and Crain, S. R.,** Innervation in cultures of fetal rodent skeletal muscle by organotypic explants of spinal cord from different animals, *Z. Zellforsch.*, 106, 1, 1970.
33. **Peterson, E. R. and Crain, S. M.,** Regeneration and innervation in cultures of adult mammalian skeletal muscle coupled with fetal rodent spinal cord, *Exp. Neurol.*, 36, 136, 1972.
34. **Hooisma, J., Slaf, D. W., Meeter, E., and Stevens, W. F.,** Trophic support by neural explants of cultured muscle fibers, *Exp. Neurol.*, 62, 628, 1978.
35. **Globus, M. and Vethamany-Globus, S.,** Transfilter mitogenic effect of dorsal root ganglia on cultured regeneration blastemata, in the newt, *Notophthalmus viridescens, Dev. Biol.*, 56, 316, 1977.
36. **Oh, T. H.,** Neurotrophic effects: characterization of the nerve extract that stimulates muscle development in culture, *Exp. Neurol.*, 46, 432, 1975.
37. **Gospodarowicz, D., Weseman, J., and Moran, J.,** Presence in brain of a mitogenic agent promoting proliferation of myoblasts in low density culture, *Nature (London)*, 256, 216, 1975.
38. **Allen, R. E., Dodson, M. V., and Luiten, L. S.,** Regulation of skeletal muscle satellite cell proliferation by bovine pituitary fibroblast growth factor, *Exp. Cell Res.*, 152, 154, 1984.
39. **Albuquerque, E. X., Warnick, J. E., Tasse, J. R., and Sansone, F. M.,** Effect of vinblastine and colchicine on neural regulation on the fast and slow skeletal muscle of the rat, *Exp. Neurol.*, 37, 607, 1972.
40. **Hofmann, W. W. and Thesleff, S.,** Studies on the trophic influence of nerve on skeletal muscle, *Eur. J. Pharmacol.*, 20, 256, 1972.
41. **Oh, T. H.,** Neurotrophic effects of sciatic nerve extract on muscle development in culture, *Exp. Neurol.*, 50, 376, 1976.
42. **Festoff, B. W., Duell, M. J., and Fernandez, H. L.,** Trophic effects of axonally transported proteins on muscle cells in cultures, in *Maturation of Neurotransmission*, Vernadakis, A., Giacobini, E., and Filogamo, G., Eds., S. Karger, Basel, 1978, 152.
43. **Markelonis, G. J. and Oh, T. H.,** A protein fraction from peripheral nerve having neurotrophic effects on skeletal muscle cells in culture, *Exp. Neurol.*, 58, 285, 1978.
44. **Popiela, H.,** Trophic effects of adult peripheral nerve extract on muscle cell growth and differentiation *in vitro, Exp. Neurol.*, 62, 405, 1978.
45. **Oh, T. H. and Markelonis, G. J.,** Dependence of *in vitro* myogenesis on a trophic protein present in chicken embryo extract, *Proc. Natl. Acad. Sci. U.S.A.*, 77, 6922, 1980.
46. **Oh, T. H., Markelonis, G. J., Reier, P. J., and Zalewski, A. A.,** Persistence in degenerating sciatic nerve of substances having a trophic influence upon cultured muscle, *Exp. Neurol.*, 67, 646, 1980.
47. **Hooisma, J., DeKrijger, J., DeGroot, D. M. G., Magchielse, T., and Muijser, H.,** Effect of extracts of denervated muscles on the morphology of cultured muscle cells, *Neurosci. Lett.*, 22, 47, 1981.
48. **Markelonis, G. J., Bradshaw, R. A., Oh, T. H., Johnson, J. L., and Bates, O. J.,** Sciatin is a transferrin-like polypeptide, *J. Neurochem.*, 39, 315, 1982.
49. **Oh, T. H. and Markelonis, G. J.,** Chicken serum transferrin duplicates the myotrophic effects of sciatin on cultured muscle cells, *J. Neurosci. Res.*, 8, 535, 1982.
50. **Goldberg, A. L.,** Protein turnover in skeletal muscle. II. Effects of denervation and cortisone on protein catabolism in skeletal muscle, *J. Biol. Chem.*, 244, 3223, 1969.
51. **Goldspink, D. F.,** The effect of denervation on protein turnover of the soleus and extensor digitorum longus muscles of adult mice, *Comp. Biochem. Physiol.*, 61B, 37, 1978.
52. **Singer, M.,** Neurotrophic control of limb regeneration in the newt, *Ann. N.Y. Acad. Sci.*, 228, 308, 1974.
53. **Singer, M.,** The influence of the nerve in regeneration of the amphibian extremity, *Q. Rev. Biol.*, 27, 169, 1952.
54. **Dresden, M. H.,** Denervation effects on newt limb regeneration: DNA, RNA and protein synthesis, *Dev. Biol.*, 19, 311, 1969.
55. **Lebowitz, P. and Singer, M.,** Neurotrophic control of protein synthesis in the regenerating limb of the newt, *Triturus, Nature (London)*, 225, 824, 1970.
56. **Singer, M. and Caston, J. D.,** Neurotrophic dependence of macromolecular synthesis in the early limb regeneration of the newt, *Triturus, J. Embryol. Exp. Morphol.*, 28, 1, 1972.
57. **Singer, M. and Llan, J.,** Nerve-dependent regulation of absolute rates of protein synthesis in newt limb regenerates, *Dev. Biol.*, 57, 174, 1977.
58. **Singer, M., Maier, C. E., and McNutt, W. S.,** Neurotrophic activity of brain extracts in forelimb regeneration of the urodele, *Triturus, J. Exp. Zool.*, 196, 131, 1976.
59. **Jabaily, J. A. and Singer, M.,** Neurotrophic stimulation of DNA synthesis in the regenerating forelimb of the newt, *Triturus, J. Exp. Zool.*, 199, 251, 1977.

60. **Choo, A. F., Logan, D. M., and Rathbone, M. P.,** Nerve trophic effects: an in vitro assay for factors involved in regulation of protein synthesis in regenerating amphibian limbs, *J. Exp. Zool.*, 206, 347, 1978.
61. **Choo, A. F., Logan, D. M., and Rathbone, M. P.,** Nerve trophic effects: partial purification from chick embryo brains of proteins that stimulate protein synthesis in cultured newt blastemata, *Exp. Neurol.*, 73, 558, 1981.
62. **Markelonis, G. J., Oh, T. H., and Derr, D.,** Stimulation of protein synthesis in cultured skeletal muscle by a trophic protein from sciatic nerves, *Exp. Neurol.*, 70, 598, 1980.
63. **Beach, R. L., Popiela, H., and Festoff, B. W.,** The identification of neurotrophic factor as a transferrin, *FEBS Lett.*, 156, 151, 1983.
64. **Buller, A. J., Eccles, J. C., and Eccles, R. M.,** Interactions between motorneurons and muscles in respect of the characteristic speeds of their responses, *J. Physiol. (London)*, 150, 417, 1960.
65. **Salmons, S. and Vrbová G.,** The influence of activity on some contractile characteristics of mammalian fast and slow muscles, *J. Physiol. (London)*, 201, 535, 1969.
66. **Samaha, F. J., Guth, L., and Albers, R. W.,** The neural regulation of gene expression in the muscle cell, *Exp. Neurol.*, 27, 276, 1970.
67. **Barany, M. and Close, R. I.,** The transformation of myosin in cross-innervated rat muscles, *J. Physiol. (London)*, 213, 455, 1971.
68. **Streter, F. A., Gergely, J., Salmons, S., and Romanul, F.,** Synthesis by fast muscle of light chains characteristic of slow muscle in response to long-term stimulation, *Nature (London) New Biol.*, 241, 17, 1973.
69. **Whalen, R. G., Sell, S. M., Butler-Browne, G. S., Schwartz, K., Bouveret, P., and Pinset-Härstöm, I.,** Three myosin heavy-chain isozymes appear sequentially in rat muscle development, *Nature (London)*, 292, 805, 1981.
70. **Butler-Browne, G. S., Bugaisky, L. B., Cuénoud, S., Schwartz, K., and Whalen, R. G.,** Denervation of newborn rat muscle does not block the appearance of adult fast myosin heavy chain, *Nature (London)*, 299, 830, 1982.
71. **Matsuda, R., Spector, D., and Strohman, R. C.,** Denervated skeletal muscle displays discoordinate regulation for the synthesis of several myofibrillar proteins, *Proc. Natl. Acad. Sci. U.S.A.*, 81, 1122, 1984.
72. **Whalen, R. G., Butler-Browne, G. S., Sell, S., and Gros, F.,** Transitions incontractile protein isozymes during muscle cell differentiation, *Biochimie*, 61, 625, 1979.
73. **Bader, D., Masaki, T., and Fischman, D. A.,** Immunochemical analysis of myosin heavy chain during avian myogenesis *in vivo* and *in vitro*, *J. Cell Biol.*, 95, 763, 1982.
74. **Bandman, E., Matsuda, R., and Strohman, R. C.,** Development appearance of myosin heavy and light chain iso-forms *in vivo* and *in vitro* in chicken skeletal muscle, *Dev. Biol.*, 93, 508, 1982.
75. **Ecob, M. S., Butler-Browne, G. S., and Whalen, R. G.,** The adult fast isozyme of myosin is present in a nerve-muscle tissue culture system, *Differentiation*, 25, 84, 1983.
76. **Thibault, M. C., Havaranis, A. S., and Heywood, S. M.,** Trophic effect of a sciatic nerve extract on fast and slow myosin heavy chain synthesis, *Am. J. Physiol.*, 241, C269, 1981.
77. **Kovács, T., Kovér, A., and Balogh, G.,** Studies on the localization of cholinesterase in various types of muscle, *J. Cell Comp. Physiol.*, 57, 63, 1961.
78. **Miledi, R.,** Electron-microscopical localization of products from histochemical reactions used to detect cholinesterase in muscle, *Nature (London)*, 204, 293, 1964.
79. **Eränko, O. and Terävätnen, H.,** Distribution of esterases in the myoneural junction of the striated muscle of the rat, *J. Histochem. Cytochem.*, 15, 399, 1967.
80. **Salpeter, M. M.,** Electron microscope radioautography as a quantitative tool in enzyme cytochemistry. I. The distribution of acetylcholinesterase at motor end plates of a vertebrate twitch muscle, *J. Cell Biol.*, 32, 379, 1967.
81. **Hall, Z. W. and Kelly, R. B.,** Enzymatic detachment of endplate acetylcholinesterase from muscle, *Nature (London) New Biol.*, 232, 62, 1971.
82. **Betz, W. and Sakman, B.,** Effects of proteolytic enzymes on function and structure of frog neuromuscular junctions, *J. Physiol. (London)*, 230, 673, 1971.
83. **McMahan, U. J., Sanes, J. R., and Marshall, L. M.,** Cholinesterase is associated with the basal lamina at the nueromuscular junction, *Nature (London)*, 271, 172, 1978.
84. **Hall, Z. W.,** Multiple forms of acetylcholinesterase and their distribution in endplate and non-endplate regions of rat diaphragm muscle, *J. Neurobiol.*, 4, 343, 1973.
85. **Mumenthaler, M. and Engel, W. K.,** Cytological localization of cholinesterase in the developing chick embryo skeletal muscle, *Acta Anat.*, 47, 274, 1961.
86. **Filogamo, G. and Gabella, G.,** The development of neuro-muscular correlations in vertebrates, *Arch. Biol. (Liege)*, 78, 9, 1967.
87. **Wake, K.,** Formation of myoneural and myotendinous junctions in the chick embryo. Role of acetylcholinesterase-rich granules in the developing muscle fibers, *Cell Tissue Res.*, 173, 383, 1976.

88. **Bennett, M. R. and Pettigrew, A. G.,** The formation of synapses in striated muscle during development, *J. Physiol. (London),* 241, 515, 1974.
89. **Bevan, S. and Steinbach, J. H.,** The distribution of α-bungarotoxin binding sites on mammalian skeletal muscle developing *in vivo, J. Physiol. (London),* 267, 195, 1977.
90. **Kullberg, R. W., Mikelberg, F. S., and Cohen, M. W.,** Contribution of cholinesterase to developmental decreases in the time course of synaptic potentials at an amphibian neuromuscular junction, *Dev. Biol.,* 75, 255, 1980.
91. **Lømo, T. and Slater, C. R.,** Control of junctional acetylcholinesterase by neural and muscular influence in the rat, *J. Physiol. (London),* 303, 191, 1980.
92. **Dennis, M. J., Ziskind-Conhaim, L., and Harris, A. J.,** Development of neuromuscular junctions in rat embryos, *Dev. Biol.,* 81, 266, 1981.
93. **Guth, L. and Brown, W. C.,** Changes in cholinesterase activity following partial denervation, colateral reinnervation, and hyperneurotization of muscle, *Exp. Neurol.,* 13, 198, 1965.
94. **Kiauta, T., Brzin, M., and Dettbarn, W.-D.,** Synthesis of neuromuscular cholinesterases in innervated and denervated rat diaphragm, *Exp. Neurol.,* 56, 281, 1977.
95. **Vigny, M., Koenig, J., and Rieger, R.,** The motor endplate specific form of acetylcholinesterase: appearance during embryogenesis and reinnervation of rat muscle, *J. Neurochem.,* 27, 1347, 1976.
96. **Fernandez, H. L., Duell, M. J., and Festoff, B. W.,** Neurotrophic control of 16S acetylcholinesterase at the vertebrate neuromuscular junction, *J. Neurobiol.,* 10, 441, 1979.
97. **Filogamo, G. and Gabella, G.,** Cholinesterase behavior in the denervated and reinnervated muscles, *Acta Anat.,* 63, 199, 1966.
98. **Rose, S. and Glow, P. H.,** Denervation effects on the presumed *de novo* synthesis of muscle cholinesterase and the effects of acetylcholine availability on retinal cholinesterase, *Exp. Neurol.,* 18, 267, 1967.
99. **Fernandez, H. L. and Inestrosa, N. C.,** Role of axoplasmic transport in neurotrophic regulation of muscle endplate acetylcholinesterase, *Nature (London),* 262, 55, 1976.
100. **Drachman, D. B.,** Neurotrophic regulation of muscle cholinesterase: effects of botulinum toxin and denervation, *J. Physiol. (London),* 226, 619, 1972.
101. **Giacobini, G., Filogamo, G., Weber, M., Boquet, P., and Changeux, J.-P.,** Effects of a snake α-neurotoxin on the development of innervated skeletal muscles in chick embryo, *Proc. Natl. Acad. Sci. U.S.A.,* 70, 1708, 1973.
102. **Gordon, T., Perry, R., Tuffery, A. R., and Vrbová, G.,** Possible mechanisms determining synapse formation in developing skeletal muscles of the chick, *Cell Tissue Res.,* 155, 13, 1974.
103. **Oppenheim, R. W., Pittman, R., Gray, M., and Maderdrut, J. L.,** Embryonic behavior, hatching and neuromuscular development in the chick following a transient reduction of spontaneous motility and sensory input by neuromuscular blocking agents, *J. Comp. Neurol.,* 179, 619, 1978.
104. **Betz, H., Bourgeois, J.-P., and Changeux, J.-P.,** Evolution of cholinergic proteins in developing slow and fast skeletal muscles in chick embryo, *J. Physiol. (London),* 302, 197, 1980.
105. **Cangiano, A., Lømo, T., Lutzemberger, L., and Sveen, O.,** Effects of chronic nerve conduction block on formation of neuromuscular junctions and junctional AChE in the rat, *Acta Physiol. Scand.,* 109, 283, 1980.
106. **Weinberg, C. B. and Hall, Z. W.,** Junctional form of acetylcholinesterase restored at nerve-free endplates, *Dev. Biol.,* 68, 631, 1979.
107. **Kato, A. C., Vrachliotis, A., Fulpius, B., and Dumant, Y.,** Molecular forms of acetylcholinesterase in chick muscle and ciliary ganglion: embryonic tissues and cultured cells, *Dev. Biol.,* 76, 222, 1980.
108. **Bornstein, M. B., Iwanami, H., Lehrer, G. M., and Breitbart, L.,** Observations on the appearance of neuromuscular relationships in cultured mouse tissues, *Z. Zellforsch.,* 92, 197, 1968.
109. **Nakai, J.,** The development of neuromuscular junctions in cultures of chick embryo tissue, *J. Exp. Zool.,* 170, 85, 1969.
110. **Veneroni, G. and Murray, M. R.,** Formation of *de novo* and development of neuromuscular junctions *in vitro, J. Embryol. Exp. Morphol.,* 21, 369, 1969.
111. **Moody-Corbett, M., Weldon, P. R., and Cohen, M. W.,** Cholinesterase localization at sites of nerve contact on embryonic amphibian muscle cells in culture, *J. Neurocytol.,* 11, 381, 1982.
112. **Rubin, L. L., Schuetze, S. M., Weil, C. L., and Fischbach, G. D.,** Regulation of acetylcholinesterase appearance at neuromuscular junctions *in vitro, Nature (London),* 283, 264, 1980.
113. **Koenig, J., Bournaud, R., and Rieger, F.,** Acetylcholinesterase and synapse formation in striated muscle, in *Ontogenesis and Functional Mechanisms of Peripheral Synapses,* Taxi, J., Ed., Elsevier/North-Holland, Amsterdam, 1980, 313.
114. **Weldon, P. R., Moody-Corbett, F., and Cohen, M. W.,** Ultrastructure of sites of cholinesterase activity on amphibian embryonic muscle cells cultured without nerve, *Dev. Biol.,* 84, 341, 1981.
115. **Oh, T. H., Johnson, D. D., and Kim, S. U.,** Neurotrophic effect on isolated chick embryo muscle in culture, *Science,* 178, 1298, 1972.

116. **Koenig, J. and Vigny, M.,** Neural induction of the 16S acetylcholinesterase in muscle cell cultures, *Nature (London),* 271, 75, 1978.
117. **Lentz, T. L.,** Nerve trophic function: *in vitro* assay of effects of nerve tissue on muscle cholinesterase activity, *Science,* 171, 187, 1971.
118. **Lentz, T. L.,** Effect of brain extracts on cholinesterase activity of cultured skeletal muscle, *Exp. Neurol.,* 45, 520, 1974.
119. **Johnson, D. D., Bailey, S., and Wenger, B. S.,** Trophic activity of sheep sciatic nerve extracts in skeletal muscle cultures from normal and dystrophic chick embryos: failure of dystrophic muscle to respond, *Exp. Neurol.,* 73, 421, 1981.
120. **Oh, T. H. and Markelonis, G. J.,** Neurotrophic protein regulates muscle acetylcholinesterase in culture, *Science,* 200, 337, 1978.
121. **Younkin, S. G., Brett, R. S., Davey, B., and Younkin, L. H.,** Substances moved by axonal transport and released by nerve stimulation have an innervation-like effect on muscle, *Science,* 200, 1292, 1978.
122. **Carlone, R. L., Ganagarajah, M., and Rathbone, M. P.,** Bovine pituitary fibroblast growth factor has neurotrophic activity on newt limb regenerates and skeletal muscles *in vitro, Exp. Cell Res.,* 132, 15, 1981.
123. **Haynes, L. W., Smith, M. E., and Smyth, D. G.,** Evidence for the neurotrophic regulation of collagen-tailed acetylcholinesterase in immature skeletal muscle by β-endorphin, *J. Neurochem.,* 42, 1542, 1984.
124. **Fertuck, H. C. and Salpeter, M. M.,** Localization of acetylcholine receptor by ^{125}I-labeled α-bungarotoxin binding at mouse motor endplates, *Proc. Natl. Acad. Sci. U.S.A.,* 71, 1376, 1974.
125. **Loring, R. H. and Salpeter, M. M.,** Denervation increases turnover rate of junctional acetylcholine receptors, *Proc. Natl. Acad. Sci. U.S.A.,* 77, 2293, 1980.
126. **Brett, R. S., Younkin, S. G., Konieczkowski, M., and Slugg, R. M.,** Accelerated degradation of junctional acetylcholine receptor α-bungarotoxin complexes in denervated rat diaphragm, *Brain Res.,* 233, 133, 1982.
127. **Miledi, R.,** Properties of regenerating neuromuscular synapses in the frog, *J. Physiol. (London),* 154, 190, 1960.
128. **Diamond, J. and Miledi, R.,** A study of foetal and newborn rat muscle fibers, *J. Physiol. (London),* 162, 393, 1962.
129. **Bevan, S. and Steinbach, J. H.,** The distribution of α-bungarotoxin binding sites on mammalian skeletal muscle developing *in vivo, J. Physiol. (London),* 267, 195, 1977.
130. **Burden, S.,** Development of the neuromuscular junction in the chick embryo: the number, distribution and stability of acetylcholine receptors, *Dev. Biol.,* 57, 317, 1977.
131. **Lømo, T. and Westgaard, R. H.,** Further studies on the control of ACh sensitivity by muscle activity in the rat, *J. Physiol. (London),* 252, 603, 1975.
132. **Reiness, C. G. and Hall, Z. W.,** Electrical stimulation of denervated muscles reduces incorporation of methionine into the acetylcholine receptor, *Nature, (London),* 268, 655, 1977.
133. **Jones, R. and Vrbová, G.,** Two factors responsible for the development of denervation hypersensitivity, *J. Physiol. (London),* 236, 517, 1974.
134. **Fambrough, D. M.,** Control of acetylcholine receptors in skeletal muscle, *Physiol. Rev.,* 59, 165, 1979.
135. **Dryden, W. G.,** Development of acetylcholine sensitivity in cultured skeletal muscle, *Experientia,* 26, 984, 1970.
136. **Kano, M. and Shimada, Y.,** Innervation and acetylcholine sensitivity of skeletal muscle cells differentiated *in vitro* from chick embryo, *J. Cell. Physiol.,* 78, 233, 1971.
137. **Fischbach, G. D. and Cohen, S. A.,** The distribution of acetylcholine sensitivity over uninnervated and innervated muscle fibers grown in cell culture, *Dev. Biol.,* 31, 147, 1973.
138. **Harris, A. J., Heinemann, S., Schubert, D., and Tarakis, H.,** Trophic interaction between cloned tissue culture lines of nerve and muscle, *Nature (London),* 231, 296, 1971.
139. **Steinbach, J. H., Harris, A. J., Patrick, J., Schubert, D., and Heinemann, S.,** Nerve-muscle interaction *in vitro:* role of acetylcholine, *J. Gen. Physiol.,* 62, 255, 1973.
140. **Sytkowski, A. J., Vogel, Z., and Nirenberg, M. W.,** Development of acetylcholine receptor clusters on cultured muscle cells, *Proc. Natl. Acad. Sci. U.S.A.,* 70, 270, 1973.
141. **Axelrod, D. E., Ravdin, P., Koppel, D. E., Schlessinger, J., Webb, W. W., Elson, E. L., and Podleski, T. R.,** Lateral motion of fluorescently labeled acetylcholine receptors in membranes of developing muscle fibers, *Proc. Natl. Acad. Sci. U.S.A.,* 73, 4594, 1976.
142. **Betz, W. and Osborne, M.,** Effects of innervation on acetylcholine sensitivity of developing muscle *in vitro, J. Physiol. (London),* 270, 75, 1977.
143. **Frank, E. and Fischbach, G. D.,** Early events in neuromuscular junction formation *in vitro, J. Cell. Biol.,* 83, 143, 1979.
144. **Anderson, M. J. and Cohen, M. W.,** Nerve-induced and spontaneous redistribution of acetylcholine receptors on cultured muscle cells, *J. Physiol. (London),* 268, 757, 1977.
145. **Anderson, M. J., Cohen, M. W., and Zorychta, E.,** Effects of innervation on the distribution of acetylcholine receptors on cultured muscle cells, *J. Physiol. (London),* 268, 731, 1977.

146. **Cohen, M. W. and Weldon, P. R.,** Localization of acetylcholine receptors and synaptic ultrastructure at nerve-muscle contacts in culture: dependence on nerve type, *J. Cell Physiol.,* 86, 388, 1980.
147. **Kuromi, H. and Kidokoro, Y.,** Denervation disperses acetylcholine receptor clusters at the neuromuscular junction in *Xenopus* cultures, *Dev. Biol.,* 104, 421, 1984.
148. **Cohen, S. A. and Fischbach, G. D.,** Clusters of acetylcholine receptors located at identified nerve-muscle synapses *in vitro, Dev. Biol.,* 59, 24, 1977.
149. **Jessell, T. M., Siegel, R. E., and Fischbach, G. D.,** Induction of acetylcholine receptors on cultured skeletal muscle by a factor extracted from brain and spinal cord, *Proc. Natl. Acad. Sci. U.S.A.,* 76, 5397, 1979.
150. **Salpeter, M. M., Spanton, S., Holley, K., and Podleski, T. R.,** Brain extract causes acetylcholine receptor redistribution which mimics some early events at developing neuromuscular junctions, *J. Cell Biol.,* 93, 417, 1982.
151. **Olek, A. J., Pudimat, P. A., and Daniels, M. P.,** Direct observation of the rapid aggregation of acetylcholine receptors on identified cultured myotubes after exposure to embryonic brain extract, *Cell,* 34, 255, 1983.
152. **Kalcheim, C., Vogel, Z., and Duksin, D.,** Embryonic brain extract induces collagen biosynthesis in cultured muscle cells: involvement in acetylcholine receptor aggregation, *Proc. Natl. Acad. Sci. U.S.A.,* 79, 3077, 1982.
153. **Knaack, D. and Podleski, T.,** Ascorbic acid mediates acetylcholine receptor increase induced by brain extract on myogenic cells, *Proc. Natl. Acad. Sci. U.S.A.,* 82, 575, 1985.
154. **Christian, C. N., Daniels, M. P., Sugiyama, H., Vogel, Z., Jacques, L., and Nelson, P. G.,** A factor from neurons increases the number of acetylcholine receptor aggregates on cultured muscle cells, *Proc. Natl. Acad. Sci. U.S.A.,* 75, 4011, 1978.
155. **Markelonis, G. J., Oh, T. H., Eldefrawi, M. E., and Guth, L.,** Sciatin: a myotrophic protein increases the number of acetylcholine receptors and receptor clusters in cultured skeletal muscle, *Dev. Biol.,* 89, 353, 1982.
156. **Nitkin, R. M., Wallace, B. B., Spira, M. E., Godfrey, E. W., and McMahan, U. J.,** Molecular components of the synaptic basal lamina that direct differentiation of regenerating neuromuscular junctions, *Cold Spring Harbor Symp. Quant. Biol.,* 49, 653, 1093.
157. **Godfrey, E. W., Nitkin, R. M., Wallace, B. G., and McMahan, U. J.,** Identifying components of extracellular matrix that organize ACh receptors at synaptic sites on regenerating muscle cells, in *Developing and Regenerating Vertebrate Nervous Systems,* Coates, P. W., Markwald, R. R., and Kenney, A. D., Eds., Alan R. Liss, New York, 1983, 61.
158. **Godfrey, E. W., Nitkin, R. M., Wallace, B. G., Rubin, L. L., and McMahan, U. J.,** With technical assistance of Marshall, R. M., Components of *Torpedo* electric organ and muscle that cause aggregation of acetylcholine receptors on cultured muscle cells, *J. Cell Biol.,* 99, 615, 1984.
159. **Connolly, J. A., St. John, P. A., and Fischbach, G. D.,** Extracts of electric lobe and electric organ from *Torpedo californica* increase the total number as well as the number of aggregates of chick myotube acetylcholine receptors, *J. Neurosci.,* 2, 1207, 1982.
160. **Redfern, P. and Thesleff, S.,** Action potential generation in denervated rat skeletal muscle. II. The action of tetrodotoxin. *Acta Physiol. Scand.,* 82, 70, 1971.
161. **Guth, L. and Albuquerque, E. X.,** The neurotrophic regulation of resting membrane potential and extrajunctional acetylcholine sensitivity in mammalian skeletal muscle, *Physiol. Bohemoslov.,* 27, 401, 1978.
162. **Harris, J. B. and Thesleff, S.,** Nerve stump length and membrane changes in denervated skeletal muscle, *Nature (London) New Biol.,* 236, 60, 1972.
163. **Bray, J. J., Hawken, M. J., Hubbard, J. I., Pockett, S., and Wilson, L.,** The membrane potential of rat diaphragm muscle fibers and the effect of denervation, *J. Physiol. (London),* 266, 651, 1976.
164. **Deshpande, S., Albuquerque, E. X., and Guth, L.,** Neurotrophic regulation of prejunctional and postjunctional membrane at the mammalian motor endplate, *Exp. Neurol.,* 53, 151, 1976.
165. **Bray, J. J., Mills, R. G., Forrest, J. W., and Hubbard, J. I.,** Partial restoration by tissue extracts of membrane properties affected by denervation of rat skeletal muscle, *Proc. Univ. Otago Med. Sch.,* 56, 39, 1978.
166. **Kuromi, H. and Hasagawa, S.,** Neurotrophic effect of spinal cord extract on membrane potentials of organ-cultured mouse skeletal muscle, *Brain Res.,* 100, 1978, 1975.
167. **Hasagawa, S. and Kuromi, H.,** Effects of spinal cord and other tissue extracts on resting and action potentials of organ-cultured mouse skeletal muscle, *Brain Res.,* 119, 133, 1977.
168. **Kuromi, H., Gonoi, T., and Hasegawa, S.,** Neurotrophic substance develops tetrodotoxin-sensitive action potential and increases curare-sensitivity of acetylcholine response in cultured rat myotubes, *Dev. Brain Res.,* 1, 369, 1981.
169. **Kano, M., Shimada, Y., and Ishikawa, K.,** Electrogenesis of embryonic chick skeletal muscle cells differentiated *in vitro, J. Cell Physiol.,* 79, 363, 1972.

170. **Kidokoro, Y.,** Development of action potentials in a clonal rat skeletal muscle cell line, *Nature (London) New Biol.*, 241, 158, 1973.
171. **Land, B. R., Sastre, A., and Podleski, T. R.,** Tetrodotoxin-sensitive and -insensitive action potentials in myotubes, *J. Cell Physiol.*, 82, 497, 1973.
172. **Harris, J. B., Marshall, M. W., and Wilson, P.,** A physiological study of chick myotubes grown in tissue culture, *J. Physiol. (London)*, 229, 751, 1973.
173. **Fukuda, J.,** Chloride spike: a third type of action potential in tissue-cultured skeletal muscle cells from the chick, *Science*, 185, 76, 1974.
174. **Obata, K.,** Transmitter sensitivities of some nerve and muscle cells in culture, *Brain Res.*, 73, 71, 1974.
175. **Kano, M., Suzuki, N., and Ojima, H.,** Neurotrophic effect of nerve extract on development of tetrodotoxin-sensitive spike potential in skeletal muscle cells in culture, *J. Cell Physiol.*, 99, 327, 1979.
176. **Engelhardt, J. K., Ishikawa, K., Mori, J., and Shimabukuro, Y.,** Passive electrical properties of cultured chick skeletal muscle: neurotrophic effect on sample distribution, *Brain Res.*, 126, 172, 1977.
177. **Markelonis, G. J., Kemerer, V. F., and Oh, T. H.,** Sciatin: purification and characterization of a myotrophic protein from chicken sciatic nerves, *J. Biol. Chem.*, 255, 8967, 1980.
178. **Oh, T. H., Sofia, C. A., Kim, Y. C., Carroll, C., Kim, H. H., Markelonis, G. J., and Reier, P. J.,** Sciatin: immunocytochemical localization of a myotrophic protein in chicken neural tissues, *J. Histochem. Cytochem.*, 29, 1205, 1981.
179. **Oh, T. H. and Markelonis, G. J.,** Sciatin (transferrin) and other muscle trophic factors, in *Growth and Maturation Factors*, Vol. 2, Guroff, G., Ed., John Wiley & Sons, New York, 1984, 55.
180. **Li, I., Kimura, I., and Ozawa, E.,** A myotrophic protein from chick embryo extract: its purification, identity to transferrin, and indispensability for avian myogenesis, *Dev. Biol.*, 94, 366, 1982.
181. **Matsuda, R., Spector, D., and Strohman, R. C.,** There is selective accumulation of a growth factor in chicken skeletal muscle, *Dev. Biol.*, 103, 267, 1984.
182. **Mescher, A. L. and Munaim, S. I.,** ''Trophic'' effect of transferrin on amphibian limb regeneration blastemas, *J. Exp. Zool.*, 230, 485, 1984.
183. **Davis, H. L. and Heinicke, E. A.,** Prevention of denervation atrophy in muscle: mammalian neurotrophic factor is not transferrin, *Brain Res.*, 309, 293, 1984.
184. **Stamatos, C., Squicciarini, J., and Fine, R. E.,** Chick embryo spinal cord neurons synthesize a transferrin-like myotrophic protein, *FEBS Lett.*, 153, 387, 1983.
185. **Levin, M. J., Tuil, D., Uzan, G., Dreyfus, J.-C., and Kahn, A.,** Expression of the transferrin gene during development of non-hepatic tissues: high level of transferrin RNA in fetal muscle and adult brain, *Biochem. Biophys. Res. Commun.*, 122, 212, 1984.
186. **Markelonis, G. J., Oh, T. H., Park, L. P., Cha, C. Y., Sofia, C. A., Kim, J. W., and Azari, P.,** Synthesis of the transferrin receptor by cultures of embryonic spinal neurons, *J. Cell Biol.*, 100, 8, 1985.
187. **Oh, T. H., Markelonis, G. J., DionGuidera, T., Hobbs, S. L., and Park, L. P.,** Stimulation of the synthesis of fructose 1,6-diphosphate aldolase by transferrin, in *Gene Expression in Muscle*, Strohman, R. C. and Wolf, S., Eds., Plenum Press, New York, 1985, 129.
188. **Neckers, L. M. and Cossman, J.,** Transferrin receptor induction in mitogen-stimulated human T lymphocytes is required for DNA synthesis and cell division and is regulated by interleukin 2, *Proc. Natl. Acad. Sci. U.S.A.*, 80, 3494, 1983.
189. **Ekblom, P., Thesleff, S., Saxén, L., Miettinen, A., and Timpl, R.,** Transferrin as a fetal growth factor: acquisition of responsiveness related to embryonic induction, *Proc. Natl. Acad. Sci. U.S.A.*, 80, 2651, 1983.

Chapter 7

TROPHIC INFLUENCES OF NEUROGENIC SUBSTANCES ON ADULT SKELETAL MUSCLES IN VIVO

Heather L. Davis

TABLE OF CONTENTS

I. INTRODUCTION

Evidence of trophic influences by substances secreted by motor neurons on skeletal muscle fibers has long existed (Chapter 2), but much of this evidence has been indirect. The research to be presented and discussed in this chapter provides direct evidence of the existence and trophic action of neurogenic substances on mature skeletal muscle. This evidence has been provided by studies carried out to examine the in vivo effects of a proteinaceous extract from mammalian peripheral nerve on denervated mammalian skeletal muscle.[1-10] Although the active trophic agent in these studies has previously been referred to as a neurotrophic substance,[1-5] then later as a myotrophic substance,[6-10] in keeping with the terminology used in this book (Chapter 1) it will henceforth be referred to as a neurogenic substance, its action being an example of an anterograde exterotrophic function.

The provisional criteria of a neurogenic substance affecting skeletal muscle, as outlined in Chapter 1, included that it (1) be present within motor neurons; (2) be synthesized in the neuronal cell body and conveyed by anterograde axoplasmic transport; (3) be released from the nerve terminal and supplied to the muscle; (4) interact with a molecular target or receptor in the postsynaptic membrane and/or within the muscle cell; and (5) exert a dose-dependent effect on one or several definable metabolic, functional, and/or structural properties of muscle. The work to be discussed in this chapter is involved almost exclusively with the fifth provisional criterion.

A. Denervated Muscle

Denervation of a skeletal muscle results in several morphological, metabolic, and physiological alterations (Chapter 2). Atrophy is the most obvious morphological change and is manifested as reduced diameter or cross-sectional area (CSA) of component fibers.[1,11] Several ultrastructural changes are associated with denervation atrophy, including loss of contractile myofilaments, mitochondria, and areas of sarcoplasmic reticular and t-tubular membranous organelles.[12] Biochemical changes include increased or decreased concentrations or activities of many of the metabolic enzymes.[11] Physiological alterations include supersensitivity to acetylcholine (ACh), decreased resting membrane potential, and alterations in the characteristics of the contractile response to direct electrical stimulation (Chapter 5). The increased sensitivity to iontophoretically applied ACh is due to the production of new ACh receptors (AChR), which are inserted over the entire sarcolemma of the denervated fiber, rather than being localized to the postsynaptic region of the motor endplate, as occurs in normally innervated muscle.[13,14] Changes of the contractile response of a denervated muscle include increased twitch tension, decreased tetanus tension, and prolongation of the time to peak and half relaxation time of an isometric twitch.[8,13,15]

B. Disuse vs. Exterotrophic Influences

It is now generally agreed that changes occurring in a denervated muscle are due to both the disuse caused by paralysis of the muscle and to the loss of exterotrophic influences by neurogenic substances (Chapter 2). The experimental design and interpretation of results are complicated by the fact that both of these influences are exerted in the anterograde direction. However, evidence for the dual neural roles has come from two basic types of studies: depletion and repletion experiments.

In depletion studies, attempts are made to delete one of the two major neural influences: (1) contractile activity secondary to the release of ACh or (2) exterotrophic effects by secreted neurogenic substances. Muscles may be experimentally inactivated by procedures such as immobilization, pharmacological blockade of neural conduction or transmission, or spinal isolation. These models of disuse attempt to eliminate the action of contractile activity from otherwise normally innervated muscle, and although each has been shown to produce several denervation-like changes, these are almost always of smaller magnitude or later onset than those occurring following denervation.[2,11,16-18] The failure of disuse to mimic denervation is presumed to be due to the continuation of exterotrophic influences by neurogenic substances provided by the structurally intact motor neurons.

When muscles are denervated, but left with a long nerve stump attached, the onset of several of the changes of denervation is later than when only a short stump is left.[19-24] Because muscle activity and evoked release of ACh cease from the moment of denervation, regardless of the length of the nerve stump, these results suggest that the axoplasm of the nerve contains a special trophic substance which, until it is depleted, continues to influence the muscle fiber.

In repletion studies, one of the two major neural influences is replaced in a denervated muscle to determine if any of the postdenervation changes are prevented or ameliorated. The production of contractile activity in denervated muscles by direct electrostimulation has been found to reduce the severity of several of the changes in denervated muscles. However, the ameliorative effects were far less than complete, even with amounts of stimulation which mimicked normal levels of activity.[25]

Muscle cells from fetal animals may be grown in tissue culture. Such explanted cells are, of course, devoid of innervation, and, as one would expect, they soon show signs of degenerative change. However, if a piece of fetal spinal cord is co-cultured with the muscle cells, motor axons innervate the latter, and the degeneration is prevented.[26,27] This effect of innervation in vitro can be mimicked by adding to the muscle culture the medium from a spinal cord culture[28] or certain tissue extracts.[29,30] Markelonis and Oh[31] found that an extract of chicken sciatic nerve promoted morphological maturation and maintenance of aneural chicken muscle cells grown in tissue culture (Chapter 6). These investigators later purified from chicken sciatic nerve an acidic glycoprotein, sciatin, which was responsible for the trophic effects.[32] It has been shown more recently that sciatin is closely related in structure to transferrin,[33] an iron-binding glycoprotein present in serum and egg white.[34] Kimura et al.[35] studied the effects of serum on cultured muscle cells and also isolated transferrin as a neurogenic substance. These investigators demonstrated that the effects of transferrin on differentiation of chick myoblasts in vitro were also obtained with inorganic ferric salts.[36]

C. Summary and Experimental Rationale

Thus, there is considerable evidence from both in vitro and in vivo studies that exterotrophic influences by neurogenic substances distinct from ACh are at least partially responsible for maintaining many morphological, biochemical, and physiological properties of skeletal muscle.

The most direct evidence for the existence of neurogenic trophic substances came from the studies of Markelonis and Oh and colleagues[31-33] showing trophic effects of nerve extract on embryonic muscle in vitro. It is difficult, however, to extrapolate their findings to the situation of trophic maintenance of mature skeletal muscle. Owing to difficulties of maintaining whole adult muscle in organ culture, it was decided to attempt replacement of neurogenic substances in denervated muscles in vivo. In this way, the denervated muscles would be in an otherwise normal environment and would be nourished by natural vasculature.

It was hypothesized that in vivo administration of neurogenic substances would ameliorate morphological, metabolic, and/or physiological alterations of the denervated muscles. Complete prevention of these changes was not expected, since it had been well established that inactivity also plays a significant role in producing changes in denervated muscles.[2,11,16-18] Furthermore, it was hypothesized that the relative roles of disuse and loss of neurogenic substances in producing a given change in a denervated muscle would vary because it is unlikely that all properties of muscle fibers are controlled in a similar manner or by the same mechanisms.

II. ATROPHY IN DENERVATED MUSCLE

Atrophy of a denervated skeletal muscle is manifested as a decrease in the total weight of the muscle, loss of total protein, reduction of CSA of the component muscle fibers, and associated ultrastructural alterations.[1,2,10] Considerable evidence indicates that atrophy in denervated muscles is due to both disuse and loss of neurogenic substances.[1,27,31,37-40]

There is more than one type of skeletal muscle fiber, and they have different morphological, metabolic, and physiological properties. For example, three types of fiber are evident in cross sections of muscle stained for ATPase activity by the routine alkaline method (pH, 9.4), preceded by alkaline preincubation (pH, 10.4) and formaldehyde fixation.[41] Type I fibers have low ATPase activity and stain the palest. Type II fibers have high ATPase activity and stain darkly, but may be subdivided into two subtypes, since the IIB fibers are partially inhibited by alkaline and/or formaldehyde pretreatment, and thus stain with an intermediate intensity (Figure 1). Type I fibers are slow twitch and use oxidative metabolic pathways. Type II fibers are fast twitch, and although it has frequently been reported that IIA fibers are oxidative-glycolytic (FOG), whereas IIB fibers are principally glycolytic (FG), this does not appear to be the case for all species. For example, it has been shown that rabbit IIB fibers can be equated with FG fibers and IIA fibers with FOG fibers, while in humans both IIA and IIB fibers appear to have equal representation of FG and FOG fibers comprising them. In the rat, while the majority of IIB fibers are FG, some are FOG, and IIA fibers are predominantly FOG, but some are FG. Muscles of mice stain with the reverse pattern of that found in those of rats, with the majority of IIB fibers having FOG characteristics.[42]

In denervated rat extensor digitorum longus (EDL) muscle, atrophy occurs earliest and is greatest in the type IIB fibers, and is latest and least in the type I fibers.[1,43]

A. Trophic Effects of Neurogenic Proteins

Initial studies were carried out to determine if the trophic properties of neurogenic sub-

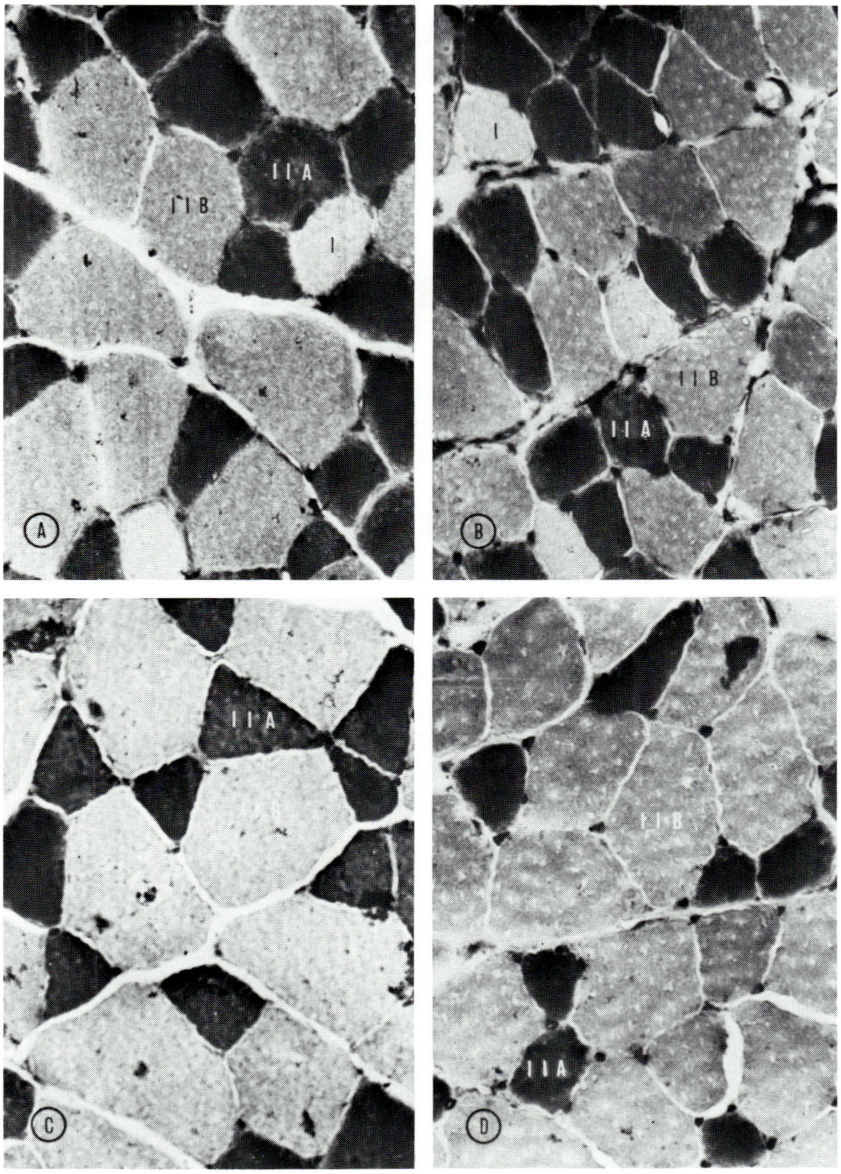

FIGURE 1. Photomicrographs of sections of adult rat EDL muscles, stained for ATPase. (A) Normal EDL. Type I, IIA, and IIB fibers are labeled. (B) Contralateral denervated EDL (7 days). (C) Normal EDL. (D) Contralateral denervated EDL muscle, injected daily with rat sciatic nerve extract (2 mg protein/mℓ for 7 days). The type IIB fibers are conspicuously smaller in B than in D. (Magnification × 390.)

stances that had previously been demonstrated in vitro on embryonic muscle cells (Chapter 6)[30-33] were also demonstrable in vivo on mature muscle fibers.

In each of 72 male Wistar rats, the right EDL muscle was denervated by removal of the portion of the sciatic nerve in the thigh region. The left EDL served as a normal contralateral control. For the following 7 days the animals received daily intramuscular injections in the denervated EDL of one of the following substances: rat sciatic nerve extract with a protein concentration of 0.25, 0.5, 1, 2, 4, 6, or 12 mg/mℓ; heat-inactivated nerve extract with a protein concentration of 2 mg/mℓ; liver extract with a protein concentration of 2 or 5 mg/

mℓ or phosphate-buffered saline. One group of animals received no injections, and there were six animals in each group.

Each substance was injected into the muscle in two volumes of 0.05 mℓ each, at points one third and two thirds of the distance between the tibial tuberosity and the anterior ankle crease. Two injection sites were used to avoid the damage which might have occurred if the entire volume of extract was injected at only one site and to ensure that a greater portion of the muscle was exposed to the injected substance.

1. Nerve Extract

Nerve extract was made from sciatic nerves dissected from anesthetized or freshly killed adult Wistar rats and frozen immediately. The nerves were ground to a powder in liquid nitrogen with a pestle and mortar. The pieces were then homogenized with an equal volume of 50 m*M* ammonium acetate (pH, 7.3) at 4°C, with a Brinkmann Polytron® PT 10-35 homogenizer. The homogenate was first centrifuged for 30 min at 17,500 *g* and then for 2 hr at 27,000 *g* at 4°C. The supernatant was filtered through a 0.45-μm Millipore® filter and was assayed for protein by the method of Lowry et al,[44] with bovine serum albumin as standard. The extract was lyophilized and then redissolved in an appropriate volume of 0.03 *M* sodium phosphate buffered saline (0.9% NaCl; pH, 7.3) to obtain the desired concentration of protein. Solutions were stored in syringes at −70°C and thawed just before use. Protein in the extract for control injections was inactivated by heating the extract for 30 min at 90°C.

2. Liver Extract

Liver extract was made from liver tissue removed from freshly killed adult Wistar rats. The procedure was the same as that described for nerve extract except that the liver was cut into pieces (about 1 cm^3) prior to homogenization.

3. Assessment of Denervation Atrophy

Muscle wet weight was determined immediately after removal of the muscles. A 5-mm segment from immediately below the midpoint of each muscle was frozen in isopentane cooled by liquid nitrogen to −150°C. The right and left EDL muscles from each animal were placed side by side on the same chuck to ensure identical treatment. Transverse sections, 10 μm thick, were cut in a cryostat at −20°C. The sections were stained for activity of ATPase by the method of Guth and Samaha.[41] The CSA of the fibers were measured with a Zeiss MOP-3® digital image analyzer from photographic prints which had been carefully standardized for magnification. The IIA and IIB fibers were measured separately, 100 fibers of each type from each muscle. There were not enough type I fibers in the sections for statistically significant comparisons to be made. Instead, the CSA of the type I fibers from all six denervated, untreated muscles were pooled and compared with those from the normal muscles of the same animals to determine the magnitude of their atrophy. Using 20 mg of muscle from the proximal remnant of the muscle, the total protein was determined by the method of Lowry et al.[44] using bovine serum albumin as standard.

4. Statistical Study

For each parameter, the value obtained from the experimental muscle was expressed as a percentage of the value for the normal contralateral muscle of the same animal. This controlled for differences in absolute values due to variation among different rats or in standard curves for protein analyses performed on different days. All data were expressed as group means plus or minus standard errors of absolute values from individual muscles or of percentage of control values from individual animals.

The significance of the differences between the groups was determined for each parameter

Table 1
EFFECT OF NERVE EXTRACT (OPTIONAL DOSAGE) AND CONTROL TREATMENTS ON ATROPHY OF DENERVATED RATS' EXTENSOR DIGITORUM LONGUS (EDL) MUSCLES

Treatment of right EDL[a]	Wet Weight[b]	Total protein[b]	CSA of IIA fibers[b]	CSA of IIB fibers[b]
Denervated (7 days)				
Untreated	80.5 ± 1.4	70.6 ± 2.1	82.8 ± 1.6	65.2 ± 1.8
Saline	83.4 ± 1.0	70.7 ± 2.1	81.0 ± 4.4	59.6 ± 2.3
Liver extract (2 mg/mℓ)[c]	81.2 ± 0.8	72.2 ± 3.2	76.8 ± 1.2	63.9 ± 3.5
Liver extract (5 mg/mℓ)[c]	82.5 ± 1.5	69.9 ± 1.9	80.3 ± 3.2	69.2 ± 3.3
Heat-inactivated nerve extract (2 mgℓmℓ)[c]	79.5 ± 1.5	65.6 ± 1.8	75.4 ± 1.0	79.5 ± 3.9
Nerve extract (2 mg/mℓ)[c]	88.0 ± 1.9	79.1 ± 2.7	79.5 ± 3.9	78.6 ± 3.0
Significance[d]	$p < 0.001$	$p < 0.05$	NS	$p < 0.025$
Innervated (7 days)				
Nerve extract (2 mg/mℓ)[c]	101.9 ± 1.0	103.1 ± 2.2	101.1 ± 1.8	100.9 ± 2.6
Significance[e]	$p < 0.50$	$p < 0.50$	$p < 0.75$	$p < 0.75$

[a] Contralateral (left) EDL served as normal control.
[b] Each value is the mean plus or minus the standard error of the mean of the percentage of control values obtained for each parameter from the six animals in each group. Percentage of control = right EDL /left EDL × 100. CSA, cross-sectional area.
[c] Extracts made from rat tissue (liver or sciatic nerve). Concentration of protein in the extract is in parentheses. Extract inactivated by heating to 90°C for 30 min.
[d] Refers to comparison of the last value (nerve extract) with any of the previous five values (control treatments) by one-factor analysis of variance and Newman-Keuls multiple-range test, based on group means of percentage of control values subjected to arcsine transformation. N.S., not significant ($p > 0.05$).
[e] Refers to comparison of the left and right muscles of each group by Student's t test based on group mean plus or minus the standard deviation of absolute values.

by one-factor analysis of variance followed by Newman-Keuls multiple-range testing. The percentages of control values were subjected to arcsine transformation before statistical comparisons were made, as recommended by Zar,[45] to validate the use of parametric statistical methods with percentages lying outside the 30 to 70% range.

5. *Effect of Denervation*

The changes that occurred in denervated and otherwise untreated muscles were similar to those described previously.[41,43,46] The CSA were reduced more in the IIB than in the IIA fibers (Table 1 and Figure 1). When the CSA of the type I fibers from all six denervated, untreated muscles were grouped and compared with those from the normal muscles of the same animals, the group means were almost identical. It is difficult, however, to state statistical significance of this finding due to the low number of fibers involved (total of 33 in six denervated muscles).[1] These observations agree with those of Niederle and Mayr[43] who found that the diameters of the IIB fibers in rat EDL were reduced the most, but that the diameters of the type I fibers remained unchanged for as long as 42 days after denervation.

6. *Effect of Nerve Extract*

In EDL muscles denervated for 1 week, the diminution of the type IIB fibers was the greatest atrophic change.[1] This parameter was also the most sensitive for assay of trophic effect of the nerve extract. Denervated muscles injected daily with 0.1 mℓ of nerve extract containing 1, 2, 4, or 6 mg protein per milliliter had significantly less diminution of the type IIB fibers than those injected with saline or those not treated ($p < 0.005$). At the optimal dosage of 2 mg protein per milliliter, almost 50% of the atrophy of IIB fibers was

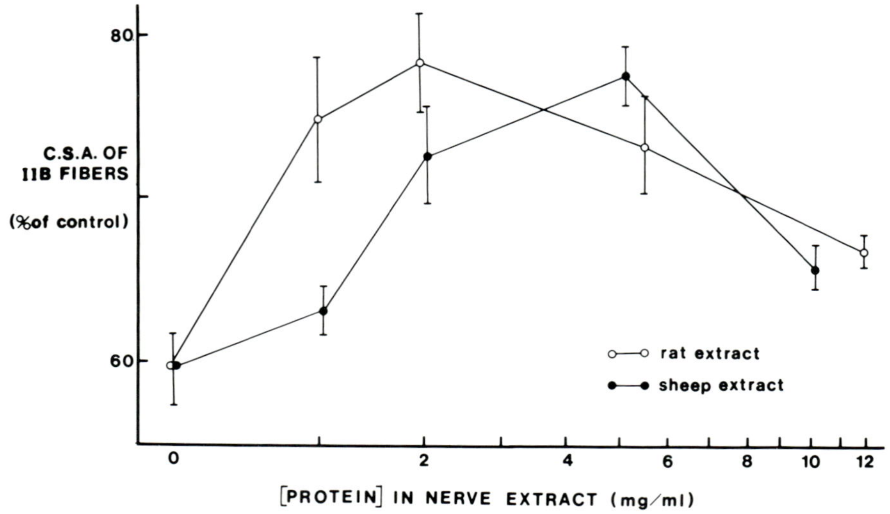

FIGURE 2. Dose-response curve for CSA of type IIB fibers in rat EDL muscles denervated and injected daily for 7 days with extract of sciatic nerves obtained from rats (○). Each point represents the mean of the percentage of normal contralateral control values (denervated/normal × 100) for six animals. Vertical lines represent standard errors of means. Extract of 0 mg/mℓ was physiological saline. Atrophy of denervated type IIB fiber was significantly reduced by injections of rat extract containing 1, 2, or 6 mg protein per milliliter or sheep extract (●) containing 2 or 5 mg protein per milliliter ($p < 0.0005$).

prevented. Extract with a protein concentration of 0.25, 0.5, or 12 mg/mℓ did not reduce the atrophy of denervated IIB fibers. These results are illustrated in Figure 2.

The type IIA fibers in denervated muscles that were injected with nerve extract at any of the concentrations tested were atrophied as much as those in controls.

Wet weight and content of total protein were decreased significantly less in denervated muscles injected with nerve extract of optimal dosage (2 mg protein per milliliter) than in those injected with saline or those not treated ($p < 0.005$). These two parameters were not affected by extracts of other concentrations tested.[3]

Thus, in vivo injection of neurogenic substances into denervated rat EDL muscle significantly ameliorated some of the atrophic changes, and this trophic effect was dose dependent. Although doses of rat nerve extract containing 1 to 6 mg protein per milliliter all significantly reduced denervation atrophy, smaller or large concentrations were ineffective.[3] Thus, the smaller doses of nerve provided insufficient amounts of the trophic substances to exert detectable ameliorative effects. The lack of trophic effect with highly concentrated extract may have been due to a high concentration of the trophic factor or to some inhibitory substances in the extract. Other investigators have demonstrated decreased effectiveness of high concentrations of crude[40] or purified[31] chick nerve extracts on chick myoblasts in vitro.

At optimal dosage the most marked trophic effect of the nerve extract on denervated muscle in vivo was the prevention of almost 50% of the atrophy in IIB fibers. The effects on total protein and on muscle weight support the validity of these findings. Complete prevention of the shrinkage of the muscle fibers would not be expected, even at an optimal concentration of the neurogenic substance, because it is well established that disuse alone causes considerable atrophy[27,37-39]

The absence of an exterotrophic effect on the CSA of the IIA fibers may be due to lack of sufficient atrophy to detect an effect. The untreated IIB fibers had a 35% decrease in area, whereas the IIA had less than a 20% decrease. The observation of a selective effect of the nerve extract on only the IIB type of fiber in the rat's EDL muscle is in agreement with the results of Riley and Allin[38] that type IIB fibers are the ones most dependent on exterotrophic factors. The possibility exists that different neurogenic substances might exist for different types of muscle fiber. However, it is also possible that an effect on the type IIA fibers would have been found if postoperative survival had been longer.

The smaller loss of total protein in the denervated muscles injected with nerve extract is probably due largely to decreased loss of myofibrillar proteins in the IIB fibers. An even greater effect on total protein might have been obtained if the sample of muscle used for protein determination was taken from the distal, rather than the proximal, end of the muscle. The distal end of the rat's EDL contains relatively more IIB fibers than the proximal end.[43,47]

The failure of heat-inactivated extract to reduce the loss of total protein rules out the possibility that the protein within the extract accumulated within the muscle and was later assayed. The same amount of protein was injected with the inactivated extract as with the active extract.

7. Systemic Administration of Neurogenic Substances

The effect of systemic administration (i.e., intraperitoneal injection) of neurogenic substances on denervation atrophy was explored because it was desirable to avoid the direct damage caused by intramuscular injection in order to carry out ultrastructural and physiological studies. Mice were used for this purpose because the quantity of nerve extract required for systemic administration into rats was prohibitively great.

The right hindlimbs of 8-week-old male C57BL/6J mice were denervated as described above for rats. The innervated left hindlimb muscles served as contralateral innervated controls. The animals either were not treated ($n = 6$) or received daily intraperitoneal injections of nerve extract (250 mg protein per kilogram of body weight; $n = 4$). The nerve extract was made from rat sciatic nerves, as described above. It was used at a protein concentration of 10 mg/mℓ, and the volume injected was approximately 0.5 mℓ per animal per day. This dosage was calculated to produce the same concentration of extract in the body tissues of the mouse as was estimated to have been produced in EDL muscles of rats which received intramuscular injections of the optimal dosage of rat nerve extract.[3] After 7 days of denervation, the mice were killed by an overdose of ether vapor. The right (denervated) and left (normal) tibialis anterior (TA), EDL, and soleus (SOL), together with gastrocnemius (G), muscles were removed from the animals and weighed. Each TA was thoroughly homogenized in 0.1 *M* NaOH (6 mg wet weight per milliliter), and the content of total protein was determined by the method of Lowry et al.[44] using bovine serum albumin as standard. The EDL and SOL plus G muscles were sectioned, stained, and used for determination of mean CSA of each type of fiber, as was described above for the rat EDL, except that the mouse muscles were embedded in pieces of liver prior to freezing because of their small size. Type I fibers were measured in SOL, type IIA in EDL and SOL, and type IIB in EDL and G. Data were expressed as group means plus or minus standard deviations of absolute values obtained from individual muscles or absolute values normalized to body weight. Student's *t*-test for unpaired samples was used for statistical comparison of normal and denervated muscles within the same group (absolute data) or of treated and untreated groups (normalized absolute data). One-factor analysis of variance followed by Newman-Keuls multiple-range testing was used for comparison of three or more values (normalized absolute data).

Unilateral denervation of hindlimbs of adult mice for 7 days produced atrophic changes similar to those found in the rat.[1] Absolute values obtained for denervated muscles were significantly smaller than those for contralateral normal muscles for all parameters (Student's one-tailed *t*-test; $p < 0.0005$).[7] Daily intraperitoneal injections of nerve extract into mice reduced the atrophy of the denervated muscles.

Absolute values were compared between groups to determine if the effect of the injected

Table 2
EFFECT OF SYSTEMIC ADMINISTRATION OF NERVE EXTRACT ON ATROPHY OF DENERVATED HINDLIMB MUSCLES OF C57BL/6J MICE

		Treatment groups[b]				Significance among groups[c]
Parameter	**Muscle[a]**	**Group 1 (NOR-UnRx)**	**Group 2 (NOR-Rx)**	**Group 3 (DN-Rx)**	**Group 4 (DN-UnRx)**	**(α = 0.05)**
Wet weight (mg/g)[d]	TA	1.83 ± 0.05	1.70 ± 0.07	1.53 ± 0.03	1.40 ± 0.02	1 2 3 4
	EDL	0.37 ± 0.01	0.35 ± 0.01	0.33 ± 0.01	0.31 ± 0.01	1 2 3 4
Protein (mg/g)[d]	TA	0.36 ± 0.01	0.30 ± 0.02	0.28 ± 0.01	0.25 ± 0.01	1 2 3 4
CSA of fibers ($\mu m^2/g$)[d]						
Type I	SOL	52.5 ± 3.7	59.3 ± 2.0	51.8 ± 2.5	39.7 ± 3.7	1 2 3 4
Type IIA	EDL	25.5 ± 0.4	29.8 ± 3.0	27.8 ± 1.9	21.0 ± 0.8	1 2 3 4
	SOL	47.5 ± 3.7	46.3 + 2.1	38.0 ± 2.0	32.5 ± 2.4	1 2 3 4
Type IIB	EDL	65.3 ± 2.4	65.2 ± 2.5	53.8 ± 1.6	43.2 ± 2.5	1 2 3 4

[a] TA, tibialis anterior; EDL, extensor digitorum longus; SOL, soleus.
[b] Each animal was subjected to unilateral hindlimb denervation. Each muscle was normally innervated (NOR) or denervated for 7 days (DN) and was obtained from a mouse which had received no other treatment (UnRx; n = 6) or was injected daily with rat sciatic nerve extract (Rx; n = 4)
[c] Significance refers to comparison of the four groups of muscle treatments by one-factor analysis of variance followed by Newman-Keuls multiple-range testing, based on absolute values normalized to body weight. Sets of similarities are underlined together ($p > 0.05$). Values not underlined by a common line were found to be significantly different ($p < 0.05$). CSA, cross-sectional area.
[d] Each value is the mean plus or minus the standard error of the mean of individual absolute values (x) normalized to body weight (x/weight animals in grams).

nerve extract was on the denervated muscle (as expected), the contralateral innervated muscle, or both. It was necessary to normalize absolute values to body weight, since the treated animals were smaller than the control animals (18.6 ± 2.5 vs. 23.1 ± 1.0 g, respectively) and their muscles were proportionately smaller. The differences in animal weights were not due to the injection of extract, since there was not a significant change in weight over the 7-day period of denervation for either group (student's two-tailed *t*-test; $p > 0.2$). Four treatments of muscle were compared by one-factor analysis of variance: innervated (normal) untreated (NOR-UnRx); innervated from injected animals (NOR-Rx); denervated untreated (DN-UnRx); and denervated from injected animals (DN-Rx). When pairwise differences between means were compared by Newman-Keuls multiple-range testing (Table 2), three types of results were found:

1. No significant difference was found between NOR-Rx and NOR-UnRx, but DN-Rx was significantly less than NOR-Rx or NOR-UnRx, and DN-UnRx was significantly less than DN-Rx. This indicates that the extract had no effect on innervated muscle, but partially ameliorated changes in denervated muscle. This was found for wet weight of TA and CSA of type IIB fibers of EDL.
2. No significant differences were found between NOR-Rx, NOR-UnRx, and DN-Rx, but DN-UnRx was significantly smaller than the other three groups, indicating no effect of the extract on innervated muscle but an almost complete amelioration of the changes in denervated muscle. This was true for CSA of type I fibers in SOL and IIA fibers in EDL.
3. No difference was found between NOR-Rx and DN-Rx, but NOR-UnRx was significantly greater and DN-UnRx was significantly smaller than the treated groups. This

was found for total protein of TA and indicates that the extract partially ameliorated the postdenervation loss and either had a deleterious effect on innervated muscle or diminished the hypertrophy that is known to occur in muscles of the contralateral unoperated side after unilateral denervation. For some parameters the results were ambiguous because of overlapping sets of similarities (Table 2).

The amount of atrophy in EDL muscles denervated for 7 days was found to be the same in the mouse as that found previously in the rat.[1] However, systemic injection of nerve extract in mice ameliorated more denervation atrophy than did intramuscular injection of extract in rats.[1] This was true for all parameters used to assess atrophy, but especially so for the CSA of type IIA fibers (57% of the shrinkage in the mouse EDL was prevented vs. 5% in the rat EDL).[2]

Thus, administration of nerve extract by intraperitoneal injection into the mouse provided a superior system for assay of exterotrophic influence of neurogenic substances than did intramuscular injection in the rat. There are two major disadvantages of the assay system in the mouse. First, a greater quantity of nerve extract is required for systemic than for local injection. Second, it is not possible to rule out an effect of the systemically injected extract on the contralateral normal control muscle. It has been shown previously, however, that direct injection of normally innervated EDL muscles in rats, at the dosage found to be optimal for amelioration of denervated atrophy, failed to change the wet weight, total protein, or areas of fiber relative to contralateral uninjected normal controls.[3]

An important advantage of assay of the extract by systemic administration is the opportunity to study different muscles in the same animal. Muscles in denervated hindlimbs of mice were not all affected equally. For example, there were significant differences between the decreases of wet weight (SOL + G > TA > EDL; one-factor analysis of variance; $p < 0.001$) and areas of fiber [IIB (G) > IIB (EDL) = IIA (SOL) > IIA (EDL); one-factor analysis of variance; $p < 0.05$]. It is interesting to note that the postdenervation loss of CSA for a given type of fiber was greater in posterior than anterior crural muscles, but the ameliorative effect of the extract was less on fibers of posterior than anterior muscles.[7] The anterior crural muscles are phasic in action, whereas the posterior crural muscles are principally tonic. Intrinsic or extrinsic factors responsible for these properties may also determine the relative roles of activity and exterotrophic influences in normal maintenance of skeletal muscles.

B. Specificity of Exterotrophic Effects

1. Species Specificity

Injection of ovine nerve extract (prepared as the rat nerve extract) reduced the atrophy of type IIB fibers in denervated rat EDL muscles. This effect was also dose dependent (Figure 2). The dose-response curve for effect of ovine nerve extract on the CSA of type IIB fibers was similar to that obtained for extract of rat sciatic nerves. However, larger doses of sheep than of rat nerve extract were required to produce significant trophic effects (Figure 2). The greatest effect obtained with ovine nerve extract was equivalent to that obtained with the optimal concentration of extract of rat nerves. Heat-inactivated ovine nerve extract had no effect.[3]

Thus, the trophic effects of neurogenic proteins on denervated mature skeletal muscle in vivo are not species specific. Other investigators have demonstrated similar lack of species specificity of neural explants or extracts on differentiation of embryonic muscle in vitro. Trophic effects have been obtained with co-cultures of mouse spinal cord and chick myotubes,[27] extracts of rat brain on newt muscle,[48] extracts of quail or hamster brain on chick muscle, and extract of mouse spinal cord on rat myotubes.[30] Our findings are valuable because the trophic substance(s) can be purified from ovine nerve extract by biochemical purification procedures and then assayed on denervated rat muscles. The sciatic nerves from one sheep provide a neurogenic extract equivalent to that from the sciatic nerves of several hundred rats.

The requirement of more concentrated extract of sheep vs. rat nerves to produce significant trophic effects may be due to a smaller proportion of trophic factor in ovine than in rat sciatic nerves relative to other soluble proteins. Alternatively, the neurogenic trophic substances of the two species may be of different potencies. A third possibility is that of greater quantities of inhibitory substances, which oppose the action of the trophic substance(s), in extracts of ovine than of rat nerves. As mentioned, such substances may be responsible for the reduced effects of concentrated extract.

Exterotrophic effects have also been obtained on denervated rat muscle with neurogenic substances from cat sciatic nerves[106] and on denervated mouse muscle with extracts from sciatic nerves of rats[7,10] or sheep.[8]

2. Specificity of Exterotrophic Phenomena

The EDL muscles denervated and injected daily for 1 week with saline, liver extract, or heat-inactivated extract atrophied as much as those denervated and not treated (Table 1), indicating that the trophic effects of the active nerve extract were not due to inflammatory edema secondary to injection of a foreign protein or to trauma. Injection of heat-inactivated extract or liver extract, both containing foreign proteins, failed to produce an effect.[1,3] If inflammatory processes were responsible for decreasing the rate of atrophy in denervated muscle fibers, they would be expected to have equivalent effects on all types of fiber. However, the ameliorative effect of the nerve extract was greatest on the type IIB fibers and small or absent on the IIA fibers.[1] Furthermore, atrophy of immobilized muscles was not reduced by injection of nerve extract.[2]

3. Source of Neurogenic Substances

The trophic effects were likely due to substances derived from neuronal rather than vascular or connective tissue components of peripheral nerves because extracts of liver, a nonneural tissue, did not produce results significantly different from injections of saline or from no treatment.[3] Similar results have been obtained in studies of effects of neural and nonneural extracts on embryonic muscle cells in vitro.[28,48,49] Although our results do not rule out the possibility that Schwann cells are the source of the active principle, findings of other researchers suggest otherwise. Oh et al.[50] demonstrated by immunohistochemical methods that the trophic substance sciatin, purified from chick sciatic nerves, was localized in neuronal perikarya and processes, but was absent from Schwann cells.

4. Mechanisms of Action

Normal EDL muscles injected with nerve extract for 7 days did not differ from uninjected contralateral controls (Table 1). This again indicated that the active nerve extract did not produce inflammatory edema.

It is unlikely that the trophic effects of the nerve extract are due to production of contractile activity in the muscle fibers. The extract does not contain ACh, which is too unstable to survive the extraction procedure.[51] Neither does it contain sufficient choline acetyltransferase to produce ACh *in situ*. The nerve extract was made from axonal segments not including nerve terminals, which contain 98% of neuronal choline acetyltransferase.[52] Furthermore, intramuscular electromyographic recording electrodes failed to record any signs of activity from denervated EDL muscles following injection with nerve extract. The same muscles exhibited short bursts of electrical activity after injection of ACh.

5. Effect of Early or Delayed Injections

Rat EDL muscles denervated for 14 days exhibited significantly greater reductions in wet

Table 3
EFFECT OF EARLY AND DELAYED INJECTIONS OF RAT NERVE EXTRACT INTO RATS' MUSCLES DENERVATED FOR 14 DAYS

Parameter	Untreated	Injected all 14 days	Injected days 1—7	Injected days 8—14	Significance[a]
No. of animals	6	6	6	6	
Wet weight[b]	56.2 ± 1.3	62.4 ± 1.5	58.6 ± 1.5	60.4 ± 0.7	NS
Total protein[b]	44.9 ± 2.9	48.3 ± 2.7	53.3 ± 3.1	54.4 ± 3.9	NS
CSA[c] of IIA fibers[b]	67.0 ± 2.7	65.9 ± 2.4	71.0 ± 2.9	64.9 ± 2.2	NS
CSA[c] of IIB fibers[b]	36.8 ± 1.6	46.1 ± 1.3	51.4 ± 2.9	43.8 ± 1.3	$p < 0.025$
CSA[c] of I fibers (μm^2)[d]					
Normal EDL (68)	751 ± 175				
Denervated EDL (53)	719 ± 176				

[a] Refers to comparison of the first column with any of the last three columns by one-factor analysis of variance and Dunnett's multiple-range test, based on group means of percentage of control values subjected to arcsine transformation. NS, not significant.

[b] Each value is the mean plus or minus the standard error based on raw percentage of control values. These figures have only descriptive value and were not used for testing the significances of differences among means.

[c] Cross-sectional area.

[d] Each value is the mean plus or minus the standard deviation based on absolute values from all six muscles in the group. Units are μm^2. Number of fibers measured is noted in parentheses. There is no significant difference between values from normal and contralateral denervated muscles $p > 0.25$; one-tailed Student's *t* test).

weight, total protein, and CSA of types IIA and IIB fibers than those denervated for 7 days (Table 3). When the CSA of the type I fibers from all six denervated untreated muscles were compared to those from the contralateral normal muscles of the same animals, no significant differences were found (Table 2). These observations are similar to those of other investigators.[43]

The mean CSA of the type IIB fibers was reduced by 38% after 1 week of denervation[1] and by 63% after 2 weeks.[3] Type IIB fibers of muscles denervated for 2 weeks shrank significantly less (54%) if injected daily with rat nerve extract (2 mg protein per milliliter) than if not treated ($p < 0.01$). Amelioration of atrophy of IIB fibers was also obtained with injections on only the first or the last 7 days of a 14-day period of denervation ($p < 0.01$; Table 3). There were no significant differences between areas of type IIA fibers in groups of animals in which the right EDL muscles were denervated for 14 days and were either not treated or were injected daily with nerve extract for 7 or 14 days (Table 3).[3]

The wet weights of EDL muscles denervated for 14 days were reduced significantly less in those injected daily with nerve extract than in those not treated ($p < 0.01$). Muscles injected with nerve extract for either the first or the last 7 days had mean wet weights between those of muscles that were not treated and those that were injected on all 14 days. However, when the percentage of normal control values for the four groups were subjected to arcsine transformation and compared by one-factor analysis of variance, no significant differences were detected among the groups. The discordance of the results of the two statistical tests is probably due to different degrees of conservatism of the tests.[3]

Injections of nerve extract did not significantly diminish the loss of total protein in muscles that were denervated for 14 days (Table 3). The lack of statistical significance may be attributed to the large standard deviations encountered.

Thus, muscles denervated for 14 days and injected with extract on all 14 days, or only on the first or last 7 days, all exhibited significantly less atrophy than those not treated. This indicated that the trophic effects of the injected extract persisted for at least 1 week and were still demonstrable when administration of the extract was delayed.

C. Relative Roles of Disuse and Loss of Neurogenic Substances

Even injection of denervated EDL muscles with the optimal dosage of nerve extract failed to completely prevent the atrophy. For example, the type IIB fibers in denervated muscles were still about 20% smaller than those in normal contralateral controls.[1] The remaining atrophy may have been due to incomplete action of the extract, or, alternatively, it may have been more likely due to disuse.

To determine the relative roles of disuse and loss of neurogenic substances in producing denervation atrophy, comparisons were made among six groups of animals, of six rats each, which received one of the following treatments.

Group 1 — Disuse of the right EDL muscle was produced by internal fixation of the right ankle at 90°. This was accomplished by inserting under anesthesia a 26-gauge needle through the calcaneus and talus into the marrow cavity of the tibia. The excess length of needle was cut off flush with the plantar surface of the calcaneus. Placement of the needle was checked by anteroposterior and lateral view radiography (Figure 3). The contralateral (left) EDL muscles served as normal controls.

Group 2 — The right EDL muscle was denervated, and the left EDL served as a normal control.

Group 3 — In this group, the animals had their right EDL muscles denervated and their left normally innervated muscles immobilized.

Group 4 — The right EDL muscle was denervated and received daily intramuscular injections of the optimal dosage of nerve extract (0.1 mℓ of protein concentration of 2 mg/mℓ). The extract was made from rat sciatic nerves and injected as described above. The contralateral EDL muscle was left intact.

Group 5 — The right EDL muscle was denervated and received daily intramuscular injections of nerve extract, also containing 2 mg protein per milliliter. The left EDL muscle was immobilized.

Group 6 — Both ankles were immobilized, and the right EDL muscle received daily injections of nerve extract containing 2 mg protein per milliliter.

After 7 days of treatment the animals were killed by an overdose of ether vapor. The right and left EDL muscles were removed, and the degree of atrophy was assessed as described above.

The amount of atrophy produced by denervation in the EDL muscle of the rat was greater than that produced by immobilization. The decreases in wet weight, total protein, and CSA of the IIB fibers were significantly greater for denervated muscles (group 2) than for immobilized muscles (group 1) compared with their respective normal control muscles. In addition, when the effects of denervation were compared with those of immobilization in the same animal (group 3), the denervated EDL muscles had significantly greater atrophy, as determined by all four parameters, than did the immobilized muscles (Table 4).

These findings may be criticized because immobilization of a muscle does not produce complete disuse. Isometric contractions of the muscle are still possible, and it has been found by electromyographic recordings that 5 to 15% of the normal number of spikes still occur in immobilized muscles.[53] However, if the amount of atrophy due to disuse is proportional to the degree of inactivity so that "total disuse" produces 5 to 15% more atrophy than immobilization, the atrophy of "total disuse" also would be significantly less than that obtained by denervation. Immobilization was used as a model of disuse in this study because

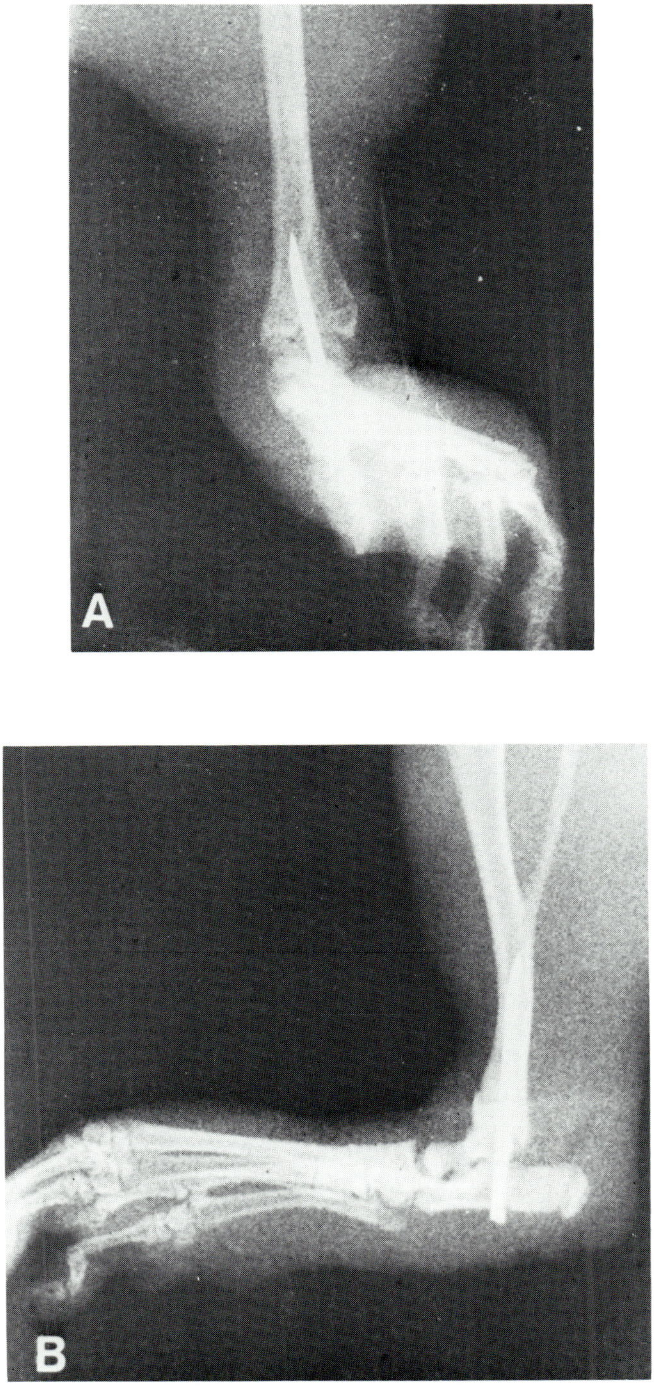

FIGURE 3. Radiograph of a rat ankle which has been immobilized at 90° by passing a 26-gauge needle through the calcaneus and talus into the marrow cavity of the tibia. (A) Anteroposterior view; (B) lateral view.

Table 4
EFFECT OF DENERVATION, IMMOBILIZATION, AND INJECTION OF NERVE EXTRACT OF SKELETAL MUSCLE

	Group 1	Group 2	Group 3	Group 4	Group 5	Group 6
Treatment						
Right EDL	Immobilized	Denervated	Denervated	Denervated, nerve extract	Denervated, nerve extract	Immobilized nerve extract
Left EDL	Normal	Normal	Immobilized	Normal	Immobilized	Immobilized
Parameter						
Wet weight[a]	87.8 ± 1.7	80.5 ± 1.4	86.6 ± 3.1	88.0 ± 1.9	99.8 ± 2.9	113.1 ± 8.1
Significance[b]	$p < 0.025$	$p < 0.01$	$p < 0.01$	NS	NS	NS
Total protein[a]	86.6 ± 3.4	72.6 ± 3.5	76.0 ± 4.8	79.1 ± 2.7	87.0 ± 6.8	110.9 ± 7.9
Significance[b]	$p < 0.025$	$p < 0.025$	$p < 0.025$	NS	NS	NS
CSA[c] of IIA fibers[a]	80.4 ± 3.6	82.8 ± 1.6	89.5 ± 3.1	79.5 ± 3.9	92.5 ± 6.7	100.1 ± 3.8
Significance[b]	$p < 0.005$	$p < 0.025$	$p < 0.01$	$p < 0.001$	NS	NS
CSA[c] of IIB fibers[a]	77.6 ± 1.4	65.2 ± 1.8	76.7 ± 4.6	78.6 ± 3.0	93.3 ± 6.4	102.1 ± 8.2
Significance[b]	$p < 0.0005$	$p < 0.0005$	$p < 0.001$	$p < 0.0025$	NS	NS

[a] Expressed as the plus or minus the standard error based on untransformed percentage right/left × 100. These figures have descriptive value only. For statistical comparisons of the differences between groups, all individual percentages were subjected to arcsine transformation.

[b] Refers to comparison of right with left EDL muscles of rats in each group. Determined by Student's *t* test using group means plus or minus standard deviations of absolute values from individual muscles. NS, not significant.

[c] Cross-sectional area.

it produced a significant and adequate amount of atrophy to test the hypothesis. Other methods of producing disuse have various disadvantages. Flaccid paralysis by spinal isolation[16] is difficult to carry out in the rat, botulinum toxin interferes with delivery of neurogenic proteins to the muscle,[54,55] local anesthetic agents block axoplasmic transport[56] and have direct toxic effects on muscle tissue,[57,58] and tetrodotoxin produces relatively brief periods of paralysis.[18,59]

The effects of denervation were significantly suppressed by daily injections of nerve extract (compare groups 2 and 4 in Table 4). The results indicated that the nerve extract did not affect atrophy due to disuse because immobilized normally innervated muscles injected with the extract did not differ significantly from contralateral, noninjected immobilized muscles (group 6). Therefore, the effects of the nerve extract are specific to that component of the atrophy which is due solely to deprivation of neurogenic substances to the muscle.

Although denervated muscles were more atrophic than their contralateral immobilized muscles (group 3), denervated muscles which received nerve extract were not significantly different from contralateral immobilized muscles (group 5). This indicated that the nerve extract was able to prevent completely that component of the atrophy which was due to loss of neurogenic substances.

The relative roles of inactivity and loss of exterotrophic influence in producing denervation atrophy in a skeletal muscle appear to be different for the various types of fiber. Denervated muscles had 35% smaller CSA of IIB fibers than their normal control muscles (group 2), however, immobilized muscles had only 22% smaller CSA of IIB fibers than their normal contralateral muscles (group 1). This indicated that 63% of the atrophy of the denervated IIB fibers was due to disuse, and the remaining 37% was due to loss of other neural influences. Injection of nerve extract into denervated muscles reduced the atrophy such that the CSA of the IIB fibers was only 21% smaller than in normal control muscles (group 4). Therefore, substances extracted from nerves could prevent 40% of the atrophy in denervated IIB fibers, suggesting that the remaining 60% of the atrophy was due to disuse (see Figure 4).

Disuse appears to play a more dominant role in producing atrophy during the first week after denervation of type IIA fibers. Both denervation (group 2) and disuse (group 1) produce about a 20% decrease in CSA of the IIA fibers compared with their respective normal controls, suggesting that purely exterotrophic influences may not regulate the type IIA fibers at all. However, when denervation was compared with immobilization in contralateral muscles of the same animals (group 3), the areas of the denervated IIA fibers were significantly smaller than those of the immobilized fibers ($p < 0.01$). This difference was abolished by injecting the denervated muscles with nerve extract, indicating that trophic influences on IIA fibers are demonstrable. If so, the loss of trophic influences accounts for only about 5% of the atrophy of type IIA fibers after 7 days of denervation (see Figure 4).

By similar comparisons of groups 1, 2, and 3, as above, it is apparent that disuse is responsible for about 60% of the losses of wet weight and total protein during 7 days of denervation, and loss of trophic influence is responsible for the remaining 40% (see Figure 4). Changes in wet weight and total protein reflect alterations in the whole muscle, not just those in the IIA and IIB fibers.

In summary, the results show that the atrophy which occurs in denervated skeletal muscle fibers is due to the combined effects of inactivity and loss of exterotrophic influence. These two factors influence the various types of fiber differently. The component of denervation atrophy due to loss of trophic influence can be ''completely'' prevented by injection of neurogenic substances.

D. Ultrastructural Changes

As discussed above, administration of neurogenic substances locally (intramuscular injection) or systemically (intraperitoneal injection) to denervated muscles of rats or mice was found to ameliorate atrophic changes manifested at the light-microscopic level (i.e, dimi-

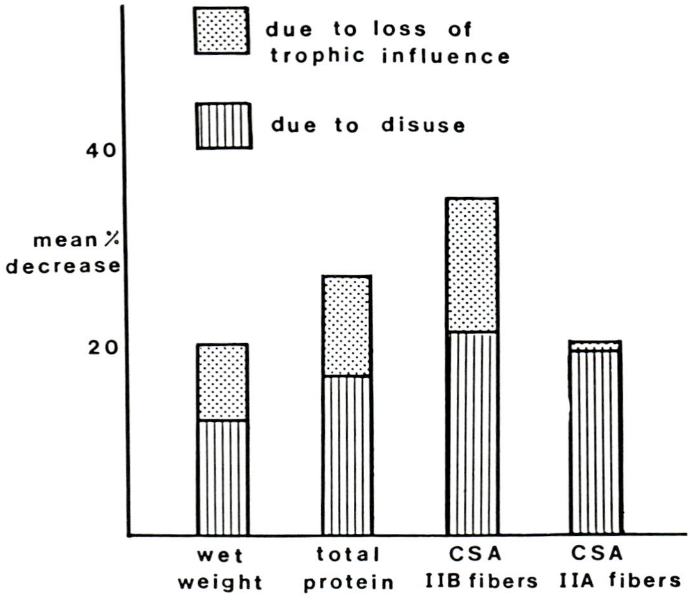

FIGURE 4. Relative effects of disuse and loss of exterotrophic influence in producing denervation atrophy. Bars represent the mean percentage decreases in wet weight, total protein, and CSA of the IIA and IIB fibers in the rat EDL muscle after 7 days of denervation. The dotted upper portion of each bar represents the mean decrease due to loss of trophic influence. The striped lower portion represents the decrease due to disuse. The values were obtained by comparison of group 2 and groups 1 and 4 (see Table 4). Loss of trophic influence is responsible for 40% of the decreases in wet weight, total protein, and CSA of the IIB fibers, and for about 5% of the atrophy of the IIA fibers.

nution of CSA of fiber). It is well known that several ultrastructural changes are associated with denervation atrophy.[12] For example, by 1 week of denervation, the volumes of myofibrils and mitochondria, and the areas of membranous organelles (sarcoplasmic reticulum and t tubules) are all significantly decreased.[12,60-64] We examined these postdenervation changes in the different types of fibers (as classified by activity of ATPase)[41] in mouse SOL (types I and IIA fibers) and G muscles (superficial portion of medial head; type IIB fibers),[65] and assessed the exterotrophic effects of systemically administered neurogenic substances on these changes.[10]

1. Preparation of Tissue

In two groups of C57BL/6J mice, the muscles of the right hindlimbs were denervated for 7 days, while those of the left hindlimbs served as innervated controls. One group of animals received no further treatment (n = 6), while the other group received daily systemic injections of neurogenic substances extracted from rat sciatic nerves (as discussed in Section II.A). The animals were killed by intracardiac perfusion with 2.5% glutaraldehyde; then pieces of the fixed SOL and G muscles were prepared for electron microscopy. This involved further fixation with 2.5% glutaraldehyde, washing with 0.1 *M* sodium cacodylate, postfixation in reduced 2% osmium tetroxide, dehydration through graded alcohols to propylene oxide, and infiltration followed by embedding in Epon. Sections of 70-nm thickness, of both longitudinal and transverse orientation, were mounted on grids and then stained with uranyl acetate and counterstained with lead citrate prior to examination by transmission electron microscopy.

Types I, IIA, and IIB fibers[41] were assayed separately. The superficial portion of the medial head of G muscle was used for type IIB fibers, since it is homogeneous for this type of fiber.[47] In the SOL muscle, fibers were classified as either type I or IIA on the basis of the width of their Z lines (those in type I being approximately twice as great as those in type IIA). This method of classifying was validated by correlation with immunocytochemical staining for the presence of type II (heavy-chain) myosin.

For immunocytochemistry, SOL muscles were dissected from freshly killed normal mice, fixed in formaldehyde, and prepared for infiltration with Lowicryl K4M. Sections were incubated in phosphate buffered saline containing 1% bovine serum albumin to block non-specific binding and then incubated in an appropriate dilution of antibody to type II (heavy-chain) myosin. Control grids were incubated in nonimmunized rabbit serum. After washing, sections were incubated with a secondary antibody (rabbit antimouse IgG), washed again, treated with a solution of 14-nm protein A-gold complex,[66,67] washed, stained with uranyl acetate, and counterstained with lead citrate.

We found a correlation between the labeling pattern for the presence of type II myosin and the width of Z lines. Fibers which had significant labeling (approximately 12 gold particles per square micron over the A band region) had narrow (70 to 90 nm) Z lines (type IIA), whereas those with wide (110 to 125 nm) Z lines (type I) did not show significant antibody labeling (1 to 2 particles per square micron; coefficient of correlation, 0.95).

2. Stereological Analysis

Fields of sections to be measured were chosen randomly as described by Engel and Stonnington.[12] CSA of organelles were measured from photographs standardized for magnification and data were expressed as (1) mean CSA for each type of organelle in each fiber; (2) total area per myofiber occupied by the organelle, as determined by taking the sum of the individual areas of the organelle in a complete cross section of fiber; and (3) percentage volume, which was estimated to be equivalent to percentage area[68] (ratio of total area of organelle to that of nonorganelle per myofiber). The ratio of inner to outer mitochondrial membrane was calculated from the measurements of their perimeters. Thicknesses of Z lines were also measured, and a mean width of a minimum of ten individual Z lines was calculated for each fiber.

3. Statistical Analysis

Data were expressed as group means plus or minus standard deviations of absolute values normalized to body weight. The significance of differences between groups was determined from group means of normalized absolute data for each parameter using one-factor analysis of variance, followed by Newman-Keuls or Dunnett's multiple-range testing.[45] Although there were only two groups of animals (those injected with nerve extract and those untreated), comparisons were made among four groups of muscles: (1) innervated (normal) untreated (NOR-UnRx); (2) denervated untreated (DN-UnRx); (3) innervated treated (NOR-Rx); and (4) denervated treated (DN-Rx).

4. Effects of Denervation

After 1 week of denervation, there were significant reductions of mean size, total area per myofiber, and the percent volume of mitochondria for both perinuclear mitochondria of type I fibers and intermyofibrillar mitochondria in all three types of fiber, with losses being greater in type I than IIA or IIB fibers (Table 5). Mitochondria did not appear abnormal in structure, nor was there a significant change of the ratio of inner to outer mitochondrial membrane. These findings are in agreement with those of other investigators.[69] Denervated fibers also exhibited reductions in areas of sarcoplasmic reticulum and transverse tubules that were greater in type II than I fibers (Table 5). These membranous systems are normally about twice as extensive in type II than I fibers.[12] Thicknesses of Z lines were reduced by denervation in type I more than IIB fibers, but were not significantly thinner in denervated than normal type IIA fibers (Table 5).

Table 5
ULTRASTRUCTURAL CHANGES IN FIBERS OF MOUSE SOLEI AND GASTROCNEMEI MUSCLES DEHYDRATED FOR 7 DAYS

	Type of fiber[a]		
Cell element	**I[b]**	**IIA[b]**	**IIB[c]**
Perinuclear mitochondria			
Mean size (μm²)	61.8 ± 6.8		
Total area/fiber (μm²)	46.8 ± 4.9		
% Volume[d]	46.8 ± 4.9		
Ratio inner/outer mitochondrial membrane	71.0 ± 14.2		
Intermyofibrillar mitochondria			
Mean size (μm²)	39.6 ± 7.3	44.9 ± 4.0	49.4 ± 1.9
Total area/fiber (μm²)	59.8 ± 5.4	49.4 ± 3.4	63.6 ± 3.3
% Volume[d]	59.8 ± 5.4	49.4 ± 3.4	63.6 ± 3.3
Ratio inner/outer mitochondrial membrane	71.0 ± 14.2	84.4 ± 5.6	75.6 ± 3.8
Sarcoplasmic reticulum			
Total area/fiber (μm²)	77.7 ± 7.0	40.8 ± 8.2	46.8 ± 7.4
% Volume[d]	77.8 ± 7.1	40.8 ± 8.2	46.8 ± 7.4
Transverse tubules			
Total area/fiber (μm²)	74.3 ± 5.1	53.1 ± 6.4	69.0 ± 5.1
% Volume[d]	74.5 ± 5.2	53.1 ± 6.3	69.0 ± 5.2
Z line thickness (nm)	60.8 ± 3.7	93.3 ± 4.0	71.5 ± 6.2

[a] Each value represents the mean plus or minus the standard deviation (n = 6) of percentage of control value (percentage of control = denervated/normal × 100; denervated and normal muscles from same animals).
[b] From soleus muscle.
[c] From gastrocnemius muscle (superficial portion of medial head).

5. *Trophic Effects of Neurogenic Proteins*

Although we had demonstrated trophic effects by neurogenic substances on CSA of denervated muscle fibers several years ago, an important finding from our recent work was that administration of nerve extract also ameliorated ultrastructural changes that occurred in denervated fibers.

The bulk of a muscle fiber is made up of myofilaments, but some of the CSA is also represented by nuclei (which are located peripherally) and sarcoplasmic elements such as the membrane systems (sarcoplasmic reticulum and t tubules) and discrete organelles (predominantly mitochondria). Results from the present study indicate that the decreased loss of CSA of denervated fibers treated with nerve extract is due to decreased losses of both myofilaments and other sarcoplasmic elements.

While the nerve extract had been found to prevent a large portion of the postdenervation losses that occurred at the light microscopic level, its ameliorative effects were even greater when assessed at the ultrastructural level. Administration of neurogenic substances prevented from 80 to 98% of the postdenervation losses for all of the parameters studied (Table 6 and Figure 5).

When the four treatments of muscles (innervated untreated, NOR-UnRx; denervated untreated, DN-UnRx; innervated treated, NOR-Rx; and denervated treated, DN-Rx) were compared by one-factor analysis of variance and Newman-Keuls multiple-range testing, four types of results were noted. These results are discussed below and are summarized in Table 6. Although absolute data are presented, statistical analyses were carried out on data normalized to body weight.

Table 6
EFFECT OF NERVE EXTRACT ON ULTRASTRUCTURE OF DENERVATED SOLEUS AND GASTROCNEMIUS MUSCLES OF C57BL/6J MICE

Cell element	Muscle	Treatment groups[a] Group 1 (DN-UnRx)	Group 2 (DN-Rx)	Group 3 (NOR-Rx)	Group 4 (NOR-UnRx)	% Atrophy[b] prevented	Significance among groups[c] ($\alpha = 0.05$)
Perinuclear Mitochondria							
Type I							
Mean size	SOL	0.71 ± 0.14	1.00 ± 0.15	1.12 ± 0.06	1.17 ± 0.02	63.1	1 2 3 4
% Volume[d]	SOL	10.50 ± 1.17	21.87 ± 1.33	22.71 ± 1.48	22.50 ± 1.70	94.8	1 2 3 4
Ratio inner/outer mitochondrial membranes	SOL	2.30 ± 0.26	3.10 ± 0.40	3.35 ± 0.43	3.30 ± 0.40	—	1 2 3 4
Intermyofibrillar Mitochondria							
Mean size (μm^2)							
I[e]	SOL	0.43 ± 0.08	0.85 ± 0.09	0.90 ± 0.09	0.97 ± 0.10	77.8	1 2 3 4
IIA[e]	SOL	0.36 ± 0.10	0.56 ± 0.16	0.69 ± 0.18	0.73 ± 0.12	54.1	1 2 3 4
IIB[e]	G	0.34 ± 0.17	0.50 ± 0.14	0.61 ± 0.29	0.71 ± 0.05	70.3	1 2 3 4
Ratio inner/outer mitochondrial membranes							
I[e]	SOL	1.98 ± 0.32	2.28 ± 0.60	2.35 ± 0.28	2.48 ± 0.19	—	1 2 3 4
IIA[e]	SOL	1.90 ± 0.10	2.02 ± 0.01	2.10 ± 0.13	2.22 ± 0.14	—	1 2 3 4
IIB[e]	G	1.63 ± 0.07	1.87 ± 0.10	1.96 ± 0.17	2.10 ± 0.15	—	1 2 3 4
SR Total area (μm^2)							
I[e]	SOL	0.83 ± 0.03	1.01 ± 0.06	1.04 ± 0.05	1.07 ± 0.10	96.0	1 2 3 4
IIA[e]	SOL	0.94 ± 0.15	1.55 ± 0.25	2.00 ± 0.18	2.28 ± 0.24	98.2	1 2 3 4
IIB[e]	G	1.14 ± 0.22	2.22 ± 0.28	2.23 ± 0.10	2.45 ± 0.12	94.3	1 2 3 4

Table 6 (continued)
EFFECT OF NERVE EXTRACT ON ULTRASTRUCTURE OF DENERVATED SOLEUS AND GASTROCNEMIUS MUSCLES OF C57BL/6J MICE

Cell element	Muscle	Treatment groups[a] Group 1 (DN-UnRx)	Group 2 (DN-Rx)	Group 3 (NOR-Rx)	Group 4 (NOR-UnRx)	% Atrophy[b] prevented	Significance among groups[c] ($\alpha = 0.05$)
T Tubule Total Area (μg^2)							
I[e]	SOL	0.38 ± 0.04	0.51 ± 0.02	0.52 ± 0.03	0.52 ± 0.02	86.0	1 2 3 4
IIA[e]	SOL	0.46 ± 0.06	0.69 ± 0.07	0.72 ± 0.12	0.83 ± 0.06	97.4	1 2 3 4
IIB[e]	G	0.49 ± 0.08	0.71 ± 0.18	0.75 ± 0.21	0.88 ± 0.16	97.7	1 2 3 4
Z Line Thickness (nm)							
I[e]	SOL	93.5 ± 5.7	121.0 ± 7.5	123.0 ± 10.0	122.0 ± 3.5	84.4	1 2 3 4
IIA[e]	SOL	74.0 ± 7.0	78.0 ± 8.6	80.5 ± 6.2	81.7 ± 9.6	—	1 2 3 4
IIB[e]	G	46.3 ± 3.6	60.5 ± 9.0	61.6 ± 7.1	65.5 ± 6.0	96.2	1 2 3 4

[a] Each animal was subjected to unilateral hindlimb denervation. Each muscle was normally innervated (NOR) or denervated for 7 days (DN) and was obtained from a mouse which had received no other treatment (UnRx; n = 6) or was injected daily with rat sciatic nerve extract (Rx; n = 4).

[b] $$\% \text{ Atrophy prevented} = 100 - \left(\frac{\text{NOR-UnRx} - \text{DN-Rx}}{\text{NOR-UnRx} - \text{DN-UnRx}}\right) \times 100.$$

[c] Significance refers to comparison of the four groups of muscle treatments by one-factor analysis of variance followed by Newman-Keuls multiple-range testing. Sets of similarities are underlined together ($p > 0.05$). Values not connected by a common line were found to be significantly different ($p < 0.05$). Absolute data normalized by body weight were used for statistical comparisons.

[d] % Volume = cell element as a percentage of cell volume.

[e] Types of fibers as classified by activity of ATPase.[41]

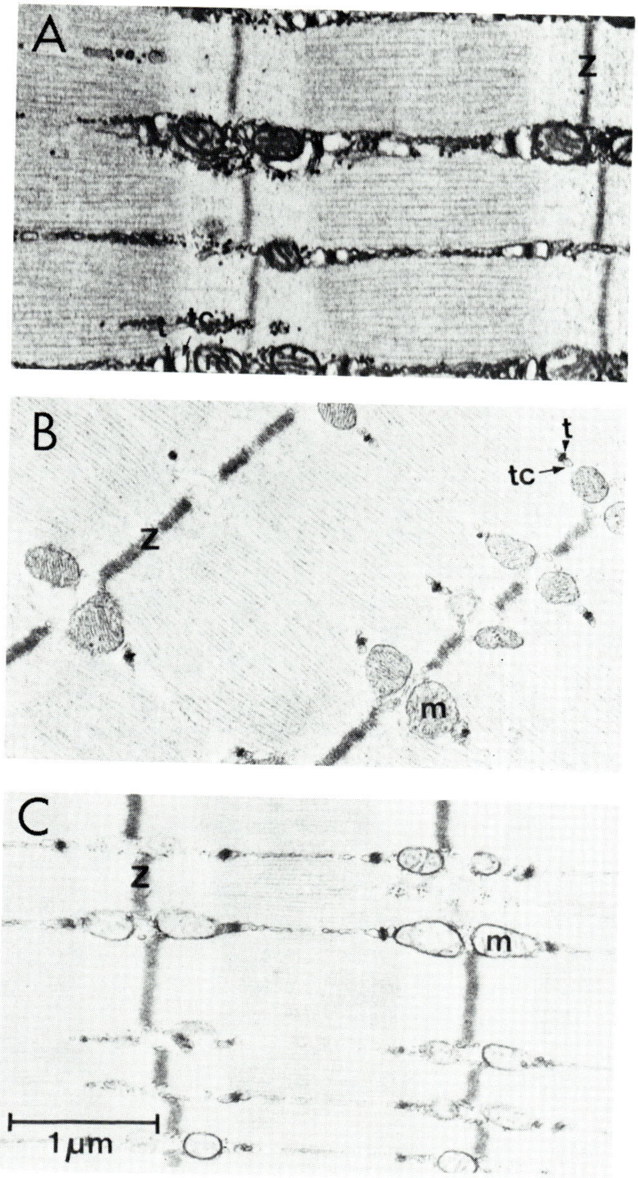

FIGURE 5. Electron micrographs showing the effect of nerve extract on ultrastructural changes in type I fibers of SOL denervated for 7 days. (A) Denervated SOL from untreated animal (DN-UnRx); (B) denervated SOL from mouse injected daily with rat nerve extract (250 mg protein per kilogram of body weight per day intraperitoneally; DN-Rx); (C) normal SOL from untreated animal (NOR-UnRx). The nerve extract prevented from 80 to 95% of the losses of total areas per myofiber occupied by sarcoplasmic reticulum and t tubules; mean size, total area, and percentage of volume of mitochondria; and width of Z lines. (m, mitochondria; sr, sarcoplasmic reticulum; t, t tubule; tc, terminal cisternae of sr; Z, Z line.)

1. There were no significant differences among NOR-UnRx, NOR-Rx, and DN-Rx groups, while DN-UnRx was significantly smaller than the other three groups. This indicated that the nerve extract had no effect on innervated muscle and almost completely ameliorated the changes in denervated muscle. This was true for mean size, total area occupied by mitochondria, and volume of mitochondria as a percentage of total cell volume for perinuclear and intermyofibrillar mitochondria of all three types of fiber, and of the total area of myofiber occupied by sarcoplasmic reticulum and t tubules in type I fibers.
2. There was no significant difference between NOR-Rx and DN-Rx while NOR-UnRx was significantly greater and DN-UnRx significantly smaller than either treated group. This indicated that the extract partially ameliorated postdenervation losses and either had a deleterious effect on innervated muscles or diminished the hypertrophy that normally occurs in muscles of the contralateral unoperated side after unilateral denervation. This situation was found for total area occupied by sarcoplasmic reticulum in IIB fibers and total area occupied by t tubules in IIA and IIB fibers.
3. All groups were significantly different from each other, with DN-UnRx less than DN-Rx, which was less than NOR-Rx, which was less than NOR-UnRx. This indicated that nerve extract prevented atrophy in DN-Rx group but may have had a deleterious effect on the normal innervated group (NOR-Rx). This was true for total area occupied by sarcoplasmic reticulum in IIA fibers.
4. There were no significant differences among any of the groups. This indicated that the nerve extract had no effect on denervated or normal muscles. This was true for the ratio of the inner to outer mitochondrial membranes and for thickness of Z lines of type IIA fibers.

The exterotrophic effects of neurogenic substances on ultrastructural changes in denervated muscle fibers were of different magnitude for the different types of fibers. For example, the nerve extract had a greater ameliorative effect on the membrane systems of types I and IIB than IIA fibers, whereas the loss of total area of intermyofibrillar mitochondria per cross section of muscle fiber was reduced more in type II than I fibers. The study presented above is the first to demonstrate trophic effects by neurogenic substances on ultrastructural alterations in denervated skeletal muscle fibers.

III. BIOCHEMICAL CHANGES IN DENERVATED MUSCLE

The metabolic changes produced in muscle fibers following denervation largely involve altered concentrations or activities of various enzymes. Many enzymes are decreased in activity following denervation due to the reduced metabolic demands; however, some are increased in activity.[11]

A. Acetylcholinesterase

Acetylcholinesterase (AChE) is an enzyme found in skeletal muscle which is responsible for inactivating ACh after it has been released by the motor nerve and bound to the postsynaptic AChR to initiate the excitation-contractile coupling response. It exists in several different molecular forms, which are found free in the sarcoplasm or are bound to the sarcolemma. The distribution of the different forms of AChE may be selective; for example, 16S AChE is found to be highly concentrated at the postsynaptic region of the motor endplate (see Chapter 4).

In rat EDL muscle denervated for 7 days the total AChE activity, as determined quantitatively,[70] was reduced to approximately 40% of that found in normal contralateral control muscles.[1] This postdenervation loss was not ameliorated by daily injections of rat sciatic nerve extract of the dosage found optimal to reduce atrophy of denervated type IIB fibers.[1]

The absence of an exterotrophic effect on the activity of AChE agrees with previously described results which suggest that the activity of this enzyme is at least partly regulated by its substrate, ACh (see Chapter 3). When ACh or its analogue, methacholine, was added to denervated muscle in vitro, the AChE activity decreased less than in control cultures. This effect was blocked by puromycin, suggesting a mechanism of enzyme induction.[71] When muscles were poisoned with botulinum toxin, which inhibits the release of ACh from presynaptic terminals, the activity of AChE decreased to values similar to those found after denervation.[72] Ranish and Dettbarn[73] found the decrease in total AChE activity after denervation to be unrelated to length of nerve stump. However, Davey and Younkin[21] found that nerve stump length affected activity of AChE at the endplate, but not elsewhere in the muscle. Other authors reported little or no effect on AChE activity after tenotomy[46] or immobilization of joints;[74] however, these are poor models of disuse as both allow ACh to reach the muscle. The addition of extracts of peripheral nerve to embryonic mammalian muscle tissue[30,40] enhanced AChE activity. Perhaps the extracts of peripheral nerves can only exert an effect on AChE in immature muscles or on the endplate specific form of the enzyme (see Chapter 11).

B. Parvalbumin

Parvalbumin is a low-molecular-weight (12,000) soluble protein present in high concentrations in the sarcoplasm of fast-twitch muscles of lower vertebrates.[75] Each molecule of parvalbumin binds two calcium ions with an affinity constant similar to that of troponin. It is hypothesized that parvalbumin plays the role of a soluble relaxing factor and thus controls the rates of contraction and relaxation by removing calcium from the myofibrils (where calcium interacts with troponin to initiate cross-bridge formation and contraction) and transferring it to the sarcoplasmic reticulum (where it is sequestered until it is released by membrane depolarization to initiate the contractile response again).[75-78] Evidence for this will be discussed subsequently with the effects of nerve extract on physiological changes in denervated muscle (Section IV.B).

The right hindlimbs of male C57BL/6J mice were denervated at 2 weeks of postnatal age and denervated again 2 weeks later, to prevent reinnervation. During a 4-week period of denervation, animals were either untreated (UnRx, control group) or received daily injections of neurogenic proteins derived from sheep sciatic nerves (Rx, treated group). The nerve extract was injected intraperitoneally with 500 mg protein per kilogram of body weight per day. It had previously been shown that systemic administration of neurogenic substances to adult mice ameliorated atrophy of hindlimb muscles denervated for 7 days.[7]

The mice were killed at 6 weeks of age by cervical dislocation, and fast-twitch muscles from both denervated (DN) and contralateral normally innervated (NOR) hindlimbs were immediately removed. The muscles examined were the TA, EDL, and G. The muscles were homogenized and prepared for isoelectric focusing (IEF) as described by Jasch and Moase.[79] Proteins separated by IEF were quantitated by densitometric scanning of the stained bands on the gels, and the relative quantity of parvalbumin was expressed as a ratio of total myosin light chain 2-fast (LC2f + LC2f-p).[80] Denervation for 4 weeks produced a 40% decrease in the quantity of parvalbumin in the fast-twitch muscles relative to normal contralateral controls ($p < 0.001$).

Systemic treatment of mice with nerve extract did not affect the levels of parvalbumin in normally innervated muscles (NOR-Rx), but it completely prevented the postdenervation losses of parvalbumin in fast-twitch muscles of the denervated hindlimbs (DN-Rx). There was no significant difference between the amounts of parvalbumin in NOR-UnRx and DN-Rx muscles (Figure 6).[8]

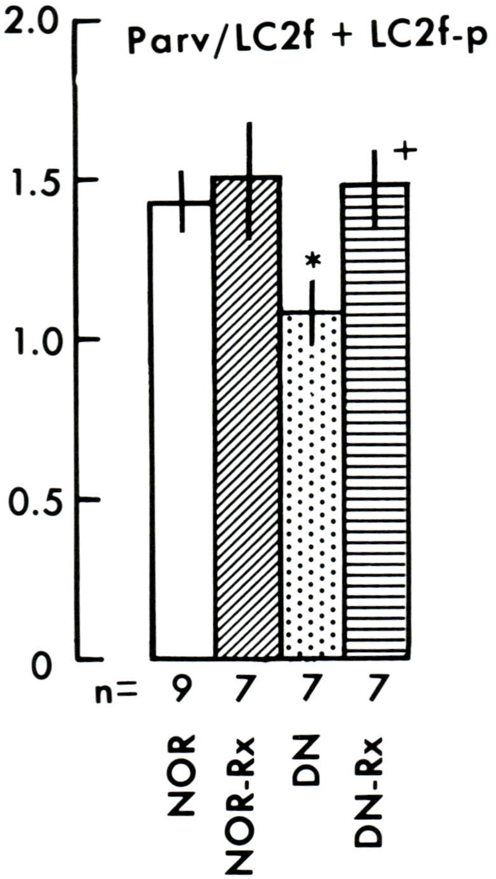

FIGURE 6. Relative amounts of parvalbumin in muscles of C57BL/6J mice. Animals received unilateral denervation of hindlimbs for 4 weeks and were otherwise untreated, or were injected with sheep nerve extract (500 mg protein per kilogram of body weight per day intraperitoneally; Rx), following which normally innervated (NOR) and denervated (DN) fast-twitch muscles were removed. Proteins were separated by isoelectric focusing and quantified by densitometric scanning of stained gel. Denervation produced significant decrease in amount of parvalbumin ($p < 0.05$; asterisk). Nerve extract had no effect on the amount of parvalbumin in NOR-Rx muscles, but completely ameliorated the loss in DN-Rx muscles. There was not a significant difference between NOR and DN-Rx (cross). (LC2f, myosin light chain 2-fast; p, phosphorylated.)

C. Other Soluble Proteins

Trophic effects by neurogenic substances have been demonstrated on postdenervation losses of total protein,[1-3,7] structural proteins,[10] and certain soluble proteins or enzymes[8,80] of skeletal muscle. It is conceivable that in vivo administration of nerve extract could also ameliorate the postdenervation changes of many other soluble proteins (e.g., enzymes).

Right EDL rat muscles were denervated, and after 7 days soluble proteins were extracted from denervated and contralateral control muscles and fractionated into 32 bands by electrophoresis on polyacrylamide gels.[81,82] The distribution of protein among the bands was calculated for each muscle by densitometric scanning of stained gels. In denervated muscles the proportion of the total protein was increased in 11 bands and decreased in 6, relative to control muscles. Daily injection of denervated muscles with nerve extracts prevented the

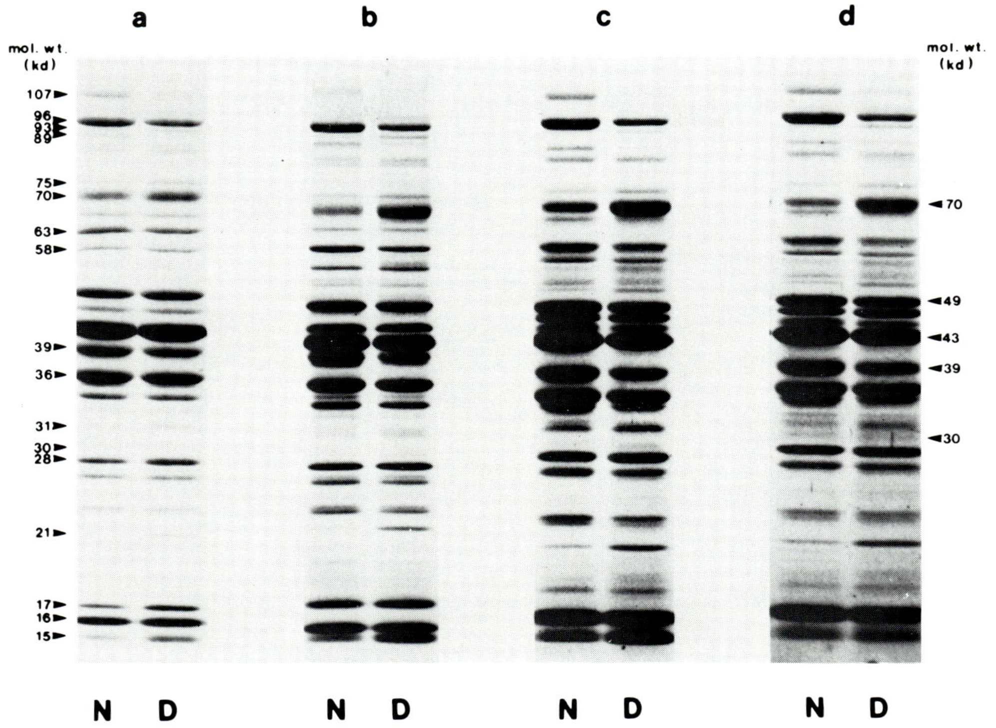

FIGURE 7. Electrophoresis of soluble proteins from normal and denervated EDL muscles run on 6 to 16% gradient polyacrylamide gels and stained with Coomassie brilliant blue R-250. Each lane contains proteins from one muscle. (N, normal muscles; D, contralateral muscles denervated for 1 week; a, untreated; b, denervated muscle injected with saline; c, denervated muscle injected with inactivated extract of sheep sciatic nerve; d, denervated muscle injected with active extract of sheep sciatic nerves.) Arrows on the left indicate bands affected by denervation. Arrows on the right show bands affected by treatment with extract.

increased level of one cytosolic protein and exaggerated the change in two others. Two bands, not affected by denervation, were decreased by extract (Figure 7).[6]

The effect of the extract on individual protein bands suggests that its action is on specific metabolic processes within the sarcoplasm rather than a general net increase in protein synthesis. As discussed above, it has previously been shown that the extract acts specifically on parvalbumin by completely preventing the loss of this protein in denervated mouse muscle.[8]

The identity of the 30,000-dalton protein which is restored to normal proportion by the extract has not been determined. It may be noted that troponin T, part of the troponin complex which interacts with parvalbumin, also has a molecular weight of 30,000.[83,84] Troponin forms part of the myofibril but also exists as a cytosolic protein,[84] and it may be speculated that it is under a similar exterotrophic influence to that of parvalbumin.

The anomalous effect of the extract in further increasing the denervation-caused changes in proportion of the 39,000- and 70,000-dalton bands was unexpected. The changes in these two bands and in the two bands that were not affected by denervation, mol wt 43,000 and 49,000, may represent an intermediate step in the response of the sarcoplasm to the extract (see Chapter 11). The balance between synthesis and breakdown of the soluble components may not be stable after 1 week of treatment. It may also be that these changes were not a direct response to the trophic agents. They may be secondary to the other possible responses of the muscles to the extract, such as increased synthesis or decreased degradation of myofibrillar proteins.

Table 7
EXPERIMENTAL PLAN FOR STUDY OF EFFECT OF NERVE EXTRACT ON ACETYLCHOLINE RECEPTORS IN NORMAL, DISUSED, AND DENERVATED RAT EDL MUSCLES

	Duration of treatment (days)			
Treatment of right EDL[a]	**1**	**3**	**5**	**7**
Denervated				
No treatment	6	6	6	6
Nerve extract[b] (2 mg/mℓ)[c]	6 (2×)	6 (1×)	6 (1×)	6 (1×)
Nerve extract[b] (4 mg/mℓ)[c]		6 (2×)		
Saline[b]	6 (2×)	6 (1×)		6 (1×)
Immobilized	6	6	6	6
Innervated (normal)				
Nerve extract[b] (2 mg/mℓ)[c]	6 (2×)	6 (1×)		
Nerve extract[b] (4 mg/mℓ)[c]	6 (2×)	6 (2×)		
Saline[b]	6 (2×)	6 (2×)		

[a] In each rat the left EDL received no treatment and served as a normal control.
[b] Nerve extract and saline (0.9% NaCl) were injected intramuscularly into the right EDL once (1 ×) or twice (2 ×) daily.
[c] Indicates the concentration of protein in the nerve extract.

Treatment with extract did not suppress the changes in the levels of the other 14 bands which were affected by denervation, indicating that these changes may be more the result of disuse than of loss of neurogenic trophic substances. This would be similar to the decrease in CSA of the type IIA fibers in denervated EDL muscles of rats which was attributed predominantly to disuse.[2] It is also possible that the changes in distribution of the soluble proteins resulted from the loss of a neurogenic substance which is required in a greater amount than that in the extract, or which may either not be extractable from the nerves under the conditions used or may have been inactivated during the extractions.

IV. PHYSIOLOGICAL CHANGES IN DENERVATED MUSCLE

A. Acetylcholine Receptors

Denervation causes a marked increase in sensitivity of the muscle to ACh that is due to a 50- to 100-fold increase in density of AChR at sites other than the neuromuscular junctions.[13,14] That change is the results of increased synthesis of new receptor molecules.[85-87] A study was carried out to determine whether or not, in addition to reducing atrophy, the nerve extract could suppress the increase in number of AChR in denervated muscles.[5]

1. Animals

The right EDL muscles of adult male Wistar rats received one of several treatments as outlined in Table 7, while the left EDL muscles served as normal controls. The procedures of denervation, immobilization, and/or intramuscular injection of nerve extract were carried out as described previously.[1,2] After the assigned period of treatment (Table 7), the animals were killed by an overdose of ether vapor, both EDL muscles were removed and used immediately to assay the number of AChR.

2. Acetylcholine Receptor Assay

The assay was based on the binding of α-bungarotoxin (α-BTX) which binds specifically and irreversibly to nicotinic AChR.[88-90] Labeling of α-BTX with a radioisotope diminishes the potency of the toxin by about 40%, but does not alter the specific binding to the receptor. The amount of toxin bound is directly proportional to the number of receptors.[91] Binding of α-BTX to AChR in vitro is affected by concentration of the toxin, the period of incubation, and the temperature.[92] Those factors were therefore kept constant.

The muscles were incubated 4 hr at 24°C in Krebs solution containing [^{125}]-α-BTX (1.0 μCi/mℓ; approximately 0.1 μg/mℓ). This concentration of toxin is similar to that found by other investigators[93] to be sufficient for saturation. The Krebs solution[5] was aerated with 5% CO_2 and 95% O_2. The left and right muscles from each animal were contained in the same perforated stainless steel capsule to ensure identical treatment. After incubation, the muscles were washed to remove all the unbound radioactivity.

The quantity of [^{125}I]-α-BTX bound to each muscle was estimated by gamma emission counting with a Micromedic Systems 4/200 Automatic Gamma Counter® with a counting efficiency of 80%. Data were expressed as disintegrations per minute per muscle and as disintegrations per minute per milligram of muscle tissue. The values calculated from the treated muscles were all expressed as percentages of those from their contralateral control muscles. Individual values for the number of disintegrations were subjected to logarithmic transformation $X' = \log (X - 1)$ to validate the use of parametric statistical methods with heteroscedastic data in which the standard deviations varied in proportion to the means.[45]

3. *Effect of Immobilization and Denervation*

The number of AChR in the denervated EDL remained normal for 1 day and then increased steadily. This observation agrees with those of other investigators.[20,90-94] Immobilization of the same muscle produced a smaller and transient increase in the number of receptors (Figure 8). Such a change was probably responsible for the transient increase in sensitivity to ACh found by Fischbach and Robbins[95] in the immobilized rat SOL muscle.

Our findings with immobilized and denervated muscles indicate that the method used for the assay of AChR was capable of detecting and quantitating both small and large changes resulting from experimental manipulations.

4. *Effect of Neurogenic Substances on Normal Muscles*

A small increase in AChR was found after injection of either nerve extract or saline into otherwise normal muscles. The magnitude of the effect was even smaller, however, than the increase that followed immobilization. Substances injected into rat EDL muscles pass within the endomysium among the muscle fibers, which are forced apart. This may injure some terminal branches of the motor axons. The paralyzed muscle fibers would then be expected to produce increased numbers of AChR. The effect of injections per se does not, of course, affect deliberately denervated muscle fibers, which are already inactive and devoid of exterotrophic influences.

5. *Effect of Neurogenic Substances on Denervated Muscles*

Available evidence supports the hypothesis that the increased number of AChR in denervated muscles results from both disuse and loss of exterotrophic influences.[14,96,97] In other investigations we demonstrated that injection of nerve extract into denervated muscles prevented some 45% of the decrease in CSA of the type IIB fibers,[7,8] which make up about one half the bulk of the rat EDL muscle.[43] Therefore, it was reasonable to expect that the nerve extract might also have restrained the increase of AChR in denervated muscles. However, we were surprised to find that denervated muscles injected with nerve extract bound more α-BTX than those that were not injected (Figure 9). The effects of the extract on denervated muscles were not due to the injections per se.

These results are consonant with those of other investigators. Younkin et al.[55] found that

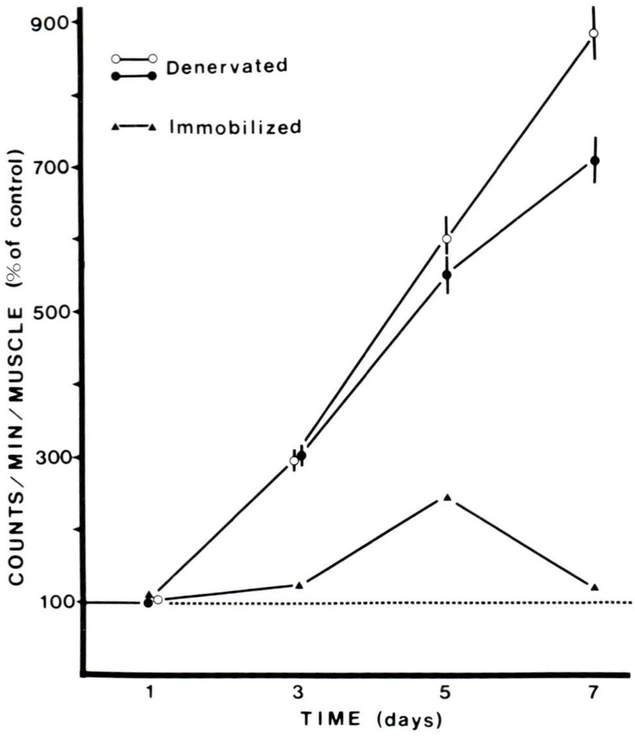

FIGURE 8. A diagrammatic representation of the amount of [^{125}I]-α-BTX bound by muscles denervated (circles) or immobilized (triangles) for 1, 3, 5, and 7 days. Each point represents the mean of percentage of control values (treated/normal × 100) from six rats. Solid circles represent ratios derived from absolute data expressed in counts per minute per muscle. Empty circles were derived from data expressed in counts per minute per milligram of muscle. Vertical lines represent the standard error. Where vertical lines are absent, the standard error was within the area covered by the point. The interrupted baseline indicates normal values.

adult rat EDL muscles denervated for 2 to 3 days and then maintained in organ culture for 3 days exhibited greater sensitivity to iontophoretically applied ACh in both junctional and extrajunctional regions, when nerve extract enriched with substances carried by axoplasmic transport (i.e., from the proximal stump of nerve ligated 4 days earlier) was added to the culture medium, than when they received extract of normal nerves (see Chapter 4). They suggested that axonally derived substances administered in vitro would promote the deployment of AChR over the whole surface of the muscle fiber. Vyskocil et al.[98] found that a free autograft sciatic nerve implanted on the surface of the denervated EDL muscle of the rat caused increased sensitivity to ACh if the graft came from an intact nerve, but not if it was from the distal portion of a nerve transected 7 days previously. This observation and our findings lend support to the suggestion made by Younkin et al.[55]

6. *Exterotrophic Regulation of Acetylcholine Receptors*

There is disagreement as to whether or not neurogenic substances normally have any effect on extrajunctional AChR.[23,96-98] In our investigation, the failure of nerve extract to suppress the increased binding of α-BTX by denervated muscles supports the contention that the additional receptors in denervated muscles, which are predominantly extrajunctional, do not arise because of the loss of exterotrophic influences. It is possible, however, that a different neurogenic substance, not present in our extract, functions together with activity

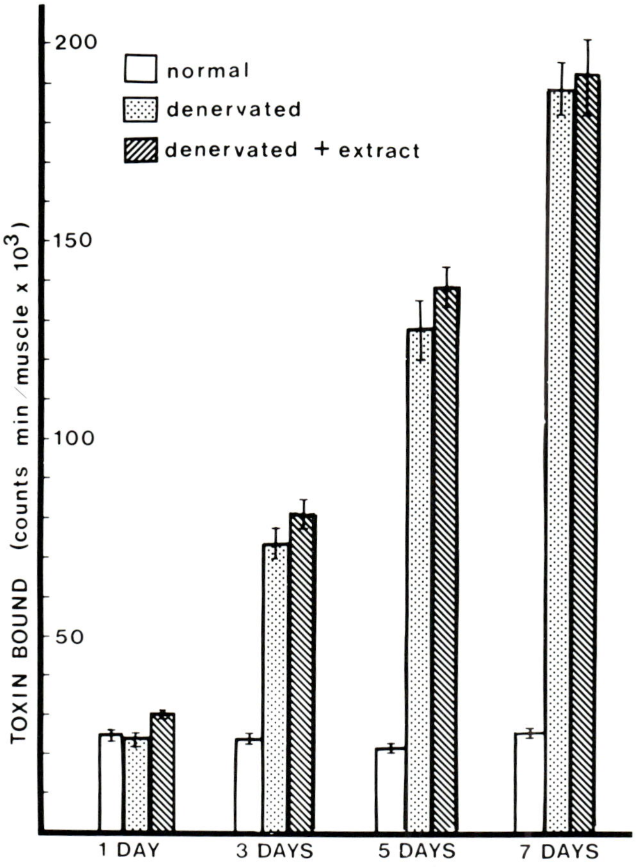

FIGURE 9. A diagrammatic representation of the amount of [^{125}I]-α-BTX bound by muscles denervated (stippled bars) or denervated and injected with nerve extract (2 mg protein per milliliter; striped bars) for 1, 3, 5, or 7 days. The empty bars represent normal muscles. The height of the bar indicates the mean disintegrations per minute (DPM) per muscle; the vertical lines represent the standard error. The amount of toxin bound was significantly greater in denervated muscles injected with nerve extract than in those denervated but not injected for the 1-day group, but not for the 3-, 5-, or 7-day groups.

to prevent the synthesis or accumulation of extrajunctional receptors in innervated muscle cells. Such an agent might be metabolically unstable or insoluble in saline and thus unable to exert a trophic effect under the conditions of the study. However, substances extracted from rat sciatic nerves are able to increase the number of AChR in the denervated EDL muscle. This apparent paradoxical response may result from an increased rate of synthesis or a prolonged life span of the new receptors. It is also possible for the trophic maintenance of a high density of AChR at the postsynaptic region of the sarcolemma. Although it contains a substance that can ameliorate the atrophic changes of denervated muscle fiber, the nerve extract does not appear to inhibit the development of extrajunctional sensitivity to ACh. Thus, different mechanisms may operate to maintain the morphological integrity of innervated muscle fibers and to restrict AChR to the region of the motor endplate.

B. Isometric Contractile Properties

From the studies discussed above, it is apparent that neurogenic substances exert exterotrophic influences on morphological and biochemical properties of mature skeletal muscle.

Administration of nerve extract intramuscularly (in rats) or intraperitoneally (in mice) has been found to ameliorate several postdenervation changes in hindlimb muscles. These include atrophic changes such as losses of weight, total protein, and CSA of fibers and associated ultrastructural alterations; and metabolic changes such as decreased levels of cytosolic parvalbumin. Of prime importance was to determine whether these ameliorative effects were of functional significance. That is to say, did they alter any physiological parameters? As discussed above, in vivo administration of nerve extract to denervated rat EDL muscle did not diminish the number of AChR and thus did not reduce the supersensitivity to ACh.[5] In the study to be discussed below, exterotrophic influences of neurogenic substances were assessed on isometric contractile properties of denervated fast-twitch muscles.

1. Animals

The right hindlimbs of male C57BL/6J mice were denervated at 2 weeks and denervated again at 4 weeks of age. During a 4-week period of denervation the mice either were untreated (UnRx) or received daily intraperitoneal injections of sheep nerve extract (500 mg protein per kilogram of body weight per day; Rx). EDL muscles from both denervated (DN) and contralateral normally innervated (NOR) hindlimbs were examined for isometric contractile properties. This is the same experimental protocol used to study the effects of nerve extract on levels of parvalbumin in denervated EDL and other fast-twitch muscles of mice (see Section III).

2. Assessment of Contractile Properties

After the 4-week period of treatment, denervated and normal muscles were carefully removed from chloral hydrate anesthetized mice and assessed for isometric twitch and tetanus tensions (Pt and Po), time to peak twitch tension (TTP), half-relaxation time (1/2 RT), and posttetanic twitch potentiation (PTP), as described by Webster and Bressler.[15] In brief, the EDL muscle was secured between the two ends of an electromagnetic puller which was part of a closed loop length servosystem, allowing examination of isometric contractile properties. The muscle was maintained in oxygenated Krebs solution at 37°C, and prior to data collection its length was set to obtain maximum isometric twitch tension. Muscles were stimulated directly with supramaximal square pulses of 10 to 16 V and of 1-msec duration. For isometric tetanus, the voltage used was the same as that for single twitches, the fusion frequency was 120 to 160 shocks per second, and the duration was 220 msec. A stimulus pattern of three twitches followed by a tetanus with a contraction once every 90 sec was used throughout the experiment to avoid fatigue. For each muscle a minimum of 10 twitches and 4 tetani were recorded. To assess posttetanic potentiation, a single twitch was recorded 1 msec after the completion of a 1 sec tetanus and compared to previous twitches.

3. Effect of Denervation

The changes in isometric contractile properties of denervated EDL muscles in mice not treated with nerve extract (UnRx) were similar to those described by other investigators in mice, rats, and cats.[15,99-102] The most noticeable changes were decreased Po and prolongation of the contraction time (evident as increased TTP and 1/2 RT). The twitch tension (Pt) per unit of muscle weight was increased, a response thought to be due to more complete activation of the contractile proteins in a twitch of a denervated muscle fiber.[100] There was also a loss of PTP (Table 8). The decreased Po was largely due to atrophy, since the tetanus tension per unit of muscle weight was not significantly different for denervated (DN-UnRx) and normally innervated (NOR-UnRx) muscles (Table 8).

4. Effect of Nerve Extract

Injections of nerve extract ameliorated some of the physiological changes in denervated

Table 8
EFFECT OF NERVE EXTRACT ON ISOMETRIC CONTRACTILE PROPERTIES OF DENERVATED EDL MUSCLES OF C57BL/6J MICE

Parameter[a]	Treatment groups[b] Group 1 (NOR-UnRx)	Group 2 (Dn-UnRx)	Group 3 (DN-Rx)	Significance[c]
n	7	6	5	
Pt (g)[d]	5.01 ± 0.41	6.04 ± 0.85	6.85 ± 0.29	$\underline{1\ 2}\ \underline{3}$
Pt/muscle wt[d]	0.66 ± 0.09	1.57 ± 0.27	1.65 ± 0.08	$\underline{1}\ \underline{2\ 3}$
Po (g)[d]	26.5 ± 2.5	8.5 ± 1.1	12.5 ± 0.7	$\underline{1}\ \underline{2}\ \underline{3}$
Po/muscle wt[d]	3.38 ± 0.46	2.67 ± 0.45	3.01 ± 0.12	$\underline{1\ 2}\ \underline{3}$
TTP (msec)[d]	8.8 ± 0.4	17.0 ± 1.2	12.6 ± 0.3	$\underline{1}\ \underline{2}\ \underline{3}$
1/2RT (msec)[d]	9.6 ± 0.6	24.2 ± 2.0	15.9 ± 0.9	$\underline{1}\ \underline{2}\ \underline{3}$
PTP[d]	1.22 ± 0.03	0.96 ± 0.02	1.00 ± 0.02	$\underline{1}\ \underline{2\ 3}$

[a] Parameters: twitch tension, Pt; tetanus tension, Po; time to peak, TTP; half-relaxation time, 1/2RT; posttetanic twitch potentiation, PTP.

[b] Each animal was subjected to unilateral hindlimb denervation. Each muscle was normally innervated (NOR) or denervated for 4 weeks (DN) and was obtained from a mouse which had received no other treatment (UnRx) or was injected daily with sheep sciatic nerve extract (Rx).

[c] Significance refers to comparison of the three groups of muscle treatments by one-factor analysis of variance followed by Newman-Keuls multiple-range testing. Sets of similarities are underlined together ($p > 0.05$). Values not underlined by a common line were found to be significantly different $p < 0.05$).

[d] Each value is the mean plus or minus the standard deviation of individual absolute values.

EDL muscles of mice: the decreased Po and the prolongation of the contraction time. There was no effect on Pt or PTP.

Denervated EDL muscles from mice treated with neurogenic substances exhibited 26% stronger tetanus tensions than those from untreated mice (Table 8 and Figure 10).[8] This was probably due to amelioration of denervation atrophy, which, although less than that found for EDL muscles denervated and treated for 1 week,[7] was nevertheless significant. EDL muscles denervated for 4 weeks exhibited 30% larger mean CSA of type II fibers in nerve extract treated versus untreated mice.

The most significant results were that administration of neurogenic substances prevented over 50% of the increases of TTP and 1/2 RT in denervated EDL muscles, compared to those from untreated mice (Table 8 and Figure 11). This is probably related to the complete prevention of the postdenervation loss of parvalbumin in fast-twitch muscles of mice treated with nerve extract (see Section III and Figure 6).

There is considerable evidence correlating concentration of cytosolic parvalbumin with speed of contraction in muscle fibers:

1. In muscles of homogenous fiber type in the same animal, those with fast-twitch fibers have much higher concentration of parvalbumin than those composed solely of slow-twitch fibers.
2. Muscles of small animals have faster relaxation times and higher concentrations of parvalbumin than those of larger animals.
3. Denervation of a fast-twitch muscle results in a marked prolongation of the contraction time and simultaneous decreased amount of parvalbumin, which returns to normal values after reinnervation.

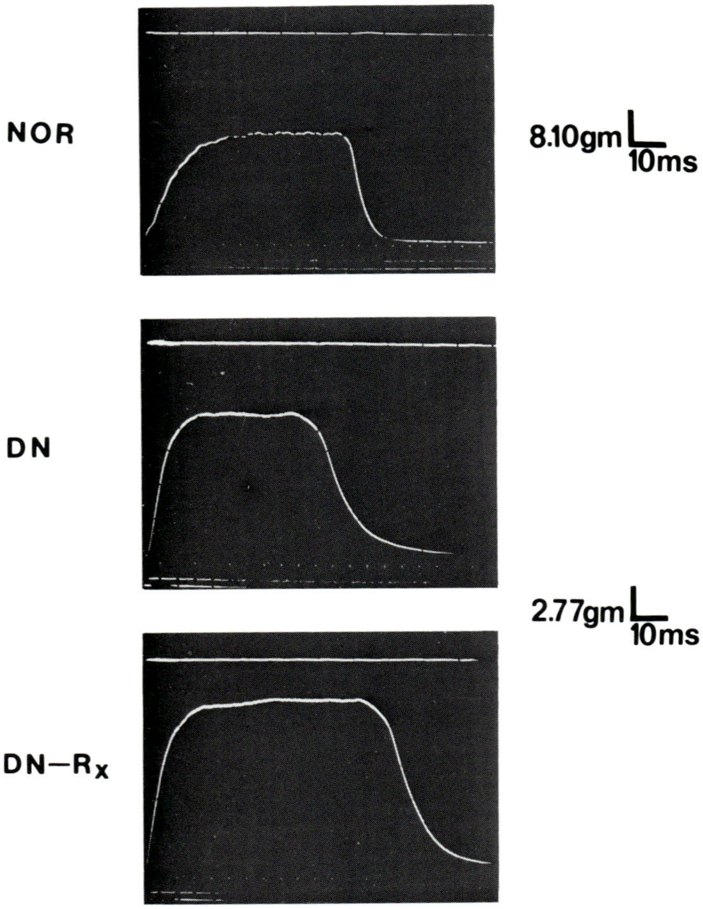

FIGURE 10. Original records of maximum isometric tetanus tension. Tension scales are different for NOR than for DN and DN-Rx records. Note the marked decrease in tetanus tension in mouse EDL denervated for 4 weeks (DN) relative to normal (NOR) and return of tetanus tension toward normal values in denervated muscles from mice injected daily with sheep nerve extract (DN-Rx).

4. Chronic indirect electrical stimulation of a fast-twitch muscle to induce a transformation from fast to slow contractile characteristics also induces a rapid decrease in concentration of parvalbumin.
5. Cross-reinnervation of a fast-twitch muscle (i.e., EDL) with a slow nerve (one that normally innervates a slow-twitch muscle, i.e., SOL) transforms the contractile characteristics of the muscle toward those of a slow-twitch muscle and also causes a marked decrease in the content of parvalbumin.[75-78]

As discussed earlier, it is believed that parvalbumin can act as a soluble relaxing factor which is able to remove calcium ions from the contracting myofibrils and transfer them to the sarcoplasmic reticulum, where they are sequestered in a relaxed muscle fiber (see Section III).

Failure of the nerve extract to restore PTP in denervated EDL muscles corresponds with the failure of the extract to affect the change in phosphorylation of myosin light chain 2-fast (LC2f). There is evidence that PTP is correlated with the proportion of LC2f in the phosphorylated form (LC2f-p).[107]

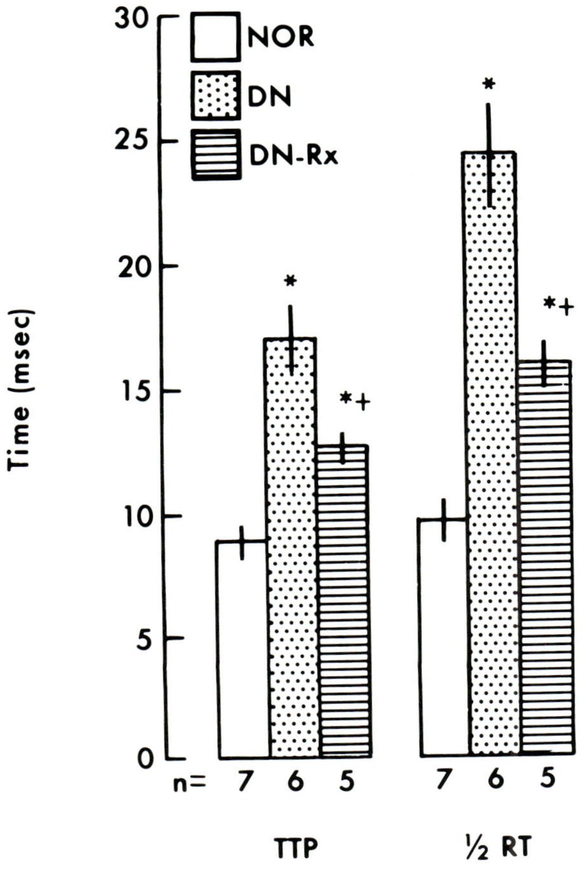

FIGURE 11. Changes in the time to peak (TTP) and half-relaxation time (1/2 RT) of normal mouse EDL (NOR) and mouse EDL denervated for 4 weeks (DN). Note the prolongation of both TTP and 1/2 RT with denervation and a significant reduction in both of these parameters in mice injected daily with nerve extract (DN-Rx).

Thus, the exterotrophic effects of neurogenic substances on denervated fast-twitch muscles in mice are greater on time- than strength-related changes in isometric contractile function.

V. PURIFICATION OF NEUROGENIC FACTOR

Administration of neurogenic substances has been found to exert trophic influences on several morphological, biochemical, and physiological alterations in denervated skeletal muscle. This trophic effect is dose dependent, but not species specific. We have partially purified the neurogenic substance which acts in vivo to ameliorate changes in denervated muscle. Trophic influences of extracts of peripheral nerve on embryonic muscle have also been demonstrated. Markelonis and Oh[31] found that an extract of chicken sciatic nerve promoted differentiation, maturation, and maintenance of aneural chicken muscle cells grown in tissue culture (see Chapter 6). Those investigators purified from chicken sciatic nerve an acidic glycoprotein, sciatin, that was responsible for the trophic effects.[32] It was later found that sciatin was the same as transferrin, an iron-binding glycoprotein present in serum.[33]

A. Neurogenic Substance Acting In Vivo is Not Transferrin

Prior to attempting purification of the active neurogenic trophic substance responsible for

Table 9
EFFECT OF TRANSFERRIN AND NERVE EXTRACT ON RATS' EDL MUSCLES DENERVATED FOR 7 DAYS

Treatment of right EDL[a]	CSA of IIB fibers[b]	Significance[c]
Controls		
No treatment	65.2 ± 1.8	
Sodium phosphate-saline	59.6 ± 2.3	
Ammonium acetate-saline	59.2 ± 2.1	
Transferrin (in sodium phosphate-saline)		
0.02 mg/mℓ[d]	57.6 ± 1.9	NS[e]
0.08 mg/mℓ[d]	59.3 ± 2.6	NS[e]
0.5 mg/mℓ[d]	65.2 ± 2.5	NS[e]
2.0 mg/mℓ[d]	57.6 ± 1.1	NS[e]
Nerve extract (2.5 mg/mℓ;[d] in ammonium acetate-saline)		
Crude	76.7 ± 4.8	$p < 0.025$[f]
Transferrin free	75.0 ± 2.8	$p < 0.05$[f]

[a] Substances were injected daily into denervated (right) EDL muscles for 7 days. Contralateral (left) EDL served as normal controls.
[b] Each value is the mean plus or minus the standard error of the mean of the percentage of control values obtained from the six animals in each group. Percentage of control = right EDL/left EDL × 100. CSA, cross-sectional area.
[c] Refers to comparison of values by one-factor analysis of variance and Newman-Keuls multiple-range test, based on group means of percentage of control values subjected to arcsine transformation.
[d] Concentration of protein in injected solution. Volumes injected into each animal were equal (0.1 mℓ/day).
[e] Refers to comparison of the CSA value from transferrin-injected muscles with the control values. NS, not significant ($p > 0.2$).
[f] Refers to comparison of the CSA value from nerve extract-injected muscles with any of the values from transferrin-injected or control muscles.

the ameliorating effects observed in our studies, we determined whether the trophic action of mammalian peripheral nerve extract on denervated muscles in vivo was due to the presence of serum transferrin in the extract.[4]

First, the effect of transferrin on atrophy of rat EDL muscles denervated for 7 days was assayed in the same manner as previous assays for dose of nerve extract or control treatments (see Sections IIA and IIB).[1-3] Muscles denervated for 7 days and injected daily with one of several doses of iron-conjugated rat transferrin exhibited a rate of atrophy equivalent to that of denervated muscles that either were not treated or were injected with saline (Table 9).[4]

Second, removal of transferrin from crude rat nerve extract by immunoaffinity chromatography did not diminish its ameliorative effects on denervated muscles (Table 9). The specificity and capacity of the antirat transferrin affinity chromatography resin were determined by passing through solutions of (1) transferrin purified from rat serum combined with horseradish peroxidase, (2) rat serum, and (3) rat nerve extract. The absence of transferrin from the nonabsorbed fraction elution directly through the column was confirmed by visualization of proteins in the eluate by polyacrylamide slab gel electrophoresis.[4,81]

Thus, the in vivo trophic action of neurogenic substances on denervated muscle was not due to the presence of serum transferrin in the extract, and purification of the trophic substance in sheep nerve extract was undertaken.

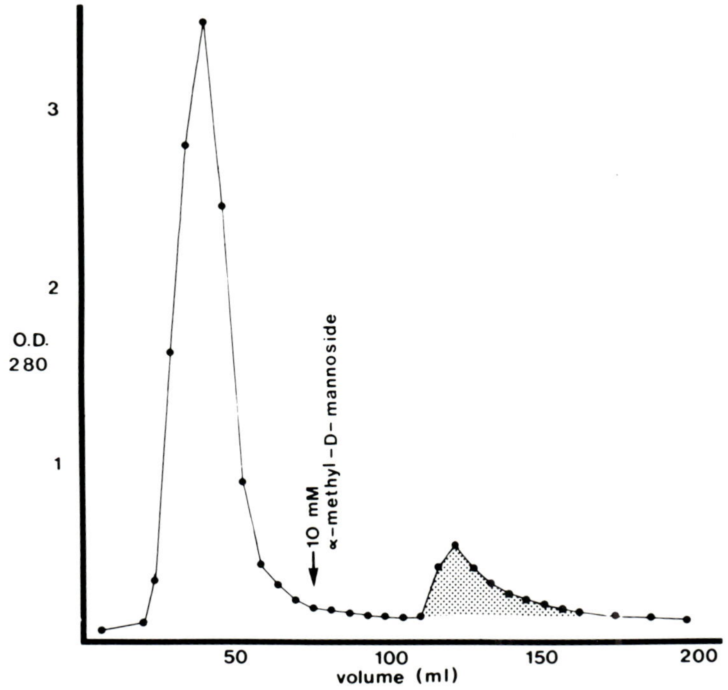

FIGURE 12. Affinity chromatography. Elution profile of sheep sciatic nerve protein from Con-A agarose. Soluble nerve protein was applied to a 1.6- × 30-cm column of Con-A agarose and eluted with Con-A buffer at a flow rate of 20 mℓ/hr. When the optical density (O.D.) 280 of the effluent returned to baseline, bound glycoproteins were desorbed with 10 m*M* α-methyl-D-mannoside in Con-A buffer. Assay of the two peaks for exterotrophic activity showed that only the second peak (stippled), which contained glycoproteins, contained trophic material.

B. Affinity Chromatography

The initial step of purification of the crude nerve extract was carried out with concanavalin A-affinity chromatography. Concanavalin-A is a lectin with specific binding affinity for carbohydrates containing α-D-glycopryranosyl or α-D-mannopyranosyl moieties,[103] which are commonly found in glycoproteins.

Concanavalin-A affinity chromatography was used to isolate the glycoproteins from other nerve-derived proteins. The bound glycoproteins were desorbed with α-methyl-D-mannoside because it has the highest binding affinity for concanavalin-A of all commonly available sugars. Using a 0 to 60 m*M* linear gradient of α-methyl-D-mannoside, it was shown that desorption started with less than 5 m*M* concentration of sugar; so in subsequent runs 10 m*M* was used. This was considered adequate, as no further protein was desorbed if 50 m*M* followed the 10 m*M* α-methyl-D-mannoside.

A typical elution profile of sciatic nerve protein from concanavalin-A Sepharose is shown in Figure 12. A large portion of the protein passed through the gel unhindered (peak 1). In 35 runs this void peak contained 54.6 ± 1.9% ($\bar{x}$ ± SE) of the total protein loaded on the column. The desorbed glycoproteins (peak 2) contained 7.6 ± 0.3% of the total protein. Approximately 25% of the protein was not recovered. This is consistent with the known behavior of concanavalin-A agarose, which is influenced by many variables.[104]

When the two peaks were assayed for trophic activity it was found that only the glycoprotein fraction (peak 2) ameliorated the atrophy of denervated muscle fibers. The mean percentage of control values for the CSA of type IIB fibers in denervated muscles injected with crude nerve extract, or peak 2, or peaks 1 and 2 recombined were not significantly

Table 10
SUMMARY OF PURIFICATION OF TROPHIC FACTOR FROM SHEEP NERVE EXTRACT

Chromatography procedure	Active fraction[a]	Dose (mg/mℓ)[b]	Purification[c]
None	Crude nerve extract	5.0	1×
Con A affinity	Glycoproteins (peak 2)	0.375	13×
G-100 gel filtration	Assay II, fraction b	0.079	66×
DEAE-ion exchange	Peak 6	0.031	250×

[a] Fractions were assayed for trophic activity on denervated EDL muscles of rats. Active fractions were those found to significantly ameliorate atrophy of denervated IIB fibers.
[b] Expressed as concentration of protein (mg/mℓ) of injected solutions. Volumes injected into muscles were the same (0.1 mℓ) for all groups.
[c] Refers to purity of the trophic factor relative to crude nerve extract.

different from each other but were all significantly greater than those for denervated muscles not treated or injected with peak 1 ($p < 0.05$). There was no significant difference between the latter two groups. These findings indicate that the trophic substance(s) is a glycoprotein. Furthermore, this one step of purification removed 93% of unwanted material (Table 10).

C. Gel Filtration

The elution profile of sciatic nerve glycoproteins from Sephadex® G-100 fine (Figure 13) consisted of two large peaks which contained approximately equal quantities of protein, followed by several smaller, poorly defined peaks. The first peak (void volume) contained molecules larger than the exclusion limit of the gel (approximately 100,000 daltons).

It was shown by assay on denervated muscles that both major peaks contained trophic material (Figure 13A). This indicated one of three possibilities: (1) two distinct trophic substances with different molecular weights; (2) one trophic protein with more than one structure (i.e., it may exist as a monomer and a dimer); or (3) one trophic substance that has a molecular weight near the exclusion limit of the gel. To determine which possibility was correct a second assay was carried out. In another gel filtration, the fractions making up the trough region between the two main peaks were pooled separately so that the two peaks were arbitrarily divided into three fractions (a, b, and c). A fourth fraction (d) consisted of all the remaining smaller peaks (Figure 13B). When assayed on denerved muscles, only fraction b ameliorated atrophy of the IIB fibers ($p < 0.05$). Comparison with the elution volumes of standard proteins indicated that the active fraction contained substances with molecular weights from 90,000 to 130,000. Polyacrylamide slab gel electrophoresis in the presence of SDS revealed the presence of 13 bands in fraction b. The simplest explanation of these results is that there is one trophic protein with a molecular weight near the exclusion limit of the gel (about 100,000 daltons).

D. Anion Exchange Chromatography

Further fractionation of glycoproteins of molecular weights 90,000 to 130,000 (approximately) on DEAE-cellulose produced six protein peaks. However, when these were assayed on denervated muscles it was found that all trophic activity had been lost. It had been necessary to pool glycoproteins from several runs of concanavalin-A affinity chromatography, which resulted in long storage times in the freezer, and therefore the trophic activity may have been lost owing to instability of the partially purified material. We subsequently

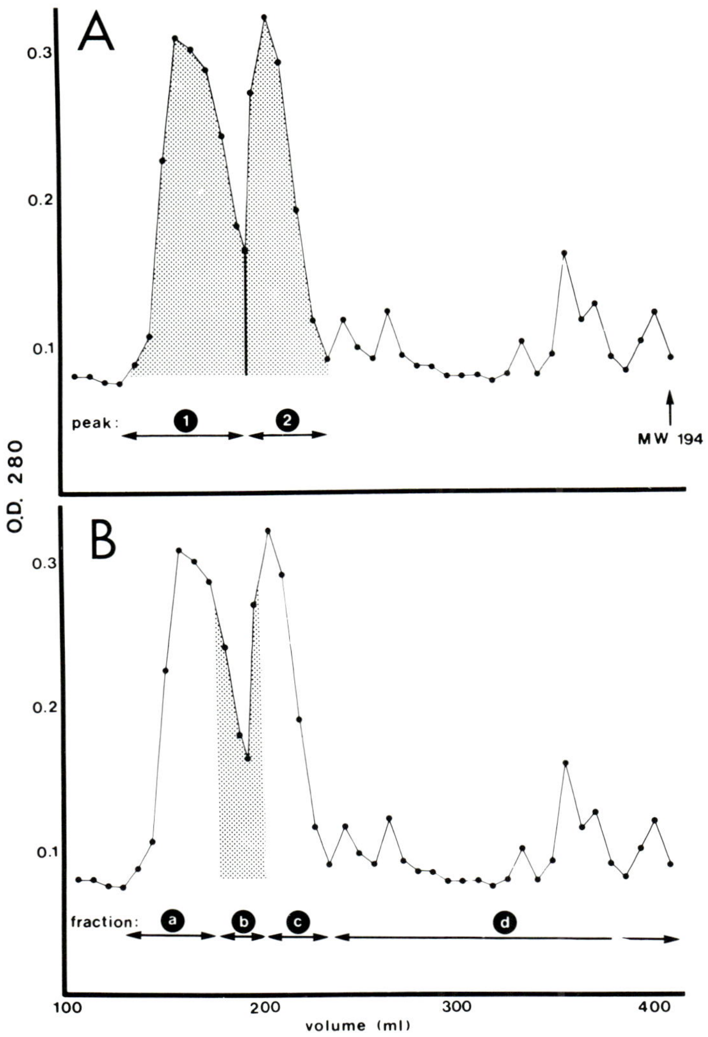

FIGURE 13. Gel filtration. Elution profile of sheep sciatic nerve glycoproteins from G-100 Sephadex.® The glycoprotein peak obtained from affinity chromatography was applied to a 2.6- × 90-cm column of G-100 Sephadex® and eluted with 50 m*M* ammonium acetate-acetic acid buffer (pH, 7.2) at a flow rate of 20 mℓ/hr. Fractions were pooled and assayed for exterotrophic activity on denervated EDL muscles of rats. (A) In the first assay both of the two main peaks were found to contain trophic material. (B) In the second assay the trophic activity was restricted to fractions making up the trough region between the two major peaks (stippled region). This fraction contained substances of molecular weights between 90,000 and 130,000 determined by passage of substances of known molecular weight through the column.

found that the trophic activity of glycoproteins purified on concanavalin-A agarose was lost after 12 weeks storage at −20°C, even if lyophilized.

The elution profile of proteins of crude sheep nerve extract from DEAE-cellulose is shown in Figure 14. Seven peaks of protein were produced, but only the sixth ameliorated denervation atrophy of type IIB fibers ($p < 0.025$). This fraction was eluted by approximately 80 m*M* NaCl in 25 m*M* sodium phosphate buffer at about pH 7.1. The trophic substance in this fraction had been purified 250-fold relative to crude nerve extract (Table 10). SDS-polyacrylamide slab gel electrophoresis revealed 21 protein bands. When the active peak

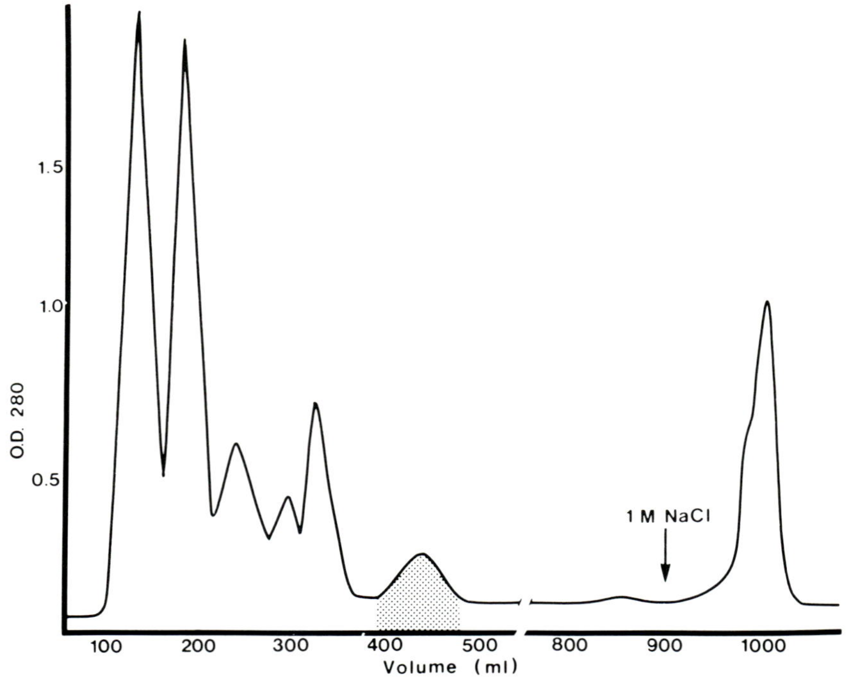

FIGURE 14. Ion exchange chromatography. Elution profile of sheep sciatic nerve proteins from DEAE-cellulose. Nerve proteins were applied to a 2.5- × 45-cm column of DEAE-cellulose and fractionated by a 800-mℓ linear gradient of 0 to 200 m*M* NaCl and pH 7.5 to 6.5 in 25 m*M* sodium phosphate buffer, followed by 1 *M* NaCl in the same buffer. The seven peaks obtained were assayed for exterotrophic activity. Only the sixth peak (stippled) ameliorated atrophy of denervated EDL muscles of rats.

was fractionated by concanavalin-A affinity chromatography prior to electrophoresis, 15 of the 21 protein bands were adsorbed to concanavalin-A; these ranged in size from 14,000 to 162,000 daltons. Two of these, with molecular weights of 91,300 and 127,400, were within the range determined by gel filtration for the active principle in its native state. If the trophic substance was not reduced to subunits under the denaturing conditions of the electrophoresis, it is likely that one of these two protein bands represents the active substance. The two bands will be cut out of stained gels and injected into rabbits for antibody production. Using immunoaffinity chromatography, the two potential trophic proteins will be isolated in pure form and assayed to determine which is the actual trophic substance. If neither is active it may be necessary to repeat this protocol for other protein bands on the gel which may correspond to subunits of the trophic molecule.

Isoelectric focusing of the active peak from the anion exchange chromatography revealed that it contained proteins with isoelectric points of 7.0 to 7.2. This provided evidence, in addition to that shown elsewhere,[4] that the mammalian neurogenic trophic substance(s) is not transferrin, which has an isoelectric point of 5.9.[105]

VI. SUMMARY AND CONCLUDING REMARKS

Numerous experiments have shown that motor nerves exert exterotrophic influences on skeletal muscle, but previous to 1980 only in vitro studies had provided direct evidence for the existence of trophic neurogenic substances. Using an in vivo assay system whereby denervated muscles receive proteinaceous extract of peripheral nerve by direct intramuscular injection (in rats) or by systemic intraperitoneal administration (in mice), it has been shown that such neurogenic substances can ameliorate several morphological, biochemical, and physiological alterations.

The reduction of atrophy in denervated muscles by nerve extract is dose dependent, but not species specific. Its trophic action is restricted to that portion of denervation atrophy not due to disuse, and furthermore its effects are not the same for different types of fiber. The amelioration of atrophy is evidenced by reduced losses of muscle weight and protein, CSA of fibers, and ultrastructural alterations of mitochondria and membranous systems.

Administration of nerve extract caused an unexpected increase in AChR, indicating that this is normally controlled by activity or that a different trophic substance, not in the extract, may act to suppress extrajunctional AChR in innervated muscle. In addition, the nerve extract may contain a substance normally responsible for maintaining high concentrations of AChR at the postsynaptic region of the sarcolemma.

Physiological changes ameliorated in denervated muscles treated with nerve extract included increased tetanus tension, which was due to prevention of atrophy, and prevention of greater than half the prolongations of time to peak and half-relaxation time of an isometric twitch. The latter effects correlated well with the complete prevention of the postdenervation loss of parvalbumin in muscles treated with extract. The active trophic neurogenic substance has been partially purified and found to be a glycoprotein of approximately 100,000 daltons.

Once the neurogenic substance responsible for exterotrophic maintenance of mature skeletal muscle is purified and antibodies to it are available, several lines of research may be pursued:

1. Immunohistochemical techniques may be used to localize the neurogenic substance in normal nerve and muscle tissue. It will be possible to determine whether the substance is present in sensory as well as motor neurons, whether it is synthesized in the perikarya and carried by fast axoplasmic transport, and whether it binds to a receptor on the sarcolemma or enters the muscle cell.
2. Administration of the purified trophic substance to denervated muscles may be used to assess many more parameters than those already examined (as discussed above).
3. Antibodies to the neurogenic substance may be administered to normally innervated muscle to determine which denervation-like changes appear and their magnitude relative to that occurring following denervation.

ACKNOWLEDGMENTS

I wish to thank Drs. J. A. Kiernan and R. A. Cook for advice; J. Wijsmann and S. L. Lui for technical assistance; and C. S. Heck, S. L. Lui, and A. D. Hines for preparation of manuscript. This work was supported by grants from the MRC of Canada and the Muscular Dystrophy Association of Canada.

REFERENCES

1. **Davis, H. L. and Kiernan, J. A.,** Neurotrophic effects of sciatic nerve extract on denervated extensor digitorum longus muscle in the rat, *Exp. Neurol.,* 69, 124, 1980.
2. **Davis, H. L. and Kiernan, J. A.,** Effect of nerve extract on atrophy of denervated or immobilized muscles, *Exp. Neurol.,* 72, 582, 1981.
3. **Davis, H. L.,** Trophic action of nerve extract on denervated skeletal muscle in vivo: dose dependency, species specificity, and timing of treatment, *Exp. Neurol.,* 80, 383, 1983.
4. **Davis, H. L. and Heinicke, E. A.,** Prevention of denervation atrophy in muscle: mammalian neurotrophic factor is not transferrin, *Brain Res.,* 309, 293, 1984.

5. **Davis, H. L. and Kiernan, J. A.,** Effect of nerve extract on number of acetylcholine receptors in denervated muscles of rats, *Exp. Neurol.*, 83, 108, 1984.
6. **Heinicke, E. A. and Davis, H. L.,** Soluble proteins of rat muscle: effect of denervation and nerve extract, *Soc. Neurosci.*, 10, 1059, 1984.
7. **Davis, H. L.,** Myotrophic effects on denervation atrophy of hindlimb muscles of mice with systemic administration of nerve extract, *Brain Res.*, 343, 176, 1985.
8. **Davis, H. L., Bressler, B. H., Jasch, L. G., and Heinicke, E. A.,** Denervated fast-twitch muscles of mice: myotrophic effects of nerve extract, *Soc. Neurosi.*, 11, 1145, 1985.
9. **Davis, H. L., Heinicke, E. A., Cook, R. A., and Kiernan, J. A.,** Partial purification from mammalian peripheral nerve of a trophic factor that ameliorates atrophy of denervated muscle, *Exp. Neurol.*, 89, 159, 1985.
10. **Heck, C. S. and Davis, H. L.,** Effect of nerve extract on ultrastructural changes in denervated skeletal muscle *in vivo*, *Soc. Neurosci.*, 12, 1108, 1986.
11. **Gutmann, E.,** *The Denervated Muscle*, Publishing House of the Czechoslovak Academy of Science, Prague, 1962, 1.
12. **Engel, A. G. and Stonnington, H. H.,** Morphological effects of denervation of muscle. A quantitative ultrastructure study, *Ann. N.Y. Acad. Sci.*, 228, 68, 1974.
13. **Thesleff, S.,** Physiological effects of denervation of muscle, *Ann. N.Y. Acad. Sci.*, 228, 89, 1974.
14. **Edwards, C.,** The effects of innervation on the properties of acetylcholine receptors in muscle, *Neuroscience*, 4, 565, 1979.
15. **Webster, D. M. S. and Bressler, B. H.,** Changes in isometric contractile properties of extensor digitorum longus and soleus muscles of C57BL/6J mice following denervation, *Can. J. Physiol. Pharmacol.*, 63, 681, 1985.
16. **Tower, S. S.,** Atrophy and degeneration in skeletal muscle, *Am. J. Anat.*, 56, 1, 1935.
17. **Van Der Meulen, J. P., Peckham, P. H., and Mortimer, J. T.,** Use and disuse of muscle, *Ann. N.Y. Acad. Sci.*, 228, 177, 1974.
18. **Stanley, E. F. and Drachman, D. B.,** Effect of disuse on the resting membrane potential of skeletal muscle, *Exp. Neurol.*, 64, 231, 1979.
19. **Albuquerque, E. X., Shuch, F. T., and Kauffman, F. C.,** Early membrane depolarization of the fast mammalian muscle after denervation, *Pfluegers Arch.*, 328, 36, 1971.
20. **Deshpande, S. S., Albuquerque, E. X., and Guth, L.,** Neurotrophic regulation of prejunctional and postjunctional membrane at the mammalian motor end-plate, *Exp. Neurol.*, 53, 151, 1976.
21. **Davey, B. and Younkin, S. G.,** Effect of nerve stump length on cholinesterase in denervated rat diaphragm, *Exp. Neurol.*, 59, 168, 1978.
22. **Fernandez, H. L., Duell, M. J., and Festoff, B. W.,** Neurotrophic control of 16S acetylcholinesterase at the vertebrate neuromuscular junction, *J. Neurobiol.*, 10, 441, 1979.
23. **Olek, A. and Robbins, N.,** A nerve stump dependent appearance of junctional and perijunctional acetylcholine receptors in organ culture, *Neuroscience*, 6, 1771, 1981.
24. **Robbins, N.,** Neurotrophic regulation of rat muscle glucose-6-phosphate dehydrogenase *in vitro*, *Brain Res.*, 225, 387, 1981.
25. **Davis, H. L.,** Is electrotherapy beneficial to denervated muscles? A review of results from basic research, *Physiol. Can.*, 35, 306, 1983.
26. **Crain, S. M. and Peterson, E. R.,** Development of neural connections in culture, *Ann. N.Y. Acad. Sci.*, 228, 6, 1974.
27. **Hooisma, J., Slaaf, D. W., Meeter, E., and Stevens, W. F.,** Trophic support by neural explants of cultured muscle fibers, *Exp. Neurol.*, 62, 628, 1978.
28. **Engelhardt, J. K., Ashikawa, K., Mori, J., and Shimbukuro, Y.,** Neurotrophic effects on the electrical properties of cultured muscle produced by conditioned medium from spinal cord explants, *Brain Res.*, 128, 243, 1977.
29. **Lentz, T. L.,** Nerve trophic function: *in vitro* assay of effects of nerve tissue on muscle cholinesterase activity, *Science*, 171, 187, 1971.
30. **Oh, T. H.,** Neurotrophic effects of sciatic nerve extracts on muscle development in culture, *Exp. Neurol.*, 50, 376, 1976.
31. **Markelonis, G. J. and Oh, T. H.,** A sciatic nerve protein has a trophic effect on development and maintenance of skeletal muscle cells in culture, *Proc. Natl. Acad. Sci. U.S.A.*, 76, 2470, 1979.
32. **Markelonis, G. J., Kemerer, V. F., and Oh, T. H.,** Sciatin: purification and characterization of a myotrophic protein from chicken sciatic nerves, *J. Biol. Chem.*, 255, 8967, 1980.
33. **Markelonis, G. J., Bradshaw, R. A., Oh, T. H., Johnson, J. L., and Bates, O. J.,** Sciatin is a transferrin-like polypeptide, *J. Neurochem.*, 39, 315, 1982.
34. **Aisen, P.,** The transferrins (siderophilims), in *Inorganic Biochemistry*, Vol. 1, Eichhorn, G. L., Ed., Elsevier, Amsterdam, 1973, 280.

35. **Kimura, I., Hasegawa, T., Miura, T., and Ozawa, E.,** Muscle trophic factor is identical to transferrin, *Proc. Jpn. Acad.*, 57(B), 200, 1981.
36. **Hasegawa, T., Saito, K., Kimura, I., and Ozawa, E.,** Fe^{3+} promotes *in vitro* growth of myoblasts and other cells from chick embryos, *Proc. Jpn. Acad.*, 57, 206, 1981.
37. **Karpati, G. and Engel, W. K.,** Correlative histochemical study of skeletal muscle after suprasegmental denervation, peripheral nerve section and skeletal fixation, *Neurology,* 681, 1968.
38. **Riley, D. A. and Allin, E. F.,** The effects of inactivity, programmed stimulation and denervation on the histochemistry of skeletal muscle fiber types, *Exp. Neurol.*, 40, 391, 1973.
39. **Tomanek, R. J. and Lund, R. D.,** Degeneration of different types of skeletal muscle fibers II. Immobilization, *J. Anat.*, 118, 531, 1974.
40. **Popiela, H.,** Trophic effects of adult peripheral nerve extract on muscle cell growth and differentiation *in vitro, Exp. Neurol.*, 62, 405, 1978.
41. **Guth, L. and Samaha, F. J.,** Procedure for the histochemical demonstration of actomyosin ATPase, *Exp. Neurol.*, 28, 365. 1970.
42. **Peter, J. B., Barnard, R. J., Edgerton, V. R., Gillespie, G. A., and Stempel, K. E.,** Metabolic profiles of three fiber types of skeletal muscle in guinea pigs and rabbits, *Biochemistry,* 11, 2627, 1972.
43. **Niederle, B. and Mayr, R.,** Course of denervation atrophy in type I and type II fibers of rat extensor digitorum longus muscle, *Anat. Embryol.*, 153, 9, 1978.
44. **Lowry, O. H., Rosebrough, N. J., Farr, A. L., and Randall, R. J.,** Protein measurement with the Folin phenol reagent, *J. Biol. Chem.*, 193, 265, 1951.
45. **Zar, J. H.,** *Biostatistical Analysis,* Prentice-Hall, Englewood Cliffs, N.J., 1984, 239.
46. **Guth, L., Albers, R. W., and Brown, W. C.,** Quantitative changes in cholinesterase activity of denervated muscle fibers and sole plates, *Exp. Neurol.*, 10, 236, 1964.
47. **Pullen, H. A.,** The distribution and relative size of fiber types in the extensor digitorum longus and soleus muscle of the adult rat, *J. Anat.*, 123, 467, 1977.
48. **Lentz, T. L.,** Effect of brain extracts on cholinesterase activity of cultured skeletal muscle, *Exp. Neurol.*, 45, 520, 1974.
49. **Hasegawa, S. and Kuromi, H.,** Effects of spinal cord and other tissue extracts on resting and action potentials of organ-cultured mouse skeletal muscle, *Brain Res.*, 119, 133, 1977.
50. **Oh, T. H., Sofia, C. A., Kim, Y. C., Carroll, C., Kim, H. H., Markelonis, G. J., and Reier, P. G.,** Sciatin: immunocytochemical localization of a myotrophic protein in chicken neural tissues, *Soc. Neurosci.*, 7, 555, 1981.
51. **Hanin, I.,** *Choline and Acetylcholine: Handbook of Chemical Assay Methods,* Raven Press, New York, 1974, 1.
52. **Giacobini, E., Pilar, G., Suszkiw, G., and Uchimura, H.,** Normal distribution and denervation changes of neurotransmitter related enzymes in cholinergic neurons, *J. Physiol. (London),* 286, 233, 1978.
53. **Fischbach, G. D. and Robbins, N.,** Changes in contractile properties of disused soleus muscles, *J. Physiol. (London),* 201, 305, 1969.
54. **Bray, J. J. and Harris, A. J.,** Dissociation between nerve-muscle transmission and nerve trophic effects on rat diaphragm using type D botulinum toxin, *J. Physiol. (London),* 253, 53, 1975.
55. **Younkin, S. G., Brett, R. S., Davey, B., and Younkin, L. H.,** Substances moved by axonal transport and released by nerve stimulation have innervation-like effects on muscle, *Science,* 200, 1292, 1978.
56. **Bisby, M. A.,** Inhibition of axonal transport in nerves chronically treated with local anesthetics, *Exp. Neurol.*, 47, 481, 1975.
57. **Benoit, P. W. and Belt, W. D.,** Some effects of local anesthetic agents on skeletal muscle, *Exp. Neurol.*, 34, 264, 1972.
58. **Schultz, E. and Lipton, B. H.,** The effect of Marcaine on muscle and non-muscle cells *in vitro, Anat. Rec.*, 191, 351, 1978.
59. **Lavoie, P. A., Collier, B., and Tenenhouse, A.,** Role of skeletal muscle activity in the control of muscle acetylcholine sensitivity, *Exp. Neurol.*, 54, 148, 1977.
60. **Sunderland, S. and Ray, L. J.,** Denervated changes in mammalian striated muscle, *J. Neurol. Neurosurg. Psychiatry,* 13, 159, 1950.
61. **Pellegrino, C. and Franzini, C.,** An electron microscope study of denervation atrophy in red and white skeletal muscle fibers, *J. Cell Biol.*, 17, 327, 1963.
62. **Gauthier, G. F.,** On the relationship of ultrastructural and cytochemical features to color in mammalian skeletal muscle, *Z. Zellforsch.*, 95, 462, 1969.
63. **Miledi, R. and Slater, C. R.,** Electron microscopic structures of denervated skeletal muscle, *Proc. R. Soc. London Ser. B,* 174, 253, 1970.
64. **Gori, Z.,** Proliferations of the sarcoplasmic reticulum and the t system in denervated muscle fibers, *Virchows Arch. B.*, 11, 147, 1972.
65. **Heck, C. S. and Davis, H. L.,** Ultrastructural changes in denervated muscle, *Can. Fed. Biol. Soc.*, 29, 164, 1986.

66. **Bendayan, M.,** Double immunocytochemical labeling applying the protein A-gold technique, *J. Histochem. Cytochem.,* 30, 81, 1982.
67. **Bendayan, M.,** Protein A-gold electron microscopic immunocytochemistry methods, applications and limitations, *J. Electron Microsc. Tech.,* 1, 243, 1984.
68. **Eisenberg, B. R.,** Quantitative ultrastructure of mammalian skeletal muscle, in *Handbook of Physiology,* Peachey, L. D. and Adrian, R. H., Eds., Williams & Wilkins, Baltimore, Md., 1983, 73.
69. **De Coster, W., De Reuck, J., and Vander Eecken, H.,** Early changes in experimental denervated rat gastrocnemius muscle, *Arch Neuropathol. Berl.,* 67, 114, 1985.
70. **Ellman, G. L., Courtney, K. D., Andres, V., and Featherstone, R. M.,** A new and rapid colorimetric determination of acetylcholinesterase activity, *Biochem. Pharmacol.,* 7, 88, 1961.
71. **Oh, T. H. and Johnson, D. D.,** Effects of acetyl-β-methylcholine on development of acetylcholinesterase and butyrylcholinesterase activities in cultured chick embryonic skeletal muscle, *Exp. Neurol.,* 37, 360, 1972.
72. **Drachman, D. B.,** Neurotrophic regulation of muscle cholinesterase: effects of botulinum toxin and denervation, *J. Physiol. (London),* 226, 619, 1972.
73. **Ranish, N. A. and Dettbarn, W. D.,** Nerve stump length and cholinesterase activity in muscle and nerve, *Exp. Neurol.,* 58, 377, 1978.
74. **Guth, L.,** Effect of immobilization on sole-plate and background cholinesterase of rat skeletal muscle, *Exp. Neurol.,* 24, 508, 1969.
75. **Gerday, C. and Gillis, J. M.,** The possible role of parvalbumins in the control of contraction, *J. Physiol. (London),* 258, 96P, 1976.
76. **Heizmann, C. W., Berchtold, M. W., and Rowlerson, A. M.,** Correlation of parvalbumin concentration with relaxation speed in mammalian muscles, *Proc. Natl. Acad. Sci., U.S.A.,* 79, 7243, 1982.
77. **Klug, G., Reichmann, H., and Pette, D.,** Rapid reduction in parvalbumin concentration during chronic stimulation of rabbit fast twitch muscle, *FEBS Lett.,* 152, 180, 1983.
78. **Muntener, M., Berchtold, M. W., and Heizmann, C. W.,** Parvalbumin in cross-reinnervated and denervated muscles, *Muscle Nerve,* 8, 132, 1985.
79. **Jasch, L. G. and Moase, E. H.,** Evidence for the identities of five proteins decreased in skeletal muscle of dystrophic mice, *Muscle Nerve,* 8, 389, 1985.
80. **Jasch, L. G., Bressler, B. H., Ovalle, W. K., and Slonecker, C. E.,** Abnormal distribution of proteins in the soleus and extensor digitorum longus of dystrophic mice, *Muscle Nerve,* 5, 462, 1982.
81. **Laemmli, U. K.,** Cleavage of structural proteins during the assembly of the head of bacteriophage T_4, *Nature (London),* 227, 680, 1970.
82. **Kelly, P. T. and Luttges, M. W.,** Electrophoretic separation of nervous system proteins on exponential gradient polyacrylamide gels, *J. Neurochem.,* 24, 1077, 1975.
83. **Obinata, T., Maruyama, K., Sugita, H., Kohama, K., and Ebashi, S.,** Dynamic aspects of structural proteins in vertebrate skeletal muscle, *Muscle Nerve,* 4, 456, 1981.
84. **Carafoli, E. and Penniston, J. T.,** The calcium signal, *Sci. Am.,* 253, 70, 1985.
85. **Fambrough, D. M.,** Acetylcholine sensitivity of muscle fiber membrane. Mechanism of regulation by motoneurons, *Science,* 168, 372, 1970.
86. **Grampp, W., Harris, J. B., and Thesleff, S.,** Inhibition of denervation changes in skeletal muscle by blockers of protein synthesis, *J. Physiol. (London),* 221, 743, 1972.
87. **Howe, P. R. C., Telfer, J. A., and Austin, L.,** Binding sites for ^{125}I-labeled α-bungarotoxin in normal and denervated mouse muscle, *Exp. Neurol.,* 52, 272, 1976.
88. **Hartzell, H. C. and Fambrough, D. M.,** Acetylcholine receptors: distribution and extrajunctional density in rat diagphragm after denervation correlated with acetylcholine sensitivity, *J. Gen. Physiol.,* 60, 258, 1972.
89. **Anderson, M. J. and Cohen, M. W.,** Fluorescent staining of acetylcholine receptors in vertebrate skeletal muscle, *J. Physiol. (London),* 237, 385, 1974.
90. **Lavoie, P. A., Collier, B., and Tenenhouse, A.,** Comparison of α-bungarotoxin binding to skeletal muscles after inactivity or denervation, *Nature (London),* 260, 349, 1976.
91. **Berg, D. K., Kelly, R. B., Sargent, P. B., Williamson, P., and Hall, Z. W.,** Binding of α-bungarotoxin to acetylcholine receptors in mammalian muscle, *Proc. Natl. Acad. Sci. U.S.A.,* 69, 147, 1972.
92. **Libelius, R.,** Binding of ^{3}H-labeled cobra neurotoxin to cholinergic receptors in fast and slow mammalian muscles, *J. Neural Transm.,* 35, 137, 1974.
93. **McConnell, M. G. and Simpson, L. L.,** The role of acetylcholine receptors and acetylcholinesterase activity in the development of denervation supersensitivity, *J. Pharmacol. Exp. Ther.,* 198, 507, 1976.
94. **Mills, R. G., Bray, J. J., and Hubbard, J. I.,** Effects of inactivity on membrane potentials in rat muscle, *Brain Res.,* 150, 607, 1978.
95. **Fischbach, G. D. and Robbins, N.,** Effect of chronic disuse of rat soleus neuromuscular junction on postsynaptic membrane, *J. Neurophysiol.,* 34, 562, 1971.

96. **Guth, L. and Albuquerque, E. X.,** The neurotrophic regulation of resting membrane potential and extrajunctional acetylcholine sensitivity in mammalian skeletal muscle, *Physiol. Bohemoslov.*, 27, 401, 1978.
97. **Fambrough, D. M.,** Control of acetylcholine receptors in skeletal muscle, *Physiol. Rev.*, 59, 165, 1979.
98. **Vyskocil, F., Syrovy, I., and Prusik, Z.,** Induction of extrajunctional sensitivity of rat EDL muscle by peptidic component of peripheral nerve, *Pflugers Arch.*, 390, 265, 1981.
99. **Gutmann, E., Melichna, J., and Syrovy, I.,** Contraction properties and ATPase activity in fast and slow muscle of the rat during denervation, *Exp. Neurol.*, 36, 488, 1972.
100. **Lewis, D. M.,** The effect of denervation on the mechanical and electrical responses of fast and slow mammalian twitch muscle, *J. Physiol. (London)*, 222, 51, 1972.
101. **Kean, C. J. C., Lewis, D. M., and McGarrick, J. D.,** Dynamic properties of denervated fast and slow twitch muscle of the cat, *J. Physiol. (London)*, 237, 103, 1974.
102. **Finol, H. J., Lewis, D. M., and Owens, R.,** The effects of denervation on contractile properties of rat skeletal muscle, *J. Physiol. (London)*, 319, 81, 1981.
103. **Goldstein, I. J., Hollerman, C. E., and Smith, E. E.,** Protein-carbohydrate interaction. II. Inhibition studies on the interaction of concanavalin A with polysaccharides, *Biochemistry*, 4, 876, 1965.
104. **Dulaney, J. T.,** Binding interactions of glycoproteins with lectins, *Mol. Cell. Biochem.*, 21, 43, 1979.
105. **Sober, H. A., Ed.,** *Handbook of Biochemistry*, 2nd Ed., CRC Press, Boca Raton, Fla., 1968, C-38.
106. **Davis, H. L., Heinicke, E. A., and Eldridge, L.,** unpublished observations.
107. **Davis, H. L., Bressler, B. H., and Jasch, L. G.,** unpublished observations.

Chapter 8

TROPHIC ROLES OF NEUROMUSCULAR JUNCTION EXTRACELLULAR MATRIX CONSTITUENTS

Nibaldo C. Inestrosa

TABLE OF CONTENTS

I. INTRODUCTION

The importance of the extracellular matrix (ECM) for normal cellular growth and development in vivo has long been recognized.[1-3] For example, epidermal basal cell layers must be in contact with the ECM in order to retain their normal orientation in vivo and to remain mitotically active in organ culture.[4,5] ECM can be produced by mesenchymal cells or epithelial cells after previous exposure to ECM produced by other tissues.[6] Studies on the role of extracellular materials at the epithelial-mesodermal interface have shown that glycosaminoglycans (GAG) at the junction of interacting tissues may be implicated in epithelial morphogenesis.[7] Cell proliferation in salivary glands is restricted to specific areas localized in the distal end of the lobules where the fastest GAG turnover is observed.[8] The ECM has also been implicated in cell migration[9] and in cellular differentiation of glial, neuronal, and muscle cells.[10-12]

An intact ECM or basal lamina scaffold is required for the maintenance of orderly tissue structure.[13] Replenishment of tissue cells following death or injury occurs in an orderly manner along the framework of the basal lamina scaffold. This process seems to be affected by the polarity of the basal lamina and enables multicellular organisms to reconstitute tissue structures and organs.[14] This is clearly illustrated by the reestablishment of neuromuscular junctions involving regenerating axons and muscle fibers, a highly ordered process which depends on a variety of cellular interactions (see Chapter 10). Regenerating axons grow through original pathways of Schwann cells and perineurium; regenerating myofibers develop within the remaining basal lamina sheaths; and new neuromuscular junctions are formed precisely at the original synaptic sites.[15-17] Sanes et al. and Burden et al.[18,19] have studied those muscle structures which influence neuromuscular junction regeneration. They demonstrated that after removal of both myofibers and motor axons the remaining synaptic ECM directs both the formation of active zones in regenerating nerve terminals[18] and the accumulation of acetylcholine receptors (AChR) in the subsynaptic membrane.[19] Their results indicate that components of the synaptic ECM can guide the differentiation of pre- and postsynaptic membranes during neuromuscular regeneration. Later, Sanes and Hall[20] demonstrated the presence of basal lamina components specific to the neuromuscular junction. Those findings led to the possibility of identifying the molecular entities responsible for the assembly of a complex structure such as the vertebrate neuromuscular junction.

In this chapter, I summarize recent work on neuromuscular junction ECM in cholinergic synapses. An overview of ECM components is followed by descriptions of synaptic and extrasynaptic basal lamina constituents; the role of heparan sulfate proteoglycan as an organizing factor for acetylcholinesterase (AChE) and AChR in the postsynaptic muscle surface; and the identification of ECM components involved in trophic functions of nerve and muscle. More extensive reviews are recommended dealing with cell surface-ECM interactions,[21] cell surface GAG,[22] and the role of ECM in neural development.[23]

II. EXTRACELLULAR MATRIX COMPONENTS

The composition of the ECM surrounding cells is closely dependent on cell type and degree of cellular differentiation. The ECM phenotype is largely retained even in cell culture, a fact that has greatly facilitated studies on the biosynthesis of ECM components and on their function at the pericellular level. The ECM contain collagenous/noncollagenous glycoproteins and sulfated proteoglycans (Table 1).

A. Collagen

One of the body's most abundant proteins, comprising more than one third of the total protein in many adult organisms, is collagen.[2,24] Collagen molecules are right-handed coiled-

Table 1
EXTRACELLULAR MATRIX COMPONENTS

Macromolecules	Molecular characteristics
Collagens[a]	
Type IV[27,28]	Two peptides, 150kdaltons: core 7s (collagenase resistant), $[\alpha 1(IV)]_3$ (procollagen-like chain)
Type V[29]	$[\alpha A]_3$, $[\alpha B]_3$, $A[\alpha B]_2$
Glycoproteins	
Fibronectin[33]	Two peptides, 220 — 250 kdaltons
Laminin[46,47]	Three chains, two A chains, 200 kdaltons; one B chain, 400 kdaltons
Entactin[54]	One peptide, 158 kdaltons (rich in sulfate)
Nidogen[56]	One peptide, 80 kdaltons (rich in galactosamine)
Thrombospondin[57]	Three subunits, 160 kdaltons (platelet derived)
Asymmetric AChE[67]	Twelve subunits, 65 kdaltons; one collagen-like tail, 55 kdaltons (*Torpedo* enzyme)
Proteoglycans[76,79]	
Heparan sulfate	D-glucuronic or L-iduronic acid plus D-glucosamine (O- and N-sulfate)
Heparin	D-glucuronic or L-iduronic acid plus D-glucosamine (O- and N-sulfate)
Keratan sulfate	D-galactose plus D-glucosamine (O-sulfate)
Dermatan sulfate	L-iduronic or D-glucuronic acid plus D-galactosamine (O-sulfate)
Chondroitin 6-sulfate	D-glucuronic acid plus D-galactosamine (O-sulfate)
Chondroitin 4-sulfate	D-glucuronic acid plus D-galactosamine (O-sulfate)
Hyaluronic acid	D-glucuronic acid plus D-glucosamine

[a] Only those types present in basement membranes are listed.

coil structures of 1.4-nm diameter, 300-nm length, and a molecular weight of 300 kdaltons.[24] Each molecule contains three α-chains with a left-handed helix of 96 kdaltons and approximately 1050 amino acid residues. About one third of the total amino acids in collagen are glycine residues, and one fifth is the sum of proline and hydroxyproline. Collagen consists of a highly repetitive amino acid sequence in which glycine occupies every third position [(Gly-X-)n]. Proline and hydroxyproline account for one third of the molecular constituents in the X and Y positions.[25] In human tissues at least six collagen isotypes have been defined which differ in primary structure and, therefore, are apparently products of different genes.[25] Collagens I, II, and III are designated interstitial to distinguish them from collagens IV and V, which are associated with basement membranes. Collagen I is an important constituent of skin, bone, tendon, blood vessels, and uterine walls, and shows large fibers with a 60-nm periodicity. Collagen III tends to accompany collagen I, whereas collagen II is mainly present in hyaline cartilage, cornea, and the vitreous body, where it forms fine 10 to 20-nm diameter fibrils. While interstitial collagen fibers display periodicity, those of collagen IV in the basement membrane do not. The failure of collagen IV to aggregate and yield cross-banding is attributed to its high (10 to 12%) carbohydrate content. Unlike the interstitial collagens, collagen IV is sensitive to pepsin digestion.[26] Also, collagen IV molecules are larger and more procollagen-like than those of interstitial collagens: they are composed of two 150-kdalton subunits and a central 7S core.[27,28] In turn, collagen V from placenta[29] has α-A and α-B chains (1:2 ratio) arranged in a pattern which is similar to that of interstitial collagens.[30] Type V collagen is also present in basement membranes, including those in skeletal muscle.[31]

B. Other Collagen-like Proteins

The sequence of Gly-X-Y is not unique to collagen. The Cl♀ subcomponents of complement[32] and the asymmetric AChE[33] have collagen-like sequences in short sections of their primary structure. In the case of AChE, such collagen-like regions may serve as recognition sites for self-aggregation processes or for anchorage in the ECM. Recent studies have shown that purified asymmetric AChE reacts with anticollagen V and to a lesser extent with anticollagen I, but not with anticollagen IV.[34] These results suggest that the collagenous tail of asymmetric AChE may have some similarities with collagen V.

C. Glycoproteins

1. Fibronectin

Fibronectin is one of the most extensively studied noncollagenous glycoproteins of the ECM.[21] It is an acidic glycoprotein which contains several domains;[35,36] is formed by 220- to 250-kdalton dimers linked by disulfide bonds near the terminal carboxyl;[33] and has binding sites for cell surface, collagen, heparin, fibrin, Cl♀, and AChE molecules.[37] Soluble fibronectin can be purified by virtue of its specific affinity to denatured collagen (gelatin).[38] It is present in large amounts in plasma (30 mg/100 mℓ),[39] where it is known as plasma fibronectin or cold-insoluble globulin.[40] Fibronectin mediates the attachment of various cell types to ECM components;[21] in neuronal cells it is apparently involved in neurite extension and survival;[41,42] and in the presence of transglutaminase clothing factor XIIIa, it can be covalently cross-linked to fibrin, collagen, and asymmetric AChE.[43-45]

2. Laminin

Laminin, a basement membrane glycoprotein, was originally isolated from mouse tumors producing large amounts of basement membrane matrix material.[46,47] This molecule is thought to be composed of three A chains (200 kdaltons each) and one B chain (400 kdaltons) covalently linked by disulfide bonds to form a molecular complex of approximately 1000 kdaltons. Purified antibodies to laminin do not cross-react with other known basement membrane antigens, including collagen IV, fibronectin, and a basement membrane proteoglycan from murine EHS sarcoma.[48] Thus far, laminin has been found in all ECM examined,[49] and in the lamina rara of glomerular basement membranes it has been localized at sites immediately adjacent to cells, suggesting that it mediates interactions between epithelial cells and the basal lamina.[48] Laminin apparently participates in the attachment of epithelial cells to collagen IV.[50] During development of mouse embryo, laminin is the first ECM component detected by immunofluorescence. It is detected in mouse embryo as early as the 16-cell preimplantation morula stage, whereas fibronectin and collagen IV are first found after implantation in the inner cell mass of the blastocyst and interstitial collagen is only detected several days later during differentiation of the mesoderm.[51,52] Laminin binds to heparin, suggesting that it may interact with ECM heparan sulfate proteoglycans in vivo.[53]

3. Entactin

Entactin is a sulfated glycoprotein (158 kdaltons) which was originally identified and isolated from a mouse teratocarcinoma-derived cell line.[54] It differs from laminin in electrophoretic mobility, cyanogen bromide peptide fragmentation pattern, immunological cross-reactivity, and incorporation of sulfate. In the kidney, entactin is found in the lamina rara interna, whereas in the basement membrane of the duodenal mucosa it is localized in the lamina densa.[55]

4. Nidogen

Nidogen is a protein which differs from other ECM components in its size, composition, and immunochemical properties.[56] It has been purified from a mouse tumor basement membrane, where it accounts for 2 to 3% ot total protein. Nidogen consists of a single polypeptide chain (80 kdaltons) which is folded into a globular structure with a small threadlike segment, stabilized by several intrachain disulfide bridges. This molecule may also aggregate into a nest like structure (250 kdaltons) and contains 10% carbohydrate, including a high percentage of galactosamine. Immunohistochemical studies have identified nidogen as a typical basement membrane protein with an ubiquitous tissue distribution.[56]

5. *Thrombospondin*

Thrombospondin is a glycoprotein composed of three large disulfide-linked subunits (160 kdaltons). It is a major platelet α-granule component which is secreted and then bound to platelet membranes.[57] Cultured human and bovine endothelial cells synthesize and secrete this protein.[58] Immunofluorescence studies have shown thrombospondin to be present in ECM of endothelial cells, smooth muscle, and fibroblasts.[59,60] This glycoprotein binds to heparin[61] and to collagen IV,[60] suggesting that it interacts with multiple components of the ECM.

6. *Acetylcholinesterase*

AChE is highly concentrated at the neuromuscular junction,[61] where it is partly associated with the basal lamina.[62,63] A specialized form of the enzyme, the asymmetric AChE, contains a collagenous subunit which is associated with ECM material of skeletal muscle.[64,65] This molecule contains a triple-helical collagenous subunit and 12 catalytic subunits linked to the nonhelical region of the collagenous portion via disulfide bridges. The four catalytic subunits are linked by disulfide bonds to each of the three collagenous chains. The collagen-like tail of the enzyme has a mass of approximately 100 kdaltons.[66] In *Electrophorus* AChE, the tail peptides appear to comprise a doublet of 30 kdaltons,[33] whereas in *Torpedo* the tail corresponds to 55-kdalton peptides.[67] The asymmetric AChE of *Torpedo* also contains a noncatalytic peptide of 100 kdaltons, which appears to be linked to terminal portions of individual chains in the tail unit and is in close apposition to the catalytic subunits.[68] The association of asymmetric AChE with various ECM components has been studied extensively, but there is conflicting evidence regarding its direct interaction with fibronectin; laminin; and collagens I, IV, and V.[34,45,69,70] Thus far, the only consistent interaction has been obtained for asymmetric AChE with heparin[71] and heparan sulfate proteoglycans.[34] We have translated poly(A)$^+$-rich mRNA prepared from *Discopyge* electric organ in a reticulocyte lysate system.[72] An anti-*Torpedo* AChE antiserum was used to precipitate a single 60- to 62-kdalton polypeptide from the products of translation obtained in vitro.[72] Recent studies by Sikorav et al.[73] have confirmed our results and showed that when *Torpedo* mRNA is used, an additional polypeptide of 65 kdaltons is synthesized. In *Electrophorus,* they found only the latter peptide. A human AChE gene has been recently identified by homology to a 2-kb DNA fragment obtained from the genetic locus controlling AChE in the *Drosophila* nervous system.[74] The gene for *Torpedo* AChE will probably be cloned soon; in fact, three cDNA clones have already been isolated.[75]

D. Proteoglycans

Proteoglycans are complex macromolecules that contain a core protein to which one or more GAG chains are covalently bound.[76] GAGs consist of unbranched polymers of disaccharide units that contain hexuronic acid and hexosamine, with a carboxylate and/or a sulfate ester, which provide the chains with linear arrays of anionic groups.[77] There are seven types of GAG, on the basis of different disaccharide units (Table 1). With the exception of hyaluronic acid, which lacks sulfate groups, GAGs are linked to proteins in different types of proteoglycans.[78] Such linkage to protein is via a neutral trisaccharide (Gal-Gal-Xyl) at the reducing end of the GAG chain, and xylose forms an O-glycosidic bond with a serine residue in the protein.[79]

In cartilage the ECM is particularly large and accounts for more than 90% of tissue volume. Proteoglycans account for 20 to 30% of the cartilage total ECM dry weight.[79] They have molecular weights ranging between 1 to 4 $\times$ 10^3 kdaltons, and each molecule has about 100 chondroitin sulfate chains (15 to 20 kdaltons), 50 keratan sulfate chains (5 to 8 kdaltons), and many O-linked and N-linked oligosaccharides; thus, carbohydrate comprises more than 90% of the molecule's weight.[80,81] The protein core has a molecular weight of approximately 250 kdaltons. Evidence from a cell-free translation system of proteoglycan mRNA has indicated that a single high-molecular-weight protein core (300 kdaltons) is synthesized from chick[82] and calf cartilage,[83] suggesting a lack of heterogeneity in the cartilage protein cores. The importance of proteoglycans for normal cellular development and growth is illustrated by experiments with certain mouse mutants. For example, brachymorphic mice have disproportionately short long bones because the proteoglycans in epiphyseal cartilage are undersulfated as a result of partially deficient synthesis of the sulfate donor 3′-phosphoadenosine 5′-phosphosulfate.[84] Undersulfated proteoglycans have a lower charge density and are less expanded than normal; hence, they are less effective in filling the space between collagen fibers of the ECM.

Heparan sulfate proteoglycan in glomerular basement membranes appear to play significant roles in the functional integrity of the filtration apparatus and in the pathogenesis of glomerular diseases.[85-87] The involvement of proteoglycans in neoplasia has been observed by a number of investigators, e.g., chondroitin sulfate is found in small quantities in most noncartilaginous tissues, but is present in higher concentrations in a number of tumors.[88,89] These findings suggest that enhanced production of chondroitin sulfate might play a direct role in malignancy.[90] One of the most consistent findings of a specific qualitative alteration in proteoglycan structure that is directly linked to transformation is a reduced degree of sulfation in the heparan sulfate molecule both in vitro[91-93] and in vivo.[94] In addition to a direct effect in the synthesis of proteoglycans, tumor cells can modify their microenvironment through degradation of ECM macromolecules. In fact, B16 melanoma cells degrade heparan sulfate synthesized by endothelial cells.[95] This enzymatic cleavage was found only when the heparan sulfate was incubated with tumor cells, indicating that melanoma cells elaborated an endoglycosidase (heparanase) specific for heparan sulfate.[96,97]

III. RECEPTORS FOR ECM COMPONENTS

The ECM markedly influences cellular differentiation, proliferation, and metabolism; therefore, it is important to know how such influences are relayed to the cells. In the nervous system we know that neurons communicate through signals which require the existence of receptors in the postsynaptic cell. Extrapolating from this system, it is likely that cell response to ECM components involve receptor molecules. Accordingly, a single cell type most probably contains receptors for certain ECM components.[21] Some plasma membrane receptors for fibronectin, laminin, and collagen have been described.[98-101] In the case of laminin a cell surface receptor has been recently isolated from several sources,[99-103] including skeletal muscle cells in culture.[102] This receptor binds to laminin with reasonably strong affinity (K_D = 2×10^{-9} *M*) and has a molecular weight of 70 kdaltons. Cell surface heparan sulfate proteoglycans are excellent candidates to be ECM receptors because they bind to several ECM components, including fibronectin, laminin, collagen, and asymmetric AChE;[22,34,53,104] behave as intercalated membrane proteins; and potentially function as integral components of ECM and plasma membranes. It has been indicated that heparan sulfate proteoglycans bind to plasma membranes via the protein core, presumably by intercalation of a hydrophobic region directly into the plasmalemma lipid moiety; and the GAG side chains through ionic interactions, and thus can be displaced by exogenous polyanionic molecules such as heparin.[105] Putative transmembrane heparan sulfate proteoglycans associate quantitatively with lipid vesicles,[106] thereby exhibiting the same chromatographic and sedimentation behavior as the vesicles. These proteoglycans are intercalated into the vesicle with their GAG-containing domains exposed to the exterior, since mild trypsin treatment removes this portion of the proteoglycan from vesicles.[106]

Another important area of current research concerns the mechanisms by which ECM components can alter cellular biosynthetic patterns and cytoskeletal organization. It has been suggested that ECM molecules, alone or in combination, affect the organization of the cytoplasm via receptors in the plasma membrane and influence cellular shape, mobility, and differentiation.[2] For example, fibronectin forms extracellular fibrous networks in patterns which correlate with intracellular patterns of actin filaments.[107] Cytochalasin, a drug which alters actin filaments, disrupts the attachment of cells to fibronectin.[108] Recent studies show that laminin and actin interact in a specific way. The addition of a membrane protein to depolymerized actin determines the formation of highly regular arrays of filaments. Such membrane protein, denominated connectin, appears to be the laminin receptor.[109] Heparan sulfate proteoglycans may also function as a link between intra- and extracellular fibers by interacting with cytoplasmic actin filaments and with ECM fribronectin, laminin, or collagen fibers. Heparan sulfate proteoglycans and actin coexist in rat fibroblast cultures;[110] and a membrane-bound heparan sulfate proteoglycan isolated from mammary cells interacts with polymerized actin.[111] ECM isolated from different sources usually contains some cytoskeletal components which co-purify with the matrix. In skeletal muscle cells both actin and myosin remain associated to ECM-like material purified according to standard procedures.[112,113] Recent evidence from our laboratory suggests that the ECM is also able to interact with intermediate filaments. In fact, when the ECM of *Discopyge* electric organ[72] was purified by extensive detergent extractions, a consistent association between the basal lamina and 10-nm filaments (subunits of 52 kdaltons) was noted.[114] Figure 1 shows typical electron micrographs of basement membranes connected to cytoplasmic filaments. In these preparations the plasma membrane was totally extracted, suggesting that either an ECM (or intermediate filament) receptor was retained or that the basal lamina interacts directly with cytoplasmic filaments. Examination of freeze and deep-etched preparations of *Torpedo* electric organ has shown what seem to be examples of direct interactions between 10-nm filaments and basal lamina components.[115] These studies clearly indicate that the ECM is not just an amorphous surface layer independent of the surrounding cells, but may be directly connected to the cytoskeleton via specific cell surface receptors.

IV. ECM CONSTITUENTS IN CHOLINERGIC SYNAPSES

The elementary unit of intercellular communication in the nervous system, the synapsis, is characterized by a well-defined spatial distribution of several macromolecular species. As previously discussed, AChE is associated with the synaptic basal lamina of cholinergic motor endplates. In turn, since the synaptic ECM participates in nerve terminal differentiation and directs the clustering of AChR in the postsynaptic membrane,[18,19] it was of interest to compare synaptic and extrasynaptic portions of the basement membrane sheath. Because the synaptic matrix represents less than 0.1% of the total surface area of muscle ECM, and no methods are available for separating this highly insoluble structure, such comparison has relied on immunohistochemical procedures. Sanes and Hall[20] obtained antibodies to immunogens which were likely to be derived from or share determinants with muscle ECM. These investigators used rhodamine (red fluorescence) coupled to α-bungarotoxin[116] to identify neuromuscular junctions by fluorescence microscopy. Each antibody was conjugated to fluorescein (green fluorescence) so that they were able to distinguish and compare the distribution of the two fluorophores on the same unfixed section of rat skeletal muscle. Under these conditions they found four antisera that distinguished synaptic from extrasynaptic regions of the muscle fiber's surface. Two of them, antilens capsule collagen and antimuscle

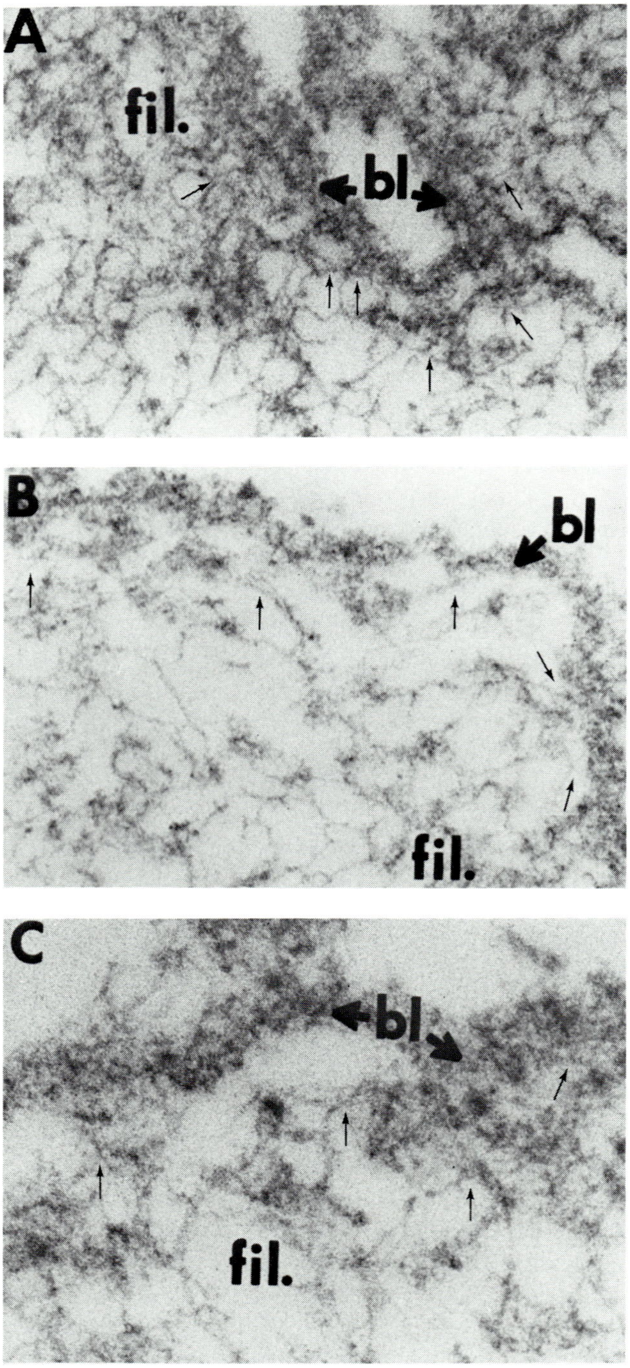

FIGURE 1. Interaction of ECM and cytoplasmic 10-nm filament detergent-insoluble ECM fractions obtained from *Discopyge* electric organ, as previously described,[104] were fixed in Karnovsky's fixative and prepared for electron microscopy.[72] (A, B, C) Pictures taken from different preparations. Arrows indicate basal lamina sheets (b1) (fil., cytoplasmic filaments.) (Magnification × 105,000.)

basement membrane collagen, were selective for endplates only after antibodies that bind to extrasynaptic areas had been removed by adsorption with connective tissue from muscle endplate-free regions. The four antibodies revealed synaptic antigens present on the muscle's external cell surface, persisted after denervation, and recognized at least three different determinants in two different molecules of the synaptic ECM. Subsequently, these immunologically and enzymatically defined markers were used to investigate the synthesis of synapse-specific ECM in muscle cells in culture.[117] Results showed that myotubes derived from a mouse muscle cell line and cultured in the absence of nerves were able to accumulate surface antigens recognized by three of these antisera.[118] Some of the synapse-specific ECM components co-localized with clusters of AChR. While in cultures examined 2 to 3 days after fusion 30 to 50% of all AChR clusters were associated with patches of the antianterior lens capsule antibody, in older cultures such association was almost complete. Patches of antimuscle basement membrane collagen that co-distributed with AChR were also occasionally observed.[119]

In addition to the aforementioned synaptic antigens,[20] other antigens have been identified which react either with synaptic and extrasynaptic ECM or only with extrasynaptic ECM components. Three well-known ECM components, laminin, fibronectin, and collagen IV, exist in the basement membrane of synaptic and extrasynaptic regions. Collagen V and a collagenous protein (high-salt soluble protein) are present in extrasynaptic, but not in synaptic basement membranes. Laminin and collagen IV are concentrated in the inner layer of the basement membrane, the high salt-soluble collagenous protein is concentrated in the outer surface and in the reticular lamina, and fibronectin is present in both basal and reticular laminae.[31] A lectin which recognized *N*-acetylgalactosamine residues of oligosaccharide chains and reveals a synapse-specific carbohydrate has been identified in rat muscles.[120] This specific carbohydrate component persists after muscle denervation and digestion of plasma membrane, suggesting that it is localized in the synaptic ECM. In this context, it is interesting to note that nidogen has a high concentration of galactosamine residues which could be responsible for the lectin binding.[56]

Another cholinergic synapse examined for the presence of ECM components is the electroplaque of the eel *Electrophorus electricus*.[121] The electroplaque of electric organ is derived from skeletal muscle fibers, but is deprived of contractile myofibrils. This system exhibits an asymmetric organization, i.e., nerve terminals contact only one side of the cell (innervated face). Immunofluorescence studies, using AChE as a marker of the innervated face, reveal that antibodies against collagen I and III stained both the innervated face as well as the cytoplasm of the electrolcytes. A more discrete concentration in the innervated membrane is observed with antibodies against collagen IV and V. Antibodies to fibronectin and laminin show a staining pattern very similar to that of AChE, suggesting that these components are localized in the synaptic region.[121]

Entactin apparently follows laminin in its distribution in the ECM of skeletal muscle, Schwann cells, and capillaries.[54,55] Antientactin antibodies can be visualized by a second antibody conjugated with horseradish peroxidase. Under these conditions, the basement membrane was entirely labeled, and no clear difference was observed between lamina rara and lamina densa. Entactin and laminin presented a similar distribution, suggesting that entactin is localized in both synaptic and extrasynaptic regions. However, the possibility still exists for entactin to be excluded from the ECM in the synaptic cleft and junctional folds. Heparan sulfate proteoglycans have also been observed in skeletal muscle by using antibodies against a proteoglycan from the EHS murine sarcoma.[122,123] These proteoglycans are not extracted by detergent and appear to be present in the basal lamina of both synaptic and extrasynaptic muscle regions. In rat muscle primary cultures, the pattern of proteoglycan immunoreactivity appears to be associated with nonmuscle cells. Most of the myotube surface shows little or no immunoreactivity, but many AChR clusters are observed in regions which exhibit proteoglycan reaction. During neuromuscular junction development heparan sulfate proteoglycans, as well as laminin, fibronectin, and collagen IV and V, are associated with the surfaces of muscle fibers as early as 16 days *in utero*. In diaphragms from 15-day-old rat fetuses (the earliest stage at which AChR aggregates are observed) the intensity of immunostaining for proteoglycans and laminin is considerably lower than that at later stages.[122]

These findings suggest the existence of both synaptic and extrasynaptic determinants. Since most of the antigens have been defined only by serological criteria, their biochemical identities remain unknown. ECM structures are composed of an array of macromolecules, many of which strongly adhere to one another. Thus, isolation of extremely pure antigens for immunization or for immunoadsorption presents a formidable problem.[124] In turn, hybridoma techniques allow one to circumvent this problem and obtain homogenous antibodies specific for single antigenic determinants.[125] Monoclonal antibodies have been generated against crude ECM fractions obtained from frog and chick muscle cells in culture[124,126] or against connective tissue-rich fractions from adult muscles[127] and chick embryos.[128-130] Immunofluorescence studies suggest that some ECM antigens are co-distributed with AChR clusters, confirming our results with crude antisera and mouse muscle cells in culture.[117-119] Subsequently, three monoclonal antibodies (C1, C4, and C10) which are IgG1 were generated.[127] C1 and C4 stain synaptic ECM more intensely than extrasynaptic ECM in adult muscle, whereas C10 stains synaptic and extrasynaptic ECM with comparable intensity. C1 and C10 were produced from a connective tissue-rich fraction from rat skeletal muscle, while C4 was raised against bovine lens capsule.[127] Also, a monoclonal antibody against heparan sulfate proteoglycan has been generated from insoluble material of *Xenopus laevis* larval tails.[126] The high-molecular-weight proteoglycan is concentrated in the basal lamina at the *Xenopus laevis* neuromuscular junction. Electron microscope autoradiography using ^{125}I-labeled monoclonal antibody has shown that the antigen is concentrated at least fivefold within the synaptic ECM. During development in culture, the muscle surface displays a nonuniform distribution of the proteoglycan. Aneural muscle cultures showed AChR clusters associated with adjacent plaques containing basal lamina heparan sulfate proteoglycan. In a parallel study on ECM in chick myogenic cultures, a similar heparan sulfate proteoglycan has been found which shows a higher correlation with AChR clusters than with several other ECM elements. In muscle cells in culture many connective tissue components, including collagen, AChR, fibronectin, proteoglycans, etc., are secreted into the culture medium, and, therefore, by metabolically labeling the cells it is possible to use the conditioned medium as a source of soluble antigen. This is convenient because protein isolation from ECM requires the use of chaotropic and denaturant agents which disrupt the normal organization of the ECM[131] and may destroy the protein's tertiary structure often important for antigen-antibody recognition. An interesting idea about the organization of ECM has emerged: the relative coexistence of different ECM components and their selective immunoprecipitation suggest that each ECM component is secreted independently.

Finally, an extracellular glycoprotein complex consisting of large disulfide-linked subunits called myotendinous antigen has been identified.[128,129] The monoclonal antibody that recognizes this antigen was prepared from crude chick collagen V isolated from pepsin-digested homogenates of 16-day-old chick embryos. The myotendinous antigen appears early during chick limb morphogenesis at sites connecting developing muscle fibers, tendon, and bone.[130] The antigen is absent from medial regions of skeletal muscle and probably corresponds to the extrasynaptic component of the ECM, as occurs with collagen V.[31] The myotendinous antigen (150 to 240 kdaltons) is resistant to bacterial collagenase and is recovered in conditioned medium of muscle cultures in combination with chondroitinase ABC-sensitive proteoglycan.[129] The appearance of myotendinous antigen during limb morphogenesis and its distribution in adult muscle and tendon are involved in attaching muscle fibers to tendon fascicles.[129] Table 2 summarizes the distribution of different ECM components in synaptic and extrasynaptic regions of cholinergic synapses.

Table 2
ECM COMPONENTS IN SYNAPTIC AND EXTRASYNAPTIC REGIONS OF CHOLINERGIC SYNAPSES

ECM components	Synaptic	Extrasynaptic
Collagen		
Rat muscle[20]	+	+
Rat muscle adsorbed[a20]	+ + +	+
Lens capsule[20]	+	+
Lens capsule adsorbed[a20]	+ + +	+
High-salt-soluble collagenous protein[31]	−	+ + +
Myotendinous antigen (derived from crude collagen type V)[128,129]	−	+ + +
Collagen type V[31]	−	+ + +
Collagen type IV[31]	+	+
Glycoproteins		
AChE[20]	+ + +	+
Laminin[31,55]	+	+
Fibronectin[31]	+	+
Entactin[b55]	+	+
Proteoglycans		
Heparan sulfate (EHS mouse tumor)[122]	+	+
Heparan sulfate (chick muscle)[124]	+ + +	+
Heparan sulfate (frog muscle)[126]	+ + +	+
Others		
Dolichos lectin[120]	+ + +	+
NaOH-treated lens capsule[20]	+ + +	+
Recognized by monoclonal antibodies C1, C4[127]	+ + +	+

Note: Most of the ECM components correspond to antigens recognized by antibodies of different sources.

[a] These antigens were synapse specific only after antibodies to extrasynaptic regions of muscle fibers were removed by adsorption with connective tissue from endplate-free regions of muscle.

[b] The presence of entactin in synaptic regions has not been demonstrated, but its distribution is similar to that of laminin in most tissues.

V. HEPARAN SULFATE PROTEOGLYCAN AS A SYNAPTIC ORGANIZER

Heparan sulfate proteoglycans interact with other ECM components, and recent evidence suggests that such interactions play a role in the structural organization and function of the ECM.[21,22] In this regard, it has been shown that in cell cultures proteoglycans and ECM components coexist;[132,133] addition of proteoglycans or GAG affects the functional properties of isolated matrix constituents;[134-136] and there is direct binding of isolated matrix components to proteoglycans or GAG.[137-139]

Studies on reconstitution of basement membrane components have shown that collagen IV, laminin, and heparan sulfate proteoglycan do not aggregate when incubated alone.[140] However, when these components are incubated together, they do undergo aggregation. This finding suggests that the formation of such a large supramolecular complex could act as a nucleation center for the assembly of the ECM. The relative amounts of laminin, collagen IV, and heparan sulfate proteoglycan in in vitro experiments, as well as in EHS murine tumors, is 1.0:1.0:0.1, suggesting that the interaction among ECM components is highly specific. The binding properties mentioned above indicate that heparan sulfate proteoglycan may play a key role in regulating the proportions of ECM components during formation of the matrix. On the other hand, recent evidence indicates that heparan sulfate proteoglycans

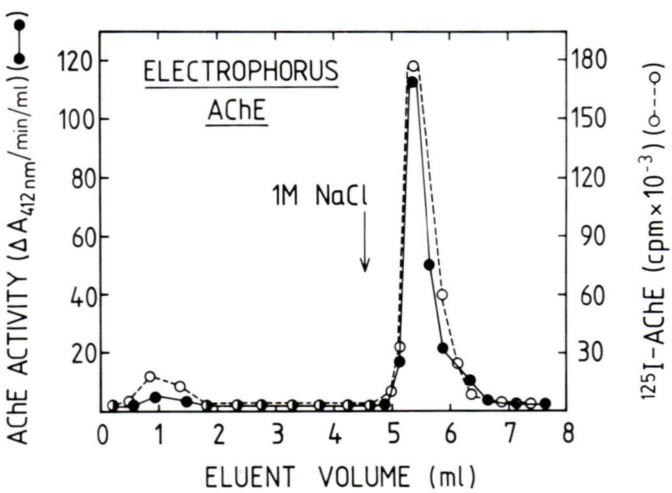

FIGURE 2. Affinity chromatography of purified *Electrophorus* AChE on a heparin-agarose resin column. Purified asymmetric 18S AChE[33] with (● — ●) and without (○----○) previous iodination[199] were loaded on a column (0.4 × 2 cm) containing 1 mg heparin per milliliter of gel. Each fraction contains 200 μℓ, and samples were assayed for AChE activity[71] or counted in a gamma counter. The arrow indicates the beginning of the elution with 1 M NaCl. The recovery of enzyme activity and total counts was approximately 90%.

are not essential for the maintenance of the structural stability of already formed pericellular matrix.[141] Since GAGs are associated with the cell surface, it is possible for these molecules to act as anchors or receptors for secreted molecules.[22] GAGs are strong polyanions, and most molecules that are positively charged at physiological pH may bind to cell surface GAG. Thus, if a macromolecule is noncovalently bound to cell-associated GAG, it should be possible to displace such a macromolecule by means of a soluble analog of the cell-associated GAG. With this in mind, we homogenized rat diaphragm muscle in the presence of heparin, hoping to solublize synaptic AChE. The result was clear-cut: heparin released asymmetric AChE from the synaptic ECM.[142] The effect of heparin was observed at low concentrations, and the half-maximal effect was obtained at 50 μg/mℓ. Hyaluronic acid and chondroitin sulfate did not release AChE from skeletal muscles. Analysis of the AChE solubilized by heparin in sucrose gradients containing 1 *M* NaCl showed the specific solubilization of the asymmetric A12 and A8 AChE molecular forms. However, if the solubilized esterase was analyzed under low ionic strength conditions, the majority of the AChE activity sedimented at the bottom of the gradient, suggesting the presence of aggregated asymmetric AChE. Further experiments showed a polydisperse molecular aggregate sedimenting at around 30S. Heparin was also able to convert the previously described low-ionic-strength AChE aggregate (obtained by dialysis of a 1 *M* NaCl muscle extract)[66] into a 30S AChE aggregate.[143] These data suggest that sulfated GAGs remove muscle surface AChE by a mechanism involving the formation of a soluble complex which contains the GAG and asymmetric AChE.

To prove the direct interaction of asymmetric AChE with heparin, we performed affinity chromatography in heparin-agarose columns and showed the specific binding of the esterase to the agarose gel.[71] This binding occurs with crude and purified enzyme. Figure 2 shows a heparin-agarose column profile obtained with *Electrophorus* 18S AChE. Note that the measured AChE activity corresponds to the pattern of radioactivity obtained with the iodinated enzyme. The interaction between AChE and heparin required an intact collagenous tail, as shown by the fact that the binding is abolished by pretreatment with collagenase.[71] These results support the idea that heparin binds to a specific domain of the collagen-like tail of the asymmetric AChE. It is well known that a large proportion (20 to 50%) of the tailed AChE species is localized in intracellular compartments of adult muscle[144,145] and cultured myotubes.[146,147] Therefore, it was important to establish whether the intracellular pool of asymmetric AChE was able to interact with heparin. Both intra- and extracellular asymmetric AChE from endplate regions of rat diaphragm muscle showed higher affinity for heparin than did the enzyme from non-endplate regions.[71]

A further level of complexity in our understanding of the biology of collagen-tailed AChE has been introduced by findings showing that two classes of asymmetric AChE forms exist in skeletal muscles.[148,149] Class A_I forms are readily brought into solution in the presence of high salt, while class A_{II} forms additionally require a chelating agent (e.g., EDTA) for solubilization. Although heparin releases only class A_I forms of AChE, both classes of collagen-tailed AChE forms bind with the same relative affinity to heparin-agarose columns.[150] These results suggest that both AChE classes possess a heparin-binding domain and that heparan sulfate proteoglycans are involved in the anchorage of both types of esterases to synaptic regions.

It is well known that synaptic AChE plays an essential role in the highly phasic transmission that occurs at the neuromuscular junction by hydrolyzing ACh released into the synaptic cleft by presynaptic nerve terminals.[151,152] To determine whether heparin detaches those AChE forms involved in neuromuscular transmission, we studied the effects of heparin on postsynaptic potentials in curarized cat skeletal muscles.[153] Results showed that heparin solubilized asymmetric AChE from muscle endplate regions, increases the amplitude of postsynaptic potentials, and decreases the time necessary to recover normal electrical activity in the curarized animals. Thus, heparin exerts an effect similar to that observed when AChE is inactivated by anticholinesterase agents.[151] Concomitantly, heparin increases total AChE activity present in blood plasma.[153] Our results are consistent with the idea that class A_I forms of AChE play a physiological role in chemical transmission at the neuromuscular junction.

The interaction of isolated asymmetric AChE and heparan sulfate proteoglycan from EHS murine tumors has also been studied.[34] AChE binds to the proteoglycan in Sepharose® 4B and in polystyrene microtiter plates. Additional results suggest that there are separate sites on the AChE collagenous subunit for the binding of heparan sulfate proteoglycan, collagen V, and laminin.[34] Another way to establish the interaction of asymmetric AChE and carbohydrate components of proteoglycans in the synaptic ECM is by using enzymes which specifically degrade GAG. In this context, we incubated ECM fractions purified from electric organs with and without heparitinase (degrades heparan sulfate) or chodroitinase ABC (digests chondroitin 4- and 6-sulfate, as well as dermatan sulfate). Increasing amounts of heparitinase, but not chondroitinase ABC, were able to release esterase into the incubation medium.[104] The heparan sulfate hydrolase liberates only collagen-tailed AChE forms. These results provide further evidence for the in vivo interaction between asymmetric AChE and heparan sulfate proteoglycans at the neuromuscular junction. Our studies on the biochemical composition of electric organ ECM showed that sulfated GAG bound to proteoglycans accounts for more than 5% of the total ECM. Using cellulose acetate electrophoresis and gel filtration on Sepharose® CL-6B columns, we established that approximately 20% of the total GAG was susceptible to heparitinase or nitrous acid oxidation (which specifically degrades heparan sulfate and heparin) and 80% was susceptible to digestion with chondroitinase ABC.[104] Our experiments have shown that asymmetric AChE and carbohydrate components of proteoglycans are associated in the ECM. They also indicate that a heparan sulfate proteoglycan is involved in the anchorage of collagen-tailed AChE at the synaptic ECM. In more recent experiments we have started to characterize the proteoglycans present

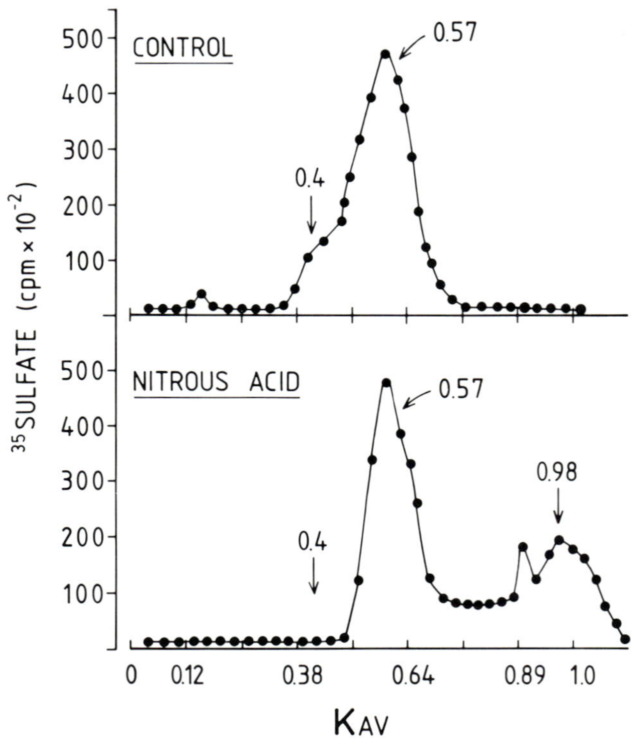

FIGURE 3. Presence of heparan sulfate in the GAG of the ECM-like material obtained from rat leg skeletal muscle. ECM-like material was prepared essentially as described for *Discopyge* ECM.[104] The material was hydrolyzed in the presence of 0.01 *M* NaOH and 1.0 *M* $NaBH_4$ during 24 hr at 45°C. Thereafter, the solution was neutralized and the free GAG precipitated with cold 70% ethanol. Aliquots of the GAG were either directly eluted in a Sepharose-4B column (100 × 1.4 cm) (top) or after nitrous acid treatment (bottom).[104] Nitrous acid released around 25% of the total ^{35}S-sulfate added to the total column volume, the treatment produces a decrease in the peak of K_{AV} = 0.4

in skeletal muscles. Figure 3A shows the profile of GAG in the ECM of rat leg muscles. Nitrous acid treatment of the sample released about 25% of the total 35S-sulfate macromolecules as low-molecular-weight material (Figure 3B); this occurs with a concomitant disappearance of the shoulder at K_{av} = 0.4 (compare Figures 3A and 3B). This result indicates the presence of heparan sulfate in the ECM of skeletal muscle.

A different line of research also provides evidence for the relation between cell surface heparan sulfate proteoglycan and asymmetric AChE. In fact, a stable variant of the rat pheochromocytoma PC12 cell line, which lacks a heparan sulfate proteoglycan on the cell surface, showed an atypical distribution of collagen-tailed AChE.[154] In normal PC12 cells virtually all the 16S AChE form is on the external cell surface, whereas in the mutant most of the enzyme (60%) is in an internal compartment. Nonetheless, the PC12 mutant cells retain their normal morphology, amount of thy-1 surface antigen, and adhesiveness to substrates such as collagen and polylysine. Furthermore, they extend neurites in response to nerve growth factor (NGF). These findings suggest that heparan sulfate proteoglycan is involved in the intracellular assembly, transport, and surface deposition of asymmetric AChE.[154] One likely explanation for the reduced extracellular 16S AChE could be that the PC12 mutant lacks one of the major anchors of this form of AChE on the cell surface, i.e., the heparan sulfate proteoglycan. To determine whether changes in proteoglycan synthesis in normal PC12 cells also result in altered accumulation of asymmetric AChE, we treated PC12 cells with β-D-xylosides, which compete with xyloside-primed core proteins as templates for GAG polymerization.[155,156] Under these conditions, only small amounts of GAG polymerized onto core proteins. Instead, large amounts of GAG polymerized on β-D-xylosides, and total heparan sulfate synthesis was dramatically reduced.[156] β-D-xylosides modify NGF-treated cells, inducing a new pattern of AChE molecular forms which is similar to that observed in heparan sulfate proteoglycan-deficient cells.[154] These results suggest that the biosynthesis and/or the cell surface deposition of asymmetric AChE is dependent on the normal synthesis and/or assembly of proteoglycans.

To study the in vivo interaction between asymmetric 16S AChE and heparan sulfate, the binding of this enzyme to intact aortic endothelial cells was evaluated. Asymmetric AChE was shown to bind to the cells in a specific, time-dependent, and saturable fashion.[157] Examination of the AChE-binding data in the form of Scatchard plots indicated the presence of one category of binding sites. At saturation, around 2.15×10^{11} molecules of 16S AChE per square centimeter were bound. The binding was reversed by exogenous heparin, but not by chondroitin sulfate or hyaluronic acid.[157] The fact that the surface of endothelial cells contains several classes of heparan sulfate proteoglycans[158,159] and that exogenous heparin detaches the 16S AChE from the endothelium suggests that the receptor site corresponds to cell-surface GAG chains of heparan sulfate.

As previously discussed, heparan sulfate proteoglycans in frog and chick muscles are highly concentrated at the neuromuscular junction, where they co-localize with AChR.[124,126] The accumulation of heparan sulfate proteoglycan has been examined in greater detail to determine whether proteoglycan plaques appear on the muscle surface before AChR clusters during myogenesis. In chick primary cultures, after only 1 day in culture the heparan sulfate proteoglycan staining was present on some, but not all, myoblast cells.[124] After 2 days, when the cells started to fuse, proteoglycan staining became visible on some myotubes. In some cases the staining material was already organized in patches, but at this stage AChR clusters were not yet apparent. Large, branching myotubes had developed after 3 days, and some of these had AChR clusters with a corresponding plaque of heparan sulfate proteoglycan. These findings indicate that specializations containing the proteoglycan are visible before AChR clustering becomes detectable. Even at the earliest stages at which AChR clusters are detected, a co-distribution between receptors and proteoglycans is evident.[126] The existence of AChR clusters along paths of nerve-muscle contact is almost invariably associated with the establishment of functional cholinergic neuromuscular junctions in *Xenopus* muscle cultures.[160] In fact, these clusters are primarily associated with plaques stained with anti-heparan sulfate proteoglycan antibody[126] and are formed *de novo* as a consequence of nerve-muscle interaction.[161,162] Accordingly, it seems reasonable to assume that their associated basal lamina-heparan sulfate proteoglycan plaques are also under neural regulation during synaptogenesis.

During development of rat muscle junctions, the aggregation of AChR beneath nerve terminals begins at embryonic day 16.[163] In turn, while focal AChR accumulation is not yet present, asymmetric AChE can be detected on the muscle cell surface at embryonic day 15.[164] This indicates that asymmetric AChE appears before AChR clustering. Available evidence suggests that at embryonic day 14 the basal lamina is in a rudimentary stage.[165] We have observed a similar situation in PC12 cells in which asymmetric AChE can be detected on the cell surface in the absence of a conventional basal lamina.[166] However, it is clear that such cells contain cell-surface heparan sulfate proteoglycans.[154] Also, our studies on the interaction of isolated asymmetric AChE with heparan sulfate proteoglycan in plasma membrane of endothelial cell indicate that potential sites for AChE anchorage are those containing the heparan sulfate proteoglycans.

These results suggest that heparan sulfate proteoglycans play a role as organizers of synaptic specializations. Studies in other systems have demonstrated that proteoglycans are implicated in the assembly and deposition of ECM components.[167] Hence, the accumulation of basal lamina proteoglycan may contribute to the establishment of normal nerve-muscle contact and participate in anchoring synaptic components. Whether the same synaptic heparan sulfate proteoglycan is involved in the anchorage of AChR and asymmetric AChE deserves further attention.

VI. ECM AND NERVE-MUSCLE TROPHIC FUNCTIONS

Until recently the search for factors implicated in nerve-target cell interactions has focused on the direct action of electromechanical activity (see Chapter 4) and soluble trophic substances (see Chapters 6 and 7). Increasing attention is now being paid to ECM constituents as another possible factor (see Chapter 10). The role of electromechanical activity can be studied in cell culture systems by comparing spontaneously active myotubes with tetrodotoxin- (TTX) paralyzed myotubes.[168,169] Soluble neural substances can be studied by supplementing culture media with neural extracts (see Chapters 3 and 6).[170-172]

The effects of activity on the accumulation of ECM components in control and TTX-paralyzed rat myotubes has been studied by immunohistochemical and ^{125}I-antibody-binding techniques with three antisera which stain the entire ECM.[173] Results showed that paralyzed myotubes reversibly accumulate less ECM than those that are active. In contrast, two synapse-specific antigens (recognized by the antilens capsule) were found to increase in inactive cultures: the mean number of AChR and antigen patches per myotube segment increased from 0.1 in control to 0.6 in TTX-treated cultures. Furthermore, AChR- and antigen-rich patches were larger in paralyzed vs. active myotubes. The number of patches stained by the ECM antibody and by rhodamine-bungarotoxin also increased in the paralyzed cultures. Clearly, synaptic ECM antigens increase in size and number under conditions that suppress the accumulation of total ECM (both shared and extrasynaptic antigens). These and previous studies[117,118] show that the presence of nerves is not entirely essential for the assembly of synaptic ECM. Other reports have shown that a rat brain extract selectively induces the accumulation of synaptic basal lamina antigens: AChE and two antigens recognized by monoclonal antibodies C1 and C4.[174] Since paralysis with TTX alone increases the number of synaptic ECM antigens and AChR-enriched patches,[173] it was important to determine whether the brain extract affected ECM by modulating the contractile activity of myotubes. The effect of soluble brain extracts was studied on paralyzed myotubes, and a further increase was detected in the accumulation of synaptic ECM antigens, as well as receptor patches. Thus, while the accumulation of ECM synaptic antigens is regulated by activity, the effects of brain extracts on synaptic ECM are largely activity independent. The effect of brain extracts on synaptic macromolecules was selective; in fact, the soluble extract did not have a significant effect on antigens shared by synaptic and extrasynaptic ECM components.[174] The substances present in adult brain extracts also induced AChE patches, which have not been observed in rat myotubes cultured without nerve.[168,175] However, such patches have been detected in rat nerve-muscle co-cultures,[175] in aneural cultures of *Xenopus* muscle,[176] and in myotubes of a mouse muscle cell line.[64] These findings suggest that both activity and soluble neural substances stimulate the accumulation of ECM components, with activity having a selective effect on shared (synaptic and extrasynaptic) antigens and brain extracts, enhancing the specific accumulation of synaptic antigens.

Myotube cultures have been also used to study the role of ECM components in the clustering of AChR. In fact, fragments of ECM from the electric organ of *Torpedo californica* have been shown to induce receptor aggregation on chick myotube cultures.[177] The insoluble ECM material caused a tenfold increase in the mean number of AChR clusters as early as 3 hr after application and reached a plateau by 20 hr. This effect was dose dependent, and the biologically active molecules could be extracted from the insoluble ECM fraction with 2 *M* $MgCl_2$. The activity of the extracts, as well as in the ECM fragments, was destroyed by boiling and by treatment with trypsin.[178] An antiserum raised against the 2 *M* $MgCl_2$ extract which stained synaptic and extrasynaptic regions of frog muscles completely blocked AChR aggregation in cultured myotubes.[179] The active molecules have been partially purified (approximately 1,000-fold) using gel filtration and ion-exchange chromatography; 80% of the clustering activity was found in the molecular-weight range of 50 to 100 kdaltons, and the most purified fraction caused AChR aggregation at a protein concentration of 0.1 n*M*.[180,181]

One possible action of the AChR clustering factors could be to enhance the accumulation of ECM components which, in turn, interact with receptors. In this regard, it has been reported that embryonic brain extracts induce a threefold increase in the number of AChR clusters and an eightfold increase in the synthesis of procollagen.[182] Both effects appear to be related, because AChR aggregates could be reduced to control levels by treating brain-extract-stimulated myotubes with Clostridial collagenase. Also, the stimulation of collagen secretion by ascorbic acid promoted a twofold increase in AChR aggregation, and the effects of the brain extract and ascorbic acid were additive. Additional evidence for the involvement of collagen in the aggregation of AChR in cultured muscle cells has been obtained by using *cis*-hydorxyproline, which inhibits collagen production by preventing secretion of the pro-collagen chain. Treatment with this compound significantly reduced the number of AChR clusters.[183] Further evidence was obtained by using purified antibodies against collagen I and IV: these antibodies blocked the aggregation of AChR induced by embryonic brain extracts.[183]

Aggregation of AChR in cultured myotubes has been evaluated in the presence of purified ECM components[184] while fibronectin, collagen IV and V, and heparan sulfate proteoglycan (purified from EHS murine tumors) had no effect. However, laminin caused a small increase in the number of AChR aggregates. Laminin enhanced the receptor-aggregating activity of medium conditioned by NG 108-15 neuroblastoma-glioma hybrid cells in a concentration-dependent manner. Furthermore, the number of aggregates formed in the presence of both laminin and nerve-conditioned medium was greater than the sum of the aggregates induced separately by each of these components. Experiments using immunofluorescence indicated that the laminin binds to myotubes within 30 min and forms patches on the cell surface over a period of hours.[184] Accordingly, the possibility exists for laminin to enhance the receptor-aggregating activity of a neuronal substance(s) released at the developing neuromuscular junction.

Other studies have provided strong evidence for a close correspondence between AChR clustering and the deposition of the ECM proteoglycan during development of the neuro-muscular junction.[185] Synaptic accumulations of AChR and heparan sulfate proteoglycan appear and progressively grow throughout a perineural zone that extends along the muscle surface on either side of the nerve process. After degeneration AChR clusters rapidly dis-appear, whereas the synaptic heparan sulfate proteoglycan survives both denervation and degeneration of muscle cells. This finding is consistent with the observed stability of the adult synaptic ECM and may explain the reappearance of junctional AChE at ectopic end-plates during reinnervation.[186] The aforementioned work[185] also indicates that presynaptic differentiation of synaptic vesicles is detectable at the developing endplates only after the formation of postsynaptic plaques containing both AChR and heparan sulfate proteoglycan. This suggests that the formation of a specialized synaptic basal lamina induced by the nerve may further stimulate the localized differentiation in both the muscle and the nerve terminals.

The mechanism by which the nerve induces muscle cells to produce a highly organized heparan sulfate proteoglycan-enriched ECM is not yet understood. One possibility is that soluble substances synthesized and secreted from nerve terminals stimulate the biosynthesis of proteoglycans. A similar role postulated for neurogenic soluble substances has been suggested for the leukocyte interleukins which induce cultured endothelial cells to produce a GAG-enriched pericellular matrix.[187] Consistent with the idea of a neuronal regulation of heparan sulfate proteoglycan are studies dealing with the effects of neonatal deafferentiation on the expression of a neuronal cell surface heparan sulfate proteoglycan in the developing rat superior cervical ganglion.[188] Following neonatal deafferentiation by section of the cervical sympathetic trunk, the expression of the core protein determinant of the heparan sulfate proteoglycan (measured by radioimmunoassay with a specific monoclonal antibody) was not altered in deafferented ganglia. In contrast, levels of side-chain determinants were significantly reduced. These results[188] provide evidence that presynaptic innervation plays a role in the regulation of the expression of heparan sulfate proteoglycan. Whether this also applies to the synaptic proteoglycan of the neuromuscular junction requires further investigation. An additional mechanism by which nerves can induce modifications in the muscle's basal lamina is through the release of specific nerve proteins supplied to the synaptic ECM.[189]

VII. SUMMARY AND CONCLUDING REMARKS

The preceding discussion illustrates that ECM structure found between the pre- and post-synaptic elements of the neuromuscular junction has become a subject of great interest. It is now clear that components of the synaptic ECM appear to induce specializations in both nerve and muscle cells during development and regeneration. An increasing number of ECM constituents are being discovered, and it is necessary to evaluate their role(s) in the neuromuscular junction, especially in relation to particular supramolecular structures, i.e., active zones in presynaptic nerve terminals or AChR-cytoskeletal interactions. The role proposed for heparan sulfate proteoglycans as organizers of synaptic specializations also deserves further investigation.

Certain diseases of muscle and peripheral nerve are likely to show alterations in some aspects of ECM biology, ECM-cell receptor interactions, and/or ECM-cytoskeleton linkages. For example, it is possible that myasthenic syndromes, characterized by asymmetric AChE deficiency and small nerve terminals,[190] involve some perturbations of the ECM. In fact, we know that heparan sulfate proteoglycans anchor AChE,[104] and both heparan sulfate proteoglycans[10] and laminin[191] are involved in neurite outgrowth mechanisms responsible for neuronal differentiation. In this context, it is interesting to note that a reduced synthesis of heparan sulfate proteoglycans was observed in the glomerular ECM of diabetic mice.[192]

Finally, it is worth mentioning that the advent of genetic engineering methods has generated enormous potential;[193] in fact, it has already revealed the arrangement of genes in neurosecretory cells[194] and transcriptional processes.[195] Considerable progress has been made in elucidating the structure of genomic or mRNA sequences coding for a number of different ECM components, including αI (IV) procollagen,[196] fibronectin,[197] and chondroitin sulfate proteoglycan core protein.[198] In the case of fibronectin, a single gene generates three different mRNAs by alternative splicing of a complex exon.[197] With the chondroitin sulfate proteoglycan, it has been found that the core protein (104 amino acids) is composed of three regions, the most striking of which is the central 49-amino-acid region composed of alternating serine and glycine residues.[198] This region functions as the acceptor site for the attachment of GAG chains.[198] Once a gene is isolated, a wide variety of questions may be addressed. Regarding the chondroitin sulfate proteoglycan core protein gene, the cDNA sequence detected provides the possibility of hybridizing unidentified genes for other core proteins, i.e., the heparan sulfate proteoglycan of synapses.

ACKNOWLEDGMENTS

This work was supported by Direccion de Investigaciones, Universidad Catolica de Chile;

Fondo Nacional de Ciencias de Chile; and Fundacion Gildemeister. I thank Miss Maria Teresa Pino for her valuable assistance with the manuscript.

REFERENCES

1. **Rudin, D. O.,** The second cell membrane: the basal lamina and the tissue cell theory of Metazoa, *J. Theor. Biol.,* 78, 143, 1979.
2. **Hay, E. D.,** Extracellular matrix, *J. Cell Biol.,* 91, 205s, 1981.
3. **Grobstein, C. J.,** Mechanisms of organogenetic tissue interaction, *Natl. Cancer Inst. Monogr.,* 26, 279, 1967.
4. **Dodson, J. W.,** On the nature of tissue interactions in embryonic skin, *Exp. Cell Res.,* 31, 233, 1963.
5. **Wessels, N. K.,** Substrate and nutrient effects upon epidermal basal cell orientation and proliferation, *Proc. Natl. Acad. Sci. U.S.A.,* 52, 252, 1964.
6. **Grobstein, C. J.,** Developmental role of intercellular matrix: retrospective and prospective, in *Extracellular Matrix Influences on Gene Expression,* Slavkin, H. C. and Grevlisch, R. C., Eds., Academic Press, New York, 1975, 9.
7. **Bernfield, M. R.,** Embryonic induction and tissue interaction during morphogenesis, in *Birth Defects, Excerpta Medica International Congress Series 432,* Littlefield, J. W. and deGrouchy, J., Eds., Excerpta Medica, Amsterdam, 1978, 111.
8. **Banerjee, S. D., Cohn, R. H., and Bernfield, M. R.,** Basal lamina of embryonic salivary epithelia, *J. Cell Biol.,* 73, 445, 1977.
9. **LeDourain, N. M.,** Extracellular factors controlling the migration and differentiation of the ganglioblasts of the autonomic nervous system, in *Extracellular Matrix Influences on Gene Expression,* Slavkin, H. C. and Grenlich, R. C., Eds., Academic Press, New York, 1975, 591.
10. **Lander, A. D., Fujii, D. K., Gospodarowicz, D., and Reichardt, L. F.,** Characterization of a factor that promotes neurite outgrowth: evidence linking activity to a heparan sulfate proteoglycan, *J. Cell Biol.,* 94, 574, 1982.
11. **Sieber-Blum, M., Sieber, F., and Yamada, K. M.,** Cellular fibronectin promotes adrenergic differentiation of quail neural crest cells *in vitro, Exp. Cell Res.,* 133, 285, 1981.
12. **Hauschka, S. D. and Konigsberg, I. R.,** The influence of collagen on the development of muscle clones, *Proc. Natl. Acad. Sci. U.S.A.,* 55, 119, 1966.
13. **Vracko, R. and Benditt, E. P.,** Basal lamina: the scaffold for orderly cell replacement, *J. Cell Biol.,* 55, 406, 1972.
14. **Vracko, R.,** Basal lamina scaffold-anatomy and significance for maintenance of orderly tissue structure, *Am. J. Pathol.,* 77, 314, 1974.
15. **Tello, F.,** Degeneración de las placas motrices después de la sección de los nervios, *Trab. Lab. Invest. Biol. Univ. Madrid,* 5, 117, 1907.
16. **Letinsky, M. S., Fishbeck, K. H., and McMahan, U. J.,** Precision of reinnervation of original postsynaptic sites frog muscle after a nerve crush, *J. Neurocytol.,* 5, 691, 1976.
17. **Marshall, L. M., Sanes, J. R., and McMahan, U. J.,** Reinnervation of original synaptic sites on muscle fiber basement membrane after disruption of the muscle cells, *Proc. Natl. Acad. Sci. U.S.A.,* 74, 3073, 1977.
18. **Sanes, J. R., Marshall, L. M., and McMahan, U. J.,** Reinnervation of muscle fiber basal lamina after removal of muscle fibers, *J. Cell Biol.,* 78, 176, 1978.
19. **Burden, S. J., Sargent, P. B., and McMahan, U. J.,** Acetylcholine receptors in regenerating muscle accumulate at original synaptic sites in the absence of the nerve, *J. Cell Biol.,* 82, 412, 1979.
20. **Sanes, J. R. and Hall, Z. W.,** Antibodies that bind specifically to synaptic sites on muscle fiber basal lamina, *J. Cell Biol.,* 83, 357, 1979.
21. **Yamada, K. M.,** Cell surface interactions with extracellular materials, *Annu. Rev. Biochem.,* 52, 761, 1983.
22. **Höök, M., Kjellén, L., Johansson, S., and Robinson, J.,** Cell-surface glycosaminoglycans, *Annu. Rev. Biochem.,* 53, 847, 1984.
23. **Sanes, J. R.,** Roles of extracellular matrix in neural development, *Annu. Rev. Physiol.,* 45, 581, 1983.
24. **Bornstein, P. and Traub, W.,** The chemistry and biology of collagen, in *The Proteins,* Neurath H. and Hill, R. L., Eds., Academic Press, New York, 1979, 411.
25. **Bornstein, P. and Sage, H.,** Structurally distinct collagen types, *Annu. Rev. Biochem.,* 49, 957, 1980.
26. **Miller, E. J.,** Biochemical characteristics and biological significance of the genetically distinct collagens, *Mol. Cell. Biochem.,* 13, 165, 1976.

27. **Alitalo, K., Vaheri, A., Krieg, T., and Timpl, R.,** Biosynthesis of two subunits of type IV procollagen and of other basement membrane proteins by a human tumor cell line, *Eur. J. Biochem.*, 109, 247, 1980.
28. **Garbisa, S., Liotta, L. A., Tryggvason, K., and Siegal, G. P.,** Antibodies to collagenase-resistant terminal regions of pro-type IV collagen recognize whole basement membrane and 7-S collagen, *FEBS Lett.*, 127, 257, 1981.
29. **Burgeson, R. T., ElAdli, F. A., Kaitila, I. I., and Hollister, D. W.,** Fetal membrane collagens: identification of two new collagen alpha chains, *Proc. Natl. Acad. Sci. U.S.A.*, 73, 2579, 1976.
30. **Kuhn, K. and Glanville, R. W.,** Molecular structure and higher organization of different collagen types, in *Biology of Collagen*, Viidik, A. and Vuust, J., Eds., Academic Press, London, 1980, 1.
31. **Sanes, J. R.,** Laminin, fibronectin, and collagen in synaptic and extrasynaptic portions of muscle fiber basement membrane, *J. Cell Biol.*, 93, 442, 1982.
32. **Porter, R. R. and Reid, K. B. M.,** The biochemistry of complement, *Nature (London)*, 275, 699, 1978.
33. **Rosenberry, T. L. and Richardson, J. M.,** Structure of 18S and 14S acetylcholinesterase. Identification of collagen-like subunits that are linked by disulfide bonds to catalytic subunits, *Biochemistry*, 16, 3550, 1977.
34. **Vigny, M., Martin, G. R., and Grotendorst, G. R.,** Interactions of asymmetric forms of acetylcholinesterase with basement membrane components, *J. Biol. Chem.*, 258, 8794, 1983.
35. **Hynes, R. O. and Yamada, K. M.,** Fibronectins: multifunctional modular glycoproteins, *J. Cell Biol.*, 95, 369, 1982.
36. **Sekiguchi, K. and Hakomori, S.,** Functional domain structure of fibronectin, *Proc. Natl. Acad. Sçi. U.S.A.*, 77, 2661, 1980.
37. **Hynes, R. O.,** Fibronectin and its relation to cellular structure and behavior, in *Cell Biology of Extracellular Matrix*, Hay, E. D., Ed., Plenum Press, New York, 1981, 295.
38. **Engvall, E. and Ruoslahti, E.,** Binding of soluble form of fibroblast surface protein, fibronectin to collagen, *Int. J. Cancer*, 20, 1, 1977.
39. **Mosesson, M. W. and Umfleet, R. A.,** The cold-insoluble globulin of human plasma. I. Purification, primary characterization, and relationship to fibrinogen and other cold-insoluble fraction components, *J. Biol. Chem.*, 245, 5728, 1970.
40. **Vuento, M. and Vaheri, A.,** Purification of fibronectin from human plasma by affinity chromatography under non-denaturing conditions, *Biochem. J.*, 183, 331, 1979.
41. **Rogers, S. L., Letourneau, P., Palm, S. L., McCarthy, J., and Furcht, L. T.,** Neurite extension by peripheral and central nervous system neurons in response to substratum-bound fibronectin and laminin, *Dev. Biol.*, 98, 212, 1983.
42. **Horie, H. and Kim, S. U.,** Improved survival and differentiation of newborn and adult mouse neurons in F12 defined medium by fibronectin, *Brain Res.*, 294, 178, 1984.
43. **Mosher, D. F.,** Cross-linking of cold insoluble globulin by fibrin-stabilizing factor, *J. Biol. Chem.*, 250, 6614, 1975.
44. **Mosher, D. F., Schad, P. E., and Vann, J. M.,** Cross-linking of collagen and fibronectin by factor XIIIa *J. Biol. Chem.*, 255, 1181, 1980.
45. **Emmerling, M. R., Johnson, C. D., Mosher, D. F., Lipton, B. H., and Lilien, J. E.,** Cross-linking and binding of fibronectin with asymmetric acetylcholinesterase, *Biochemistry*, 20, 3242, 1981.
46. **Chung, A. E., Jaffe, R., Vergnes, J. P., Freeman, I. L., Braginski, J. E., and Carlin, B.,** Properties of a basement membrane-related glycoprotein synthesized in culture by a mouse embryonal carcinoma-derived cell line, *Cell*, 16, 277, 1979.
47. **Timpl, R., Rohde, H., Robey, P. G., Rennard, S. I., Foidart, J. M., and Martin, G. R.,** Laminin — a glycoprotein from basement membranes, *J. Biol. Chem.*, 254, 9933, 1979.
48. **Foidart, J. M., Bere, E. W., Jr., Yaar, M., Rennard, S. I., Gullino, M., Martin, G. R., and Katz, S. I.,** Distribution and immunoelectron microscopic localization of laminin, a non-collagenous basement membrane glycoprotein, *Lab. Invest.*, 42, 336, 1980.
49. **Rohde, H., Wick, G., and Timpl, R.,** Immunochemical characterization of the basement membrane glycoprotein laminin, *Eur. J. Biochem.*, 102, 195, 1979.
50. **Terranova, V. P., Rohrbach, D. H., and Martin, G. R.,** Role of laminin in the attachment of PAM 212 (epithelial) cells to basement membrane collagen, *Cell*, 22, 719, 1980.
51. **Leivo, I., Vaheri, A., Timpl, R., and Wartiovaara, J.,** Appearance and distribution of collagens and laminin in the early mouse embryo, *Dev. Biol.*, 76, 100, 1980.
52. **Timpl, R. Engel, J., and Martin, G. R.,** Laminin — a multifunctional protein of basement membrane, *Trends Biochem. Sci.*, 8, 201, 1983.
53. **Sakashita, S., Engvall, E., and Ruoslahti, E.,** Basement membrane glycoprotein laminin binds to heparin, *FEBS Lett.*, 116, 243, 1980.
54. **Carlin, B., Jaffe, R., Bender, B. L., and Chung, A. E.,** Entactin, a novel basal lamina-associated sulfated glycoprotein, *J. Biol. Chem.*, 256, 5209, 1981.

55. **Martinez-Hernandez, A. and Chung, A. E.,** The ultrastructural localizational of two basement membrane components: entactin and laminin in rat tissues, *J. Histochem. Cytochem.*, 32, 289, 1984.
56. **Timpl, R., Dziaked, M., Fujiwara, S., Nowack, H., and Wick, G.,** Nidogen: a new, self-aggregating basement membrane protein, *Eur. J. Biochem.*, 137, 455, 1983.
57. **Baenziger, N. L., Brodie, G. N., and Majerus, P. W.,** Isolation and properties of a thrombin-sensitive protein of human platelets, *J. Biol. Chem.*, 247, 2723, 1972.
58. **McPherson, J., Sage, H., and Bornstein, P.,** Isolation and characterization of a glycoprotein secreted by aortic endothelial cells in culture: apparent identity with platelet thrombospondin, *J. Biol. Chem.*, 256, 11330, 1981.
59. **Jaffe, E. A., Ruggiero, J. T., Leung, L. L. K., Doyle, M. J., McKeown-Longo, P. J., and Mosher, D. F.,** Cultured human fibroblasts synthesize and secret thrombospondin and incorporate it into extracellular matrix, *Proc. Natl. Acad. Sci. U.S.A.*, 80, 998, 1983.
60. **Mumby, S. M., Rangi, G. J., and Bornstein, P.,** Interactions of thrombospondin with extracellular matrix proteins: selective binding to type V collagen, *J. Cell Biol.*, 98, 646, 1984.
61. **Dixit, V. M., Grant, G. A., Santoro, S. A., and Frazier, W. A.,** Isolation and characterization of a heparin-binding domain from the amino terminus of platelet thrombospondin, *J. Biol. Chem.*, 259, 10100, 1984.
62. **Couteaux, R.,** Localization of cholinesterase at neuromuscular junctions, *Int. Rev. Cytol.*, 4, 335, 1955.
63. **McMahan, U. J., Sanes, J. R., and Marshall, L. M.,** Cholinesterase is associated with the basal lamina at the neuromuscular junction, *Nature (London)*, 271, 172, 1978.
64. **Inestrosa, N. C., Silberstein, L., and Hall, Z. W.,** Association of the synaptic form of acetylcholinesterase with extracellular matrix in cultured mouse muscle cells, *Cell*, 29, 71, 1982.
65. **Dreyfus, P. A., Rieger, F., and Pincon-Raymond, M.,** Acetylcholinesterase of mammalian neuromuscular junctions: presence of tailed asymmetric acetylcholinesterase in synaptic basal lamina and sarcolemma, *Proc. Natl. Acad. Sci. U.S.A.*, 80, 6698, 1983.
66. **Massoulié, J. and Bon, S.,** The molecular forms of cholinesterase and acetylcholinesterase in vertebrates, *Annu. Rev. Neurosci.*, 5, 57, 1982.
67. **Lee, S. L., Heinemann, S., and Taylor, P.,** Structural characterization of the asymmetric (17S + 13S) species of acetylcholinesterase from *Torpedo*. I. Analysis of subunit composition, *J. Biol. Chem.*, 257, 12282, 1982.
68. **Lee, S. L. and Taylor, P.,** Structural characterization of the asymmetric (17S + 13S) species of acetylcholinesterase from *Torpedo*. II. Component peptides obtained by selective proteolysis and disulfide bond reduction, *J. Biol. Chem.*, 257, 12292, 1982.
69. **Anglister, L., Tarrab-Hazdai, R., Fuchs, S., and Silman, I.,** Immunological cross-reactivity between electric eel acetylcholinesterase and rat-tail-tendon collagen, *Eur. J. Biochem.*, 94, 25, 1979.
70. **Grassi, J., Massoulié, J., and Timpl, R.,** Relationship of collagen-tailed acetylcholinesterase with basal lamina components. Absence of binding with laminin, fibronectin and collagen types IV and V and lack of reactivity with different anti-collagen sera, *Eur. J. Biochem.*, 133, 31, 1983.
71. **Brandan, E. and Inestrosa, N. C.,** Binding of asymmetric forms of acetylcholinesterase to heparin, *Biochem. J.*, 221, 415, 1984.
72. **Méndez, B., Garrido, J., Maldonado, M., Jaksic, F., and Inestrosa, N. C.,** The electric organ of *Discopyge tschudii:* its innervated face and the biology of acetylcholinesterase, *Cell Mol. Neurobiol.*, 4, 125, 184.
73. **Sikorav, J.-L., Grassi, J., and Bon, S.,** Synthesis *in vitro* of precursors of the catalytic subunits of acetylcholinesterase from *Torpedo marmorata* and *Electrophorus electricus, Eur. J. Biochem.*, 145, 519, 1984.
74. **Soreq, H., Zevin-Sonkin, D., Avni, A., Hall, L. M. C., and Spierer, P.,** A human acetylcholinesterase gene identified by homology to the Ace region of *Drosophila, Proc. Natl. Acad. Sci. U.S.A.*, 82, 1827, 1985.
75. **Gaston, S. M., Tamkun, J., Burden, S., and Robbins, P. W.,** *Torpedo* acetylcholinesterase: isolation of cDNA clones with antibodies raised against an AChE active site synthetic decapeptide, *J. Cell Biochem.*, Suppl. 9B, 48, 1985.
76. **Hascall, V. C. and Hascall, G. K.,** Proteoglycans, in *Cell Biology of Extracellular Matrix,* Hay, E., Ed., Plenum Press, New York, 1981, 39.
77. **Lindahl, U. and Höök, M.,** Glycosaminoglycans and their binding to biological macromolecules, *Annu. Rev. Biochem.*, 47, 385, 1978.
78. **Hascall, V. C.,** Proteoglycans: structure and function, in *Biology of Carbohydrates,* Ginsburg, V., Ed., John Wiley & Sons, New York, 1981, 1.
79. **Hardingham, T.,** Proteoglycans: their structure, interactions and molecular organization in cartilage, *Biochem. Soc. Trans.*, 9, 489, 1981.

80. **DeLuca, S., Lohmander, L. S., Nilsson, B., Hascall, V. C., and Caplan, A. I.,** Proteoglycans from chick limb bud chondrocyte cultures, keratan sulfate and oligosaccharides which contain mannose and sialic acids, *J. Biol. Chem.*, 255, 6077, 1980.
81. **Lohmander, L. S., DeLuca, S., Nilsson, B., Hascall, V. C., Caputo, C. B., Kimura, J. H., and Heinegard, D.,** Oligosaccharides on proteoglycans from the Swarm rat chondrosarcoma, *J. Biol. Chem.*, 255, 6084, 1980.
82. **Upholt, W. B., Vertel, B. M., and Dorfman, A.,** Translation and characterization of messenger RNAs in differentiating chick cartilage, *Proc. Natl. Acad. Sci. U.S.A.*, 76, 4847, 1979.
83. **Treadwell, B. V., Mankin, D. P., Ho, P. K., and Mankin, H. J.,** Cell-free synthesis of cartilage proteins: partial identification of proteoglycan core and link proteins, *Biochemistry*, 19, 2269, 1980.
84. **Schwartz, N. B., Ostrowski, V., Brown, K. S., and Pratt, R. M.,** Defective PAPS-synthesis in epiphyseal cartilage from brachymorphic mice, *Biochem. Biophys. Res. Commun.*, 82, 173, 1978.
85. **Kanwar, Y. S. and Farquhar, M. G.,** Presence of heparan sulfate in the glomerular basement membrane, *Proc. Natl. Acad. Sci. U.S.A.*, 76, 1303, 1979.
86. **Kanwar, Y. S., Hascall, V. C., and Farquhar, M. G.,** Partial characterization of newly synthesized proteoglycans isolated from the glomerular basement membrane, *J. Cell Biol.*, 90, 527, 1981.
87. **Caulfield, J. P. and Farquhar, M. G.,** Loss of anionic sites from the glomerular basement membranes in aminonucleoside nephrosis, *Lab. Invest.*, 39, 505, 1978.
88. **Sampaio, L. O., Dietrich, C. P., and Filho, O. G.,** Changes in sulfated mucopolysaccharide composition of mammalian tissues during growth and in cancer tissues, *Biochim. Biophys. Acta*, 498, 123, 1977.
89. **Iozzo, R. V., Bolender, R. P., and Wight, T. N.,** Proteoglycan changes in the intercellular matrix of human colon carcinoma: an integrated biochemical and stereologic analysis, *Lab. Invest.*, 47, 124, 1982.
90. **Dietrich, C. P., Sampaio, L. O., and Toledo, O. M. S.,** Cell recognition and adhesiveness: a possible biological role for the sulfated mucopolysaccharides, *Biochem. Biophys. Res. Commun.*, 75, 329, 1977.
91. **Underhill, C. B. and Keller, J. M.,** A transformation-dependent difference in the heparan sulfate associated with the cell surface, *Biochem. Biophys. Res. Commun.*, 63, 448, 1975.
92. **Keller, K. L., Keller, J. M., and Joy, J. N.,** Heparan sulfates from Swiss mouse 3T3 and SV3T3 cells: O-sulfate difference, *Biochemistry*, 19, 2529, 1980.
93. **Winterbourne, D. J. and Mora, P. T.,** Cells selected for high tumorigenicity or transformed by simian virus 40 synthesize heparan sulfate with reduced degree of sulfation, *J. Biol. Chem.*, 256, 4310, 1981.
94. **Nakamura, N. and Kojima, J.,** Changes in charge density of heparan sulfate isolated from cancerous human liver tissue, *Cancer Res.*, 41, 278, 1981.
95. **Kramer, R. H., Vogel, K. G., and Nicolson, G. L.,** Solubilization and degradation of subendothelial matrix glycoproteins and proteoglycans by metastatic tumor cells, *J. Biol. Chem.*, 257, 2678, 1982.
96. **Nakajima, M., Irimura, T., DiFerrante, D., DeFerrante, N., and Nicolson, G. L.,** Heparan sulfate degradation: relation to tumor invasive and metastic properties of mouse B16 melanoma sublines, *Science*, 220, 611, 1983.
97. **Nakajima, J., Irimura, T., DiFerrante, N., and Nicolson, G. L.,** Metastic melanoma cell heparanase. Characterization of heparan sulfate degradation fragments produced by B16 melanoma endoglucuronidase, *J. Biol. Chem.*, 259, 2283, 1984.
98. **Ginsberg, M. H., Painter, R. G., Forsyth, J., Birdwell, C., and Plow, E. F.,** Thrombin increases expression of fibronectin antigen on the platelet surface, *Proc. Natl. Acad. Sci. U.S.A.*, 77, 1049, 1980.
99. **Rao, N. C., Barsky, S. H., Terranova, V. P., and Liotta, L.,** Isolation of a tumor cell laminin receptor, *Biochem. Biophys. Res. Commun.*, 111, 804, 1983.
100. **Malinoff, H. L. and Wicha, M. S.,** Isolation of a cell-surface receptor protein for laminin from murine fibrosarcoma cells, *J. Cell. Biol.*, 96, 1475, 1983.
101. **Chiang, T. M. and Kang, A. H.,** Isolation and purification of collagen 1(I) receptor from human platelet membrane, *J. Biol. Chem.*, 257, 7581, 1982.
102. **Lesot, H., Kuhl, U., and Von der Mark, K.,** A laminin receptor from skeletal muscle cells in culture, *EMBO J.*, 2, 861, 1983.
103. **Terranova, V. P., Rao, C. N., Kalebic, T., Margulies, I. M., and Liotta, L. A.,** Laminin receptor on human breast carcinoma cells, *Proc. Natl. Acad. Sci. U.S.A.*, 80, 444, 1983.
104. **Brandan, E., Maldonado, M., Garrido, J., and Inestrosa, N. C.,** Anchorage of collagen-tailed acetylcholinesterase to the extracellular matrix is mediated by heparan sulfate proteoglycans, *J. Cell Biol.*, 101, 985, 1985.
105. **Kjellén, L., Oldberg, A., and Höök, M.,** Cell surface heparan sulfate: mechanisms of proteoglycan-cell association, *J. Biol. Chem.*, 255, 10407, 1980.
106. **Rapraeger, A. C. and Bernfield, M.,** Heparan sulfate proteoglycans from mouse mammary epithelial cells. A putative membrane proteoglycan associates quantitatively with lipid vesicles, *J. Biol. Chem.*, 258, 3632, 1983.
107. **Hynes, R. O. and Destree, A. T.,** Relationship between fibronectin and actin, *Cell*, 15, 875, 1978.

108. **Kurkinen, M., Wartiovaara, J., and Vaheri, A.,** Cytochalasin-B releases a major surface-associated glycoprotein, fibronectin, from cultured fibroblasts, *Exp. Cell. Res.*, 111, 127, 1978.
109. **Brown, S. S., Malinoff, H. L., and Wicha, M. S.,** Connectin, cell surface protein that binds both laminin and actin, *Proc. Natl. Acad. Sci. U.S.A.*, 80, 5927, 1983.
110. **Woods, A., Höök, M., Kjellén, L., Smith, C. G., and Rees, D. A.,** Relationship of heparan sulfate proteoglycans to the cytoskeleton and extracellular matrix of cultured fibroblasts, *J. Cell Biol.*, 99, 1743, 1984.
111. **Rapraeger, A. C. and Bernfield, M.,** An integral membrane proteoglycan is capable of binding components of the cytoskeletal and the extracellular matrix, in *Extracellular Matrix*, Hawkes, S. and Wang, J. L., Eds., Academic Press, New York, 1982, 265.
112. **Beach, R. L., Burton, W. V., Hendricks, W. J., and Festoff, B. W.,** Extracellular matrix synthesis by skeletal muscle in culture. Proteins and effect of enzyme degradation, *J. Biol. Chem.*, 257, 11437, 1982.
113. **Hedman, K., Kurkinen, M., Alitalo, K., Vaheri, A., Johannsson, S., and Höök, M.,** Isolation of the pericellular matrix of human fibroblast cultures, *J. Cell. Biol.*, 81, 83, 1979.
114. **Inestrosa, N. C., Maldonado, M., and Garrido, J.,** Direct connection between extracellular basal lamina and cytoplasmic filaments, *J. Cell Biochem.*, Suppl. 8B, 295, 1984.
115. **Heuser, J. E. and Salpeter, S. R.,** Organization of acetylcholine receptors in quick frozen, deep-etched, and rotatory-replicated *Torpedo* postsynaptic membrane, *J. Cell Biol.*, 82, 150, 1979.
116. **Conti-Tronconi, B. M. and Raftery, M. A.,** The nicotinic cholinergic receptor: correlation of molecular structure with functional properties, *Annu. Rev. Biochem.*, 51, 491, 1982.
117. **Inestrosa, N. C., Silberstein, L., and Hall, Z. W.,** Appearance of synaptic molecules in aneural myotube cultures, *J. Supramol. Struct. Cell. Biochem.*, Suppl. 5, 297, 1981.
118. **Silberstein, L., Inestrosa, N. C., and Hall, Z. W.,** Aneural muscle cultures make synaptic basal lamina components, *Nature (London)*, 295, 143, 1982.
119. **Hall, Z. W., Silberstein, L., and Inestrosa, N. C.,** Synaptic basal lamina components made by a muscle cell line, in *Muscle Development: Molecular and Cellular Control*, Pearson, M. L. and Epstein, H. F., Eds., Cold Spring Harbor Laboratories, Cold Spring Harbor, N.Y., 1982, 459.
120. **Sanes, J. R. and Cheney, J. M.,** Lectin binding reveals a synapse-specific carbohydrate in skeletal muscle, *Nature (London)*, 300, 646, 1982.
121. **Labat-Robert J., Saitoh, T., Godeau, G., Robert, L., and Changeux, J.-P.,** Distribution of macromolecules from the intercellular matrix in the electroplaque of *Electrophorus electricus*, *FEBS Lett.*, 120, 259, 1980.
122. **Daniels, M. P., Vigny, M., Sonderegger, P., Bauer, H. C., and Vogel., Z.,** Association of laminin and other basement membrane components with regions of high acetylcholine receptor density on cultured myotubes, *Int. J. Dev. Neurosci.*, 2, 87, 1984.
123. **Hassell, J. R., Robey, P. G., Barrach, H. J., Wilczek, J., Rennard, S. I., and Martin, G. R.,** Isolation of a heparan sulfate-containing proteoglycan from basement membrane, *Proc. Natl. Acad. Sci. U.S.A.*, 77, 4494, 1980.
124. **Bayne, E. K., Anderson, M. J., and Fambrough, D. M.,** Extracellular matrix organization in developing muscle: correlation with acetylcholine receptor aggregates, *J. Cell Biol.*, 99, 1486, 1984.
125. **Kohler, G. and Milstein, C.,** Derivation of specific antibody-producing tissue culture and tumor lines by cell fusion, *Eur. J. Immunol.*, 6, 511, 1976.
126. **Anderson, M. J. and Fambrough, D. M.,** Aggregates of acetylcholine receptors are associated with plaques of a basal lamina heparan sulfate proteoglycan on the surface of skeletal muscle fibers, *J. Cell Biol.*, 97, 1396, 1983.
127. **Sanes, J. R. and Chiu, A. Y.,** The basal lamina of the neuromuscular junction, *Cold Spring Harbor Symp. Quant. Biol.*, 48, 667, 1983.
128. **Chiquet, M. and Fambrough, D. M.,** Chick myotendinous antigen. I. A monoclonal antibody as a marker for tendon and muscle morphogenesis, *J. Cell Biol.*, 98, 1926, 1984.
129. **Chiquet, M. and Fambrough, D. M.,** Chick myotendinous antigen. II. A novel extracellular glycoprotein complex consisting of large disulfide-linked subunits, *J. Cell Biol.*, 98, 1937, 1984.
130. **Bayne, E. K., Gardner, J., and Fambrough, D. M.,** Monoclonal antibodies to extracellular matrix antigens in chicken skeletal muscle, *Cold Spring Harbor Rep. Neurosci.*, 2, 259, 1981.
131. **Kefalides, N. A., Alper, R., and Clark, C. C.,** Biochemistry and metabolism of basement membrane, *Int. Rev. Cytol.*, 61, 167, 1979.
132. **Hedman, K., Johansson, S., Vartio, T., Kjellén, L., Vaheri, A., and Höök, M.,** Structure of the pericellular matrix: association of heparan and chondroitin sulfates with fibronectin-procollagen fibers, *Cell*, 28, 663, 1982.
133. **Laurie, G. W., Leblond, C. P., and Martin, G. R.,** Localization of type IV collagen, laminin, heparan sulfate proteoglycan and fibronectin to the basal lamina of basement membranes, *J. Cell Biol.*, 95, 340, 1982.

134. **Johansson, S. and Höök, M.,** Heparin enhances the rate of binding of fibronectin to collagen, *Biochem. J.,* 187, 521, 1980.
135. **Ruoslahti, E. and Engvall, E.,** Complexing of fibronectin glycosaminoglycans and collagen, *Biochim. Biophys. Acta,* 631, 350, 1980.
136. **Jilek, F. and Hörmann, H.,** Fibronectin (cold-insoluble globulin). IV. Influence of heparin and hyaluronic acid on the binding of native collagen, *Hoppe-Seyler's Z. Physiol. Chem.,* 360, 597, 1979.
137. **Perkins, M. E., Ji, T. H., and Hynes, R. O.,** Cross-linking of fibronectin to sulfated proteoglycans at the cell surface, *Cell,* 16, 941, 1979.
138. **Yamada, K. M., Kennedy, E. W., Kimata, K., and Pratt, R. M.,** Characterization of fibronectin interactions with glycosaminoglycans and identification of active proteolytic fragments, *J. Biol. Chem.,* 255, 6055, 1980.
139. **Woodley, D. T., Rao, C. N., Hassell, J. R., Liotta, L. A., Martin, G. R., and Kleinman, H. K.,** Interactions of basement membrane components, *Biochim. Biophys. Acta,* 761, 278, 1983.
140. **Kleinman, H. K., McGarvey, M. L., Hassell, J. R., and Martin, G. R.,** Formation of a supramolecular complex is involved in the reconstitution of basement membrane components, *Biochemistry,* 22, 4969, 1983.
141. **Hedman, K., Vartio, T., Johansson, S., Kjellén, L., Höök, M., Linker, A., Salonen, E. M., and Vaheri, A.,** Integrity of the pericellular fibronectin matrix of fibroblasts is independent of sulfated glycosaminoglycans, *EMBO J.,* 3, 581, 1984.
142. **Torres, J. C. and Inestrosa, N. C.,** Heparin solubilizes asymmetric acetylcholinesterase from rat neuromuscular junction, *FEBS Lett.,* 154, 265, 1983.
143. **Torres, J. C. and Inestrosa, N. C.,** Interaction of heparin with multimolecular aggregates of acetylcholinesterase, *Cell Mol. Neurobiol.,* 5, 303, 1985.
144. **Younkin, S. G., Rosenstein, C., Collins, P. L., and Rosenberry, T. L.,** Cellular localization of the molecular forms of acetylcholinesterase in rat diaphragm, *J. Biol. Chem.,* 257, 13630, 1982.
145. **Fernandez, H. L., Inestrosa, N. C., and Stiles, J. R.,** Subcellular localization of acetylcholinesterase molecular forms in endplate regions of adult mammalian skeletal muscle, *Neurochem. Res.,* 9, 1211, 1984.
146. **Inestrosa, N. C.,** 16S acetylcholinesterase of the extracellular matrix is assembled within mouse muscle cells in culture, *Biochem. J.,* 217, 377, 1984.
147. **Rotundo, R. L.,** Asymmetric acetylcholinesterase is assembled in the Golgi apparatus, *Proc. Natl. Acad. Sci. U.S.A.,* 81, 479, 1984.
148. **Ramirez, G., Gómez-Barriocanal, J., Barat, A., and Rodriguez-Borrajo, C.,** Two classes of collagen-tailed molecular forms of acetylcholinesterase, in *Cholinesterase: Fundamental and Applied Aspects,* Brzin, M., Sket, D., and Barnard, E. A., Eds., Walter de Gruyter, Berlin, 1984, 115.
149. **Barat, A., Gómez-Barriocanal, J., and Ramirez, G.,** Two classes of collagen-tailed, asymmetric molecular forms of acetylcholinesterase in skeletal muscle: differential effects of denervation, *Neurochem. Int.,* 6, 403, 1984.
150. **Brandan, E., Llona, I., and Inestrosa, N. C.,** Heparin-acetylcholinesterase interaction: specific detachment of class I-A forms, and binding of class I and II-A forms to heparin-agarose, *Neurochem. Int.,* in press, 1986.
151. **Eccles, J. C., Katz, B., and Kuffler, S. W.,** Effect of eserine on neuromuscular transmission, *J. Neurophysiol.,* 5, 211, 1942.
152. **Kuffler, S. W. and Nicholls, J. G.,** *From Neuron to Brain,* Sinauer Assoc., Sunderland, Mass., 1976, 1.
153. **Barla, R., Ramirez, B., Maldonado, M., and Inestrosa, N. C.,** Heparin solubilizes physiologically active acetylcholinesterase from the cat neuromuscular junction, *Mol. Physiol.,* in press, 1986.
154. **Inestrosa, N. C., Matthew, W. D., Reiness, C. G., Hall, Z. W., and Reichardt, L. F.,** Atypical distribution of asymmetric acetylcholinesterase in mutant PC12 pheochromocytoma cells lacking a cell surface heparan sulfate proteoglycan, *J. Neurochem.,* 45, 86, 1985.
155. **Stevens, R. L. and Austen, K. F.,** Effect of *p*-nitrophenyl-β-D-xyloside on proteoglycan and glycosaminoglycan biosynthesis in rat serosal mast cell cultures, *J. Biol. Chem.,* 257, 253, 1982.
156. **Kanwar, Y. S., Hascall, V. C., Jakubowski, M. L., and Gibbons, J. T.,** Effect of Β-D-xyloside on the glomerular proteoglycans. I. Biochemical studies, *J. Cell Biol.,* 99, 715, 1984.
157. **Brandan, E. and Inestrosa, N. C.,** The synaptic form of acetylcholinesterase binds to cell-surface heparan sulfate proteoglycans, *J. Neurosci. Res.,* 15, 185, 1986.
158. **Buonassisi, V. and Colburn, P.,** Biological significance of heparan sulfate proteoglycans, *Ann. N.Y. Acad. Sci.,* 401, 76, 1982.
159. **Buonassisi, V. and Colburn, P.,** Antibodies to the heparan sulfate proteoglycans synthesized by endothelial cell cultures, *Biochim. Biophys. Acta,* 760, 1, 1983.
160. **Kidokoro, Y., Anderson, M. J., and Gruener, R.,** Changes in synaptic potential properties during acetylcholine receptor accumulation and neurospecific interactions in *Xenopus* nerve-muscle cell culture, *Dev. Biol.,* 78, 464, 1980.

161. **Anderson, M. J. and Cohen, M. W.,** Nerve-induced and spontaneous redistribution of acetylcholine receptors on cultured muscle cells, *J. Physiol. (London),* 268, 757, 1977.
162. **Anderson, M. J., Cohen, M. W., and Zorychta, E.,** Effects of innervation on the distribution of acetylcholine receptors on cultured muscle cells, *J. Physiol. (London),* 268, 731, 1977.
163. **Ziskind-Conhaim, L. and Bennett, J. I.,** The effects of electrical inactivity and denervation on the distribution of acetylcholine receptors in developing rat muscle, *Dev. Biol.,* 90, 185, 1982.
164. **Ziskind-Conhaim, L., Inestrosa, N. C., and Hall, Z. W.,** Acetylcholinesterase is functional in embryonic rat muscle before its accumulation at the sites of nerve-muscle contact, *Dev. Biol.,* 103, 369, 1984.
165. **Kelly, A. M. and Zacks, S. I.,** The fine structure of motor endplate morphogenesis, *J. Cell Biol.,* 42, 154, 1969.
166. **Inestrosa, N. C., Reiness, C. G., Reichardt, L. F., and Hall, Z. W.,** Cellular localization of the molecular forms of acetylcholinesterase in rat pheochromocytoma PC12 cells treated with nerve growth factor, *J. Neurosci.,* 1, 1260, 1981.
167. **Heathcote, J. G., Bruns, R. R., and Orkin, R. W.,** Biosynthesis of sulphated macromolecules by rabbit lens epithelium. II. Relationship to basement membrane formation, *J. Cell Biol.,* 99, 861, 1984.
168. **Rieger, F., Koenig, J., and Vigny, M.,** Spontaneous contractile activity and the presence of the 16S form of acetylcholinesterase in rat muscle cells in culture: reversible suppressive action of tetrodotoxin, *Dev. Biol.,* 76, 358, 1980.
169. **Lawrence, J. C. and Catterall, W. A.,** Tetrodotoxin-insensitive sodium channels. Binding of polypeptide neurotoxins in primary cultures of rat muscle cells, *J. Biol. Chem.,* 256, 6223, 1981.
170. **Oh, T. H.,** Neurotrophic effects: characterization of the nerve extract that stimulates muscle development in culture, *Exp. Neurol.,* 46, 432, 1975.
171. **Jessell, T. M., Siegel, R. E., and Fischbach, G. D.,** Induction of acetylcholine receptors on cultured skeletal muscle by a factor extracted from brain and spinal cord, *Proc. Natl. Acad. Sci. U.S.A.,* 76, 5397, 1979.
172. **Salpeter, M. M., Spanton, S., Holley, K., and Podleski, T. R.,** Brain extract causes acetylcholine receptor redistribution which mimics some early events at developing neuromuscular junctions, *J. Cell Biol.,* 93, 417, 1982.
173. **Sanes, J. R. and Lawrence, J. C., Jr.,** Activity-dependent accumulation of basal lamina by cultured rat myotubes, *Dev. Biol.,* 97, 123, 1983.
174. **Sanes, J. R., Feldman, D. H., Cheney, J. M., and Lawrence, J. C., Jr.,** Brain extract induces synaptic characteristics in the basal lamina of cultured myotubes, *J. Neurosci.,* 4, 464, 1984.
175. **Koenig, J. and Vibny, M.,** Neural induction of the 16S acetylcholinesterase in muscle cell cultures, *Nature (London),* 271, 75, 1978.
176. **Moody-Corbett, F. and Cohen, M. W.,** Localization of cholinesterase at sites of high acetylcholine receptor density on embryonic amphibian muscle cells cultured without nerve, *J. Neurosci.,* 1, 596, 1981.
177. **Rubin, L. L., Gordon, A. S., and McMahan, U. J.,** Basal lamina fraction from the electric organ of *Torpedo* organizes acetylcholine receptors on cultured myotubes, *Soc. Neurosci.,* 6, 330, 1981.
178. **Rubin, L. L. and McMahan, U. J.,** Regeneration of the neuromuscular junction: steps toward defining the molecular basis of the interaction between nerve and muscle, in *Disorders of the Motor Unit,* Schotland, D. L., Ed., John Wiley & Sons, New York, 1982, 187.
179. **Godfrey, E. W., Nitkin, R. M., Wallace, B. G., Rubin, L. L., and McMahan, U. J.,** Components of *Torpedo* electric organ and muscle that cause aggregation of acetylcholine receptors on cultured muscle cells, *J. Cell Biol.,* 99, 615, 1984.
180. **Godfrey, E. W., Nitkin, R. M., Wallace, B. G., and McMahan, U. J.,** Identifying components of extracellular matrix that organize ACh receptors at synaptic sites on regenerating muscle cells, in *Developing and Regenerating Vertebrate Nervous Systems,* Coates, P. W., Markwald, R. R., and Kenny, A. D., Eds., Alan R. Liss, New York, 1983, 61.
181. **Nitkin, R. M., Wallace, B. G., Spira, M. E., Godfrey, E. W., and McMahan, U. J.,** Molecular components of the synaptic basal lamina that direct differentiation of regenerating neuromuscular junctions, *Cold Spring Harbor Symp. Quant. Biol.,* 48, 653, 1983.
182. **Kalcheim, C., Vogel, Z., and Duksin, D.,** Embryonic brain extract induces collagen biosynthesis in cultured muscle cells: involvement in acetylcholine receptor aggregation, *Proc. Natl. Acad. Sci. U.S.A.,* 79, 3077, 1982.
183. **Kalcheim, C., Duksin, D., and Vogel, Z.,** Involvement of collagen in the aggregation of acetylcholine receptors on cultured muscle cells, *J. Biol. Chem.,* 257, 12722, 1982.
184. **Vogel, Z., Christian, C. N., Vigny, M., Bauer, H. C., Sonderegger, P., and Daniels, M. P.,** Laminin induces acetylcholine receptor aggregation on cultured myotubes and enhances the receptor aggregation activity of a neuronal factor, *J. Neurosci.,* 3, 1058, 1983.
185. **Anderson, M. J., Klier, F. G., and Tanguay, K. E.,** Acetylcholine receptor aggregation parallels the deposition of a basal lamina proteoglycan during development of the neuromuscular junction, *J. Cell Biol.,* 99, 1769, 1984.

186. **Weinberg, C. B. and Hall, Z. W.**, Junctional forms of acetylcholinesterase restored at nerve-free endplates, *Dev. Biol.*, 68, 631, 1979.
187. **Montesano, R., Mossaz, A., Ryser, J. E., Orci, L., and Vassalli, P.**, Leukocyte interleukins induced cultured endothelial cells to produce a highly organized, glycosaminoglycan-rich pericellular matrix, *J. Cell Biol.*, 99, 1706, 1984.
188. **Greif, K. F. and Trenchard, H. I.**, Evidence for transsynaptic regulation of neuronal cell surface heparan sulfate proteoglycan in developing rat superior cervical ganglion, *J. Cell Biochem.*, 26, 127, 1984.
189. **Caroni, P., Carlson, S. S., Schweitzer, E., and Kelly, R. B.**, Presynaptic neurons may contribute a unique glycoprotein to the extracellular matrix at the synapse, *Nature (London)*, 314, 441, 1985.
190. **Engel, A. G., Lambert, E. H., and Gomez, M. R.**, A new myasthenic syndrome with endplate acetylcholinesterase deficiency, small nerve terminals, and reduced acetylcholine release, *Ann. Neurol.*, 1, 315, 1977.
191. **Lander, A. D., Fujii, D. K., and Reichardt, L. F.**, Laminin is associated with the neurite-outgrowth promoting factors found in conditioned media, *Proc. Natl. Acad. Sci. U.S.A.*, 82, 2183, 1985.
192. **Rorbach, D. H., Wagner, C. W., Star, V. L., Martin, G. R., Brown, K. S., and Yoon, J. W.**, Reduced synthesis of basement membrane heparan sulfate proteoglycan in streptozotocin-induced diabetic mice, *J. Biol. Chem.*, 258, 11672, 1983.
193. **Kaufman, D. L. and Tobin, A. J.**, Prospects for the isolation of genes for receptors and other proteins of pharmacological and neurobiological interest, in *Molecular and Chemical Characterization of Membrane Receptors*, Alan R. Liss, New York, 1984, 241.
194. **Scheller, R. H., Jackson, J. F., McAllister, L. B., Schwartz, J. H., Kandel, E. R., and Axel, R.**, A family of genes that codes for ELH, a neuropeptide eliciting a sterotypical pattern of behavior in Aplysia, *Cell*, 28, 707, 1982.
195. **Lewin, B. M.**, *Genes*, John Wiley & Sons, New York, 1985, 1.
196. **DosSantos, C. L. S., Villa, L. L., Sonohara, S., and Brentani, R. R.**, Construction of a cDNA clóne corresponding to mouse $\alpha 1$(IV) procollagen, *Nucleic Acid Res.*, 12, 2035, 1984.
197. **Tamkun, J. W., Schwarzbauer, J. E., and Hynes, R. O.**, A single rat fibronectin gene generates three different mRNAs by alternative splicing of a complex exon, *Proc. Natl. Acad. Sci. U.S.A.*, 81, 5140, 1984.
198. **Bourdon, M. A., Oldberg, A., Pierschbacher, M., and Ruoslahti, E.**, Molecular cloning and sequence analysis of a chondroitin sulfate proteoglycan cDNA, *Proc. Natl. Acad. Sci. U.S.A.*, 82, 1321, 1985.

Chapter 9

TROPHIC EFFECTS OF MUSCLE ON NERVE CELL SURVIVAL AND DEVELOPMENT

Guillermo R. Pilar, Stephen D. Meriney, and D. Bruce Gray

TABLE OF CONTENTS

I. INTRODUCTION

Synaptic contact and neurotransmitter interactions between a motoneuron and its target musculature are known to support cell survival and differentiation of both neurons and muscle. Anterograde effects are the influences of a neuron upon its target (e.g., muscle cells), while retrograde effects refer to target cell influences on presynaptic neuronal function (see Chapter 1). We have been interested in two areas of retrograde influences on neuronal development (1) target regulation of motoneuron survival and maturation in vivo, and (2) the use of in vitro models to examine target regulation of cholinergic neuronal differentiation. In this chapter we evaluate selected work relevant to these themes and incorporate some of our own recent observations.

First, we discuss those aspects of neuronal survival that appear to be dependent on interactions with the target. We also evaluate evidence in support of possible mechanisms underlying these interactions, including recent findings that implicate neuropeptides and possibly second messengers such as cGMP. Second, we review evidence suggesting that certain target cell factors mediate the regulation of complex characteristics of neuronal cholinergic maturation. These factors may act independently from those which are essential for neuronal survival and/or growth in culture. Results are presented implicating direct neuron-target cell contact (see Chapter 8), as well as soluble trophic substances, in the regulation of neuronal acetylcholine (ACh) chemosensitivity and synthetic responsiveness. Finally, we discuss the modulatory role that electrical activity may play in these neuron-target cell interactions.

II. RETROGRADE INFLUENCES ON NEURONAL SURVIVAL

Motor neurons send growth cones into the target region just before myotubes form from the target mesenchyme. There growth cones can release ACh either spontaneously or in response to a train of electrical stimulation.[12] Since growth cones in vivo make close contact with developing myotubes about 24 hr after myotube formation,[3,4] neurotransmitters may be available for interaction with target muscle receptors even at this early stage of development.

While target interactions do not seem to be involved in early neuronal outgrowth[5] and initial myotube development seems to be independent of neuronal innervation,[6,7] synaptic contact and functional neurotransmission seem to play an important role in many of the developmental consequences of early neuromuscular interactions. There are two developmental phases in neuromuscular synapse formation. During the first phase, many more motoneurons send processes into peripheral muscles (their targets) than survive to make mature synapses. This first period coincides with a large reduction in the number of motoneurons, and it is hypothesized that motoneurons compete among themselves for making contacts with the target cells.[8] The second phase, which follows this early neuronal competition, results in a fine-tuning of the previously formed synaptic connections as the neurons acquire mature characteristics. A conspicuous feature of the second phase is the elimination of multiple innervation at embryonic neuromuscular junctions, probably due to the retraction of axonal branches (see Chapter 10). Since synapse elimination has been reviewed elsewhere,[9] we will limit our discussion to the first phase of neuromuscular synapse formation and maturation. We describe experiments carried out in two neuromuscular systems in the chick; spinal column motor innervation of hindlimb skeletal muscles and parasympathetic ciliary ganglion innervation of smooth and skeletal muscles in the eye. Work in these preparations is complementary, and each preparation has specific advantages.

Cell death is a common occurrence in neuronal nuclei during a critical period in development. After completing migration and cell division, motoneurons have a limited time to make contact with their targets; those that do not make contact die. This was demonstrated

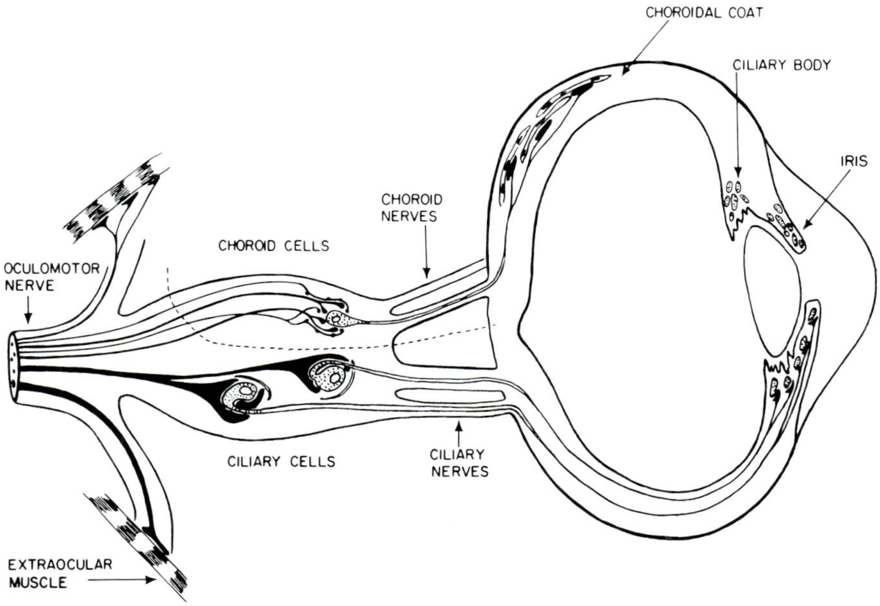

FIGURE 1. Schematic diagram of the ciliary ganglion and its peripheral targets in the chick eye: ciliary body and iris (striated muscle) and choroidal coat (smooth muscle). (From Landmesser, L. and Pilar, G., *J. Physiol. (London)*, 241, 715, 1974. With permission.)

in the avian ciliary ganglion,[10] a parasympathetic structure composed of two populations of cholinergic motoneurons: ciliary neurons project to the striated muscles of the iris and ciliary body; and choroid neurons innervate the vascular smooth muscle of the choroid coat of the eye (Figure 1). Before the period of normal cell death (day 8 of embryonic life), 6400 neurons compose the ganglion. During the subsequent formation of synapses, both populations decrease in number so that by embryonic day 14, about 3700 cells remain (1400 ciliary and 2300 choroid). Therefore, ciliary ganglion neurons decrease in number by about 50%. Quantitative estimates in the lumbosacral lateral motor column (LMC) of the spinal cord in the chick show that the number of motoneurons decreases from about 22,000 at embryonic day 6 (E6) to 13,000 at day 10 (E10). This phenomenon also occurs during the period of synapse formation in hindlimb muscles.[11,12]

A. Dependency of Neurons on Their Target Cells

Natural cell death is a physiologic process controlled by genetic, hormonal, and local factors. Some motoneurons may be preprogrammed to die. This genetic predisposition has been shown to occur in some invertebrates where cell lineage can be followed with great precision.[13] In the grasshopper, all cells in a lineage may not be necessary for every segment of the insect, and cell death may be programmed to allow for differences between segmental ganglia.[14] Hormones have been shown to influence this cell survival.[15] Presumably, cells that degenerate in one region are originally generated because they are an important part of the overall lineage, i.e., their sister cells may produce important progeny. Studies of cell lineage are difficult in higher vertebrates because of the complexity and size of the nervous system. Although we cannot rule out the possibility that vertebrate nervous systems exhibit a limited amount of programmed cell death, the experimental evidence presented in this chapter argues for a more flexible developmental strategy based on dynamic neuronal interactions, indicating that motoneurons are not genetically programmed to die. Two of the strongest pieces of evidence which argue against programmed neuronal death are that neurons

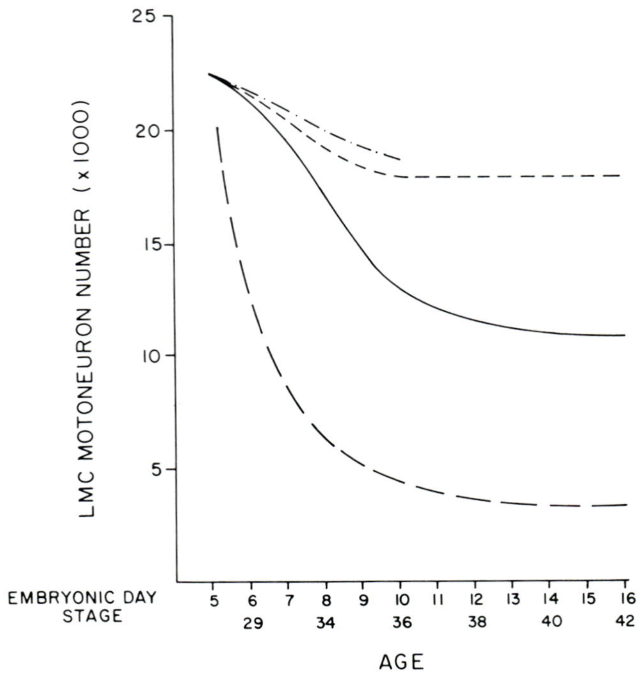

FIGURE 2. Number of motoneurons in the lumbosacral lateral motor column of the spinal cord during the normal cell death period. In control animals (—),[12] motoneuron number decreases by about 50% between stages 29 and 36. Peripherally deprived motoneurons degenerate extensively during the normal cell death period (——).[22] Embryos chronically treated with neuromuscular blockers (---)[16] or cGMP (-·-·-)[90] have significantly less motoneuron loss during the normal cell death period.

destined to die can be experimentally rescued in vivo[16] and that these same neurons can be cultured under appropriate conditions for extended periods of time.[17]

Another possibility is that during normal development, motoneurons are dependent on their target for survival. To test this hypothesis, motoneuron survival was assessed after removal of the peripheral target. Limb bud ablations in the frog *Rana pipiens* resulted in an 82% decrease in surviving LMC motoneurons.[18] Similarly, the percentage of surviving LMC motoneurons decreased dramatically in *Xenopus* tadpoles and chick embryos after similar limb bud ablations (Figure 2).[19-22] Early on investigators speculated that such decrease in LMC cell number was the result of influences on neuronal proliferation, migration, maturation, and degeneration. However, spinal motoneurons destined to die following limb bud removal in the chick develop in normal numbers, differentiate, and are indistinguishable biochemically and ultrastructurally from contralateral controls.[22] The cholinergic enzymes acetylcholinesterase (AChE) and choline acetyltransferase (CAT) were found to develop normally in peripherally deprived cells until the time of degeneration. These neurons were of normal size and sent axons out of the spinal cord which were capable of transporting horseradish peroxidase (HRP) retrogradely to their somas. Electron microscopic examination of the peripherally deprived spinal cord cells did not reveal any obvious differences when compared to control cells before the cell death period. In turn, these peripherally deprived cells do not continue to survive, presumably because they are unable to receive trophic support from their target muscles.

It is possible that the limb may exert an influence on developing motoneurons and that retrograde control of cell survival, or at least axonal guidance toward a source of trophic support, may begin even before direct neuromuscular contact. In fact, peripherally deprived neurons stop growing and do not usually reach the edge of the limb stump,[22] and may even grow in aberrant directions.[23] This dependency on the target was also demonstrated in the ciliary ganglion. Landmesser and Pilar examined ciliary ganglion development after unilateral removal of the optic vesicle before ciliary ganglion formation (St 16-20).[24,25] By the end of the cell death period, the operated side had fewer than 8% of the ganglion neurons normally present, while the contralateral ganglion was indistinguishable from ganglia of normal animals. Since the ablations were carried out before the time ganglion cells send axons into the periphery, axotomy was not a complicating factor. One of the earliest and most obvious effects of peripheral ablation was on ganglion cell migration. In most cases, two ganglia formed instead of one. Approximately half of the cells did not cease to migrate outside the remnant optic cup where the ganglion would normally form and instead formed a misplaced ganglion of both ciliary and choroid cells. Therefore, the pattern of ganglion cell migration was affected by peripheral removal. However, preganglionic axons formed normal calyciform and bouton-like synapses on both ciliary and choroid cells, indicating that information from target tissues is not required for ganglionic synapse formation. Peripherally deprived neurons proceeded to differentiate normally, sending functional axons into postganglionic nerves.

B. Selectivity of Innervation

Several theories have been formulated to explain normal cell death. Hughes[26] and recently Lamb[27] have suggested that motoneuron death occurs to eliminate those cells that have made inappropriate peripheral connections. Lamb has obtained some evidence supporting this hypothesis in his experiments on motoneurons of *Xenopus* tadpoles.[28,29] However, it was shown previously that normally developing ciliary ganglion neurons always innervate appropriate target muscles.[25] This selectivity occurs despite the fact that extraocular muscles are in close proximity. These observations have been extended to neurons innervating the chick hindlimb. If several spinal cord segments are surgically removed from the chick embryo, muscles destined to be innervated by the missing motoneurons do develop but are not innervated by the remaining motoneurons, which project only to appropriate muscles.[30] These results, together with those obtained by Landmesser,[31] suggest that chick motoneurons make peripheral projections with extreme precision. The few motoneurons that may project inappropriately and go undetected cannot account for the large number of cells that die during the normal cell death period. Moreover, when motoneurons are forced to project inappropriately, such as after large spinal cord reversals, they innervate foreign muscle but do not inevitably die.[32] It does not seem likely that motoneuron death in the ciliary ganglion and in the spinal cord functions to remove inappropriately projecting neurons. In the optic system, inappropriately projecting retinal ganglion cells refine their synaptic connections by axon collateral retraction rather than by cell death.[33]

C. Quantitative Motoneuron Matching with the Peripheral Target

Developing motoneurons may compete with each other for a limited amount of support from muscle fibers. In fact, the number of motoneurons which survive normal neuronal death in the LMC is closely correlated with the number of primary myotubes in the target musculature.[34,35] This supports the quantitative matching hypothesis for motoneuron survival, which proposes that motoneurons are overproduced to saturate the target and subsequently decrease in number to match the number of primary myotube clusters in the target musculature.[36] Since the two tissues develop independently early in embryogenesis, the initial overproduction of motoneurons may allow for variations in target size among individual animals.

Experimental manipulations that increase target size have been performed to test this hypothesis. Addition of a supernumerary limb to one side of the developing chick embryo appears to provide twice as much target for the LMC motoneurons to innervate on that side of the body.[37] Interestingly, the number of surviving motoneurons increases by only 10 to 30% rather than by 100%, which one would expect from doubling the target. This may be due to the lack of availability of the entire target tissue to innervating motoneurons. In chicks with supernumerary limbs, many motoneurons innervate foreign targets.[38,39] In addition, it is not known how many motoneurons have access to support from the supernumerary limb. Although observations in the extra limb of *Xenopus* indicate that it has been selectively innervated, the extent of innervation is less than in control limbs.[40] Therefore, the modest effect on cell survival observed after addition of an extra limb might be expected if the actual available target is less than what it seems to be.

To avoid these problems, Tanaka and Landmesser[35,41] attempted to confront developing motoneurons with a single limb which was larger or smaller than normal, so that motoneurons would have normal access to a limb of normal shape and orientation. Limbs were transplanted between two closely related avian species, the chick and the quail. Although quail have smaller limbs than chicks, their homologous limb muscles are innervated selectively by those chick motoneurons which are in a similar spinal cord position to that of the motoneurons which normally innervate that muscle in the quail.[41] Under these conditions the quantitative matching hypothesis of neuronal death was tested directly by comparing chick, quail, and chick-quail chimera LMC motoneuron numbers with primarily myotube numbers in the corresponding limbs. If motoneuron death is concerned only with achieving selective innervation,[42] the LMC cell number should not vary with limbs of different sizes. Alternatively, the quantitative matching hypothesis predicts that the number of chick motoneurons innervating a quail limb will be less than in control chick limbs, and the number of quail motoneurons innervating a chick limb will be more than in control quail limbs. In both cases, the number of motoneurons should correlate with the number of primary myotube clusters. After developmental size differences between chick and quail muscles were taken into account, the resultant motoneuron numbers agree with the predictions. In all experimental manipulations, the ratio of primary myotubes to motoneurons had a correlation coefficient of 0.996 (Figure 3). These findings support the idea that developing motoneurons may compete for trophic support in the target region and survive in proportion to the amount of that support.

D. A Test of the Competition Hypothesis

If motoneurons are not destined to die and do so only in a competitive struggle for limited target support, it should be possible to prevent cell death by decreasing the number of competing neurons. A carefully controlled test of the competition hypothesis was carried out by Pilar et al.[36] in the ciliary ganglion-target muscle system. Taking advantage of the characteristic axonal distribution in three postganglionic nerve branches, the number of competing ciliary neurons was experimentally reduced, leaving the target intact. Axotomizing all postganglionic nerves prior to the normal neuronal death period substantially reduces the number of ganglion cells. Selective axotomy of two of the three ciliary nerves allowed controlled reduction of competing ganglion neurons (Figure 4A). The number and distribution of ganglion neurons that project out of the three ciliary nerves were determined using retrogradely transported HRP after the cell death period (St 40). Cell bodies with axons projecting out of one specific nerve were labeled and counted in both control and experimental ganglia in which the other two nerves had been previously axotomized.

This axotomy-induced reduction in the number of competing motoneurons prevented about 40% of the normal cell death in ciliary neurons exiting from the third ciliary nerve. Furthermore, at least some of the motoneurons in the experimental ganglion expanded their peripheral innervation field to include target tissue normally innervated by the axotomized neurons (Figure 4B). Lamb[42] has argued that some of the axotomized nerves may have had

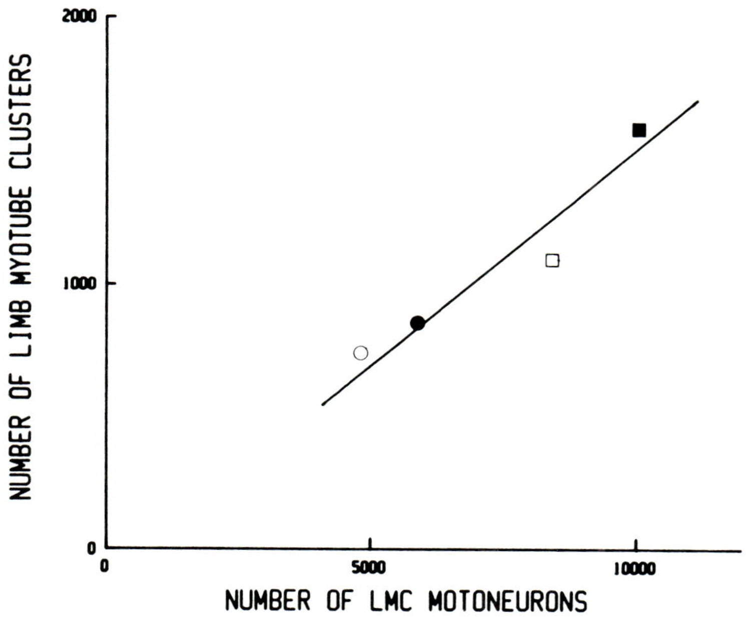

FIGURE 3. Mean number of myotube clusters during the cell death period plotted against the mean number of motoneurons after the cell death period. The regression line was calculated for quail host-chick limb (○), quail host-quail limb (●), chick host-quail limb (□), and chick host-chick limb (■). The correlation coefficient (r = 0.996) indicates a highly significant correlation.[35]

axon collaterals in more than one nerve branch which would normally retract later in development. He suggested that branch 3 neuronal survival had not increased, but contained branch 1 and 2 neurons that do not normally project out of branch 3 in the adult chick. Unfortunately, in drawing these conclusions, previous observations from electrophysiological recordings were overlooked. Those observations, made by Landmesser and Pilar,[43] demonstrate that axon collaterals cannot be detected in more than one ciliary nerve (Figure 3). This finding was later expanded by Dahm and Landmesser,[44] using retrogradely transported HRP and electrophysiological recordings. Therefore, the partial axotomy experiments provide strong evidence that competition restricts neuronal innervation to appropriate regions of the target and regulates the number of motoneurons that eventually survive. Interestingly, competition was found to decrease the rate of motoneuron maturation in the ciliary ganglion. Measurements of axon diameter, conduction velocity, and glial ensheathment revealed that axons from partially axotomized ganglia developed more rapidly than those from control ganglia.[36]

E. Neuronal Maturation

Successful competition for neuromuscular contact results in a number of biochemical changes that assure maturation and final differentiation of motoneurons. Formation of target neuromuscular junctions is known to trigger a specific increase in CAT activity in ciliary ganglion motoneurons.[45] Subsequently, these neurons acquire a high-affinity choline uptake system. In addition, iris nerve terminals of the ciliary ganglion develop the ability to respond to depolarization with accelerated ACh synthesis at hatching, when functional neuromuscular requirements are increased.[46] At the ultrastructural level these changes are apparent in the development of the newly organized rough endoplasmic reticulum (RER). This event was clearly shown in the ciliary ganglion, where a coordinated and accelerated formation of RER was synchronized with neuronal contact in the periphery.[47] RER can still develop inde-

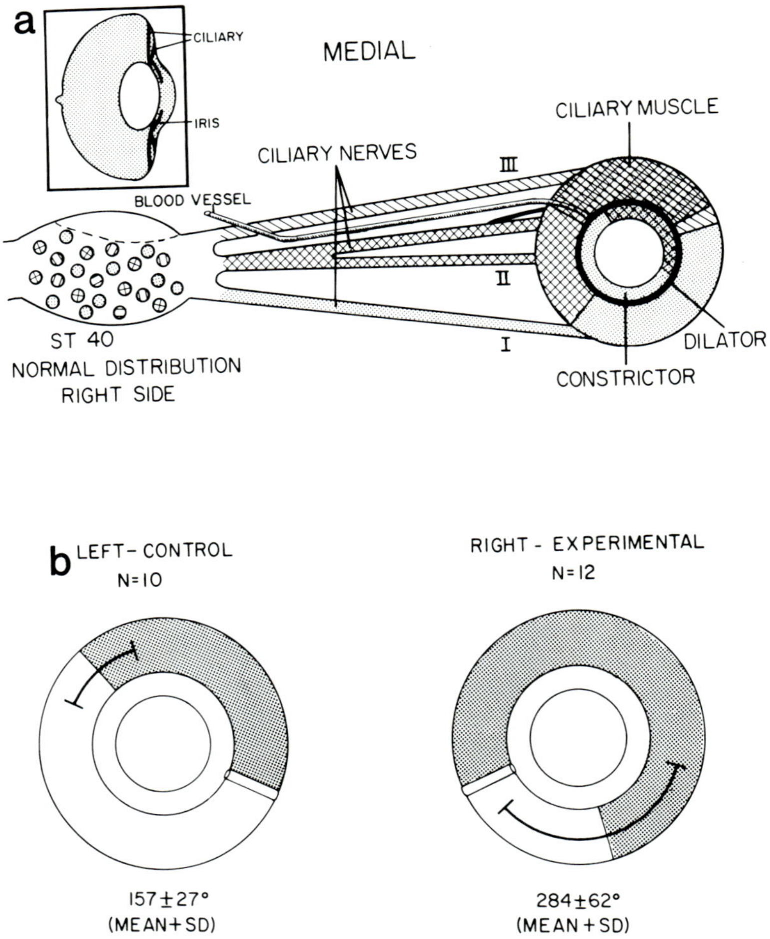

FIGURE 4. (a) Schematic diagram of the peripheral innervation pattern of the ciliary population. Each of the main ciliary nerve branches emanating from the ganglion innervates different parts of the peripheral target. Branch I (stippled) innervates the iris constrictor and the lateral portion of the ciliary muscle, branch II (cross-hatched) innervates intermediate areas of the ciliary muscle as well as the iris constrictor, and branch III (hatched) innervates only the medial portion of the ciliary muscle. A small branch (solid) emerging from branch II innervates the iris dilator. The cell bodies that project out of reach of these nerves are not localized, but distributed throughout the ciliary portion of the ganglion. (Inset) Sagittal (bisecting the lens) section of the eye showing the position of the intrinsic eye muscles. (b) Expansion of branch III innervation territory in partially axotomized preparations. The mean extent of the peripheral innervation territory of branch III is shown schematically for control and partially axotomized (experimental) preparations by the stippling of the ciliary muscle extending from the main blood vessel dorsolaterally. (From Pilar, G., Landmesser, L., and Burstein, L., *J. Neurophysiol.*, 43, 233, 1980. With permission.)

pendently of neuron-target cell contact, but at a much slower rate. The accelerated RER organization is an indication that all ciliary ganglion cells have made contact with the target before normal cell death occurs.[47]

Two types of neuronal death can be observed in the developing ciliary ganglion, depending on the presence of peripheral target. The type of cell death observed in the ciliary ganglion when the target is present is characterized by dilation of the RER (the cells look similar to a section of Swiss cheese). This type of death was termed cytoplasmic.[48] In the peripherally deprived ciliary ganglion (target tissue removed) the initial ultrastructural changes occurred in the nucleus (termed nuclear cell death). The latter phenomenon was interpreted as nuclear inactivation which subsequently results in a cessation of transcription.[48] Unlike in the ciliary ganglion, during normal motoneuron cell death in the lumbar spinal cord, both types of ultrastructural cell death are observed.[49] However, in the spinal cord not all the growing nerves reach their muscular target in time to survive;[50] consequently, both types of ultrastructural cell death would be expected.

III. ANALYSIS OF FACTORS MEDIATING THE REGULATION OF CELL SURVIVAL IN VIVO

A. Neural and Muscular Activity Blockade

Since it is known that motoneurons depend on target cells for survival, it would be of interest to know if synaptic transmission is critical or if only contact with the target is sufficient. Motoneurons may be competing for either a finite number of synaptic sites or a survival substance(s) produced in limited amounts by the target. Although we cannot differentiate between these possibilities, many researchers favor the idea that the amount of a substance(s) produced or released by the target cell directly regulates neuronal survival. This is supported by many recent studies involving nerve growth factor (NGF), which has been shown to influence sensory and sympathetic neuron survival.[51-53] However, we feel that the importance of motoneuron terminal sprouting, synaptic contact, and synapse formation has not been equally emphasized in the context of cell survival and maturation (see Chapter 10).

Since many of the characteristics of the formation of motoneuron synapses with the target are dependent on neural or muscular activity, several experiments have been performed to test the influence of functional activity on cell death. Pittman and Oppenheim pharmacologically blocked neuromuscular transmission during the neuronal death period.[16] This treatment does not prevent the formation of neuromuscular contacts,[54-56] but dramatically reduces the amount of motoneuron death (see Figure 2). The degree of prevention of motoneuron death was not uniform across all lumbosacral segments: the largest effect was observed between segments LS-3 and LS-7, a moderate effect occurred between LS-7 and LS-9, and no change was recorded between LS-1 and LS-2.[16] The implications of this seemingly selective effect are unknown, but the uneven atrophy of limb muscles observed after treatment may provide an explanation.[57] When treatment was discontinued, cell death occurred in a normal fashion. Therefore, neuromuscular blockade delays neuronal death as long as the treatment is maintained. Oppenheim has conjectured that synaptic transmission acts to regulate the availability of a putative trophic substance through the activation of muscle fibers.[12,58]

The role played by muscular activity in the regulation of neuronal survival was further investigated by Oppenheim and Nuñez.[58] They electrically stimulated the thigh region of the chick hindlimb during the normal motoneuron cell death period and were able to moderately increase neuronal death in α-bungarotoxin (α-BTX) and ditubocurarine (dTB) - paralyzed animals. However, the stimulus paradigm used was not physiologic (suprathreshold stimulation every second for 12 to 16 hr), and a large number of embryos had various degrees of limb necrosis which may have been caused by excessive manipulations or electrolytic damage. Additional controls are certainly necessary to rule out artifactual results due to damaged motor axons, which could have a direct influence on cell survival. While these experiments support the hypothesis for neuromuscular activity, a causal relationship between cell death and a specific change in nerve or muscle characteristics does not clearly emerge. Subsequently, there have been a number of modifications of this hypothesis.

One of the most obvious effects of neuromuscular blockade is atrophy of the target tissue.[59,60] It is not known if atrophied muscle fibers produce more or less survival-promoting substances, and it has already been shown that using hormones to increase the size of muscle fibers does not cause an increase in neuronal survival.[61] Additionally, atrophied muscle fibers have increased numbers of ACh receptors (AChR), and a causal relationship between AChR number and neuronal survival has been proposed.[56,62] Muscle paralysis is also known to cause nerve terminal sprouting.[63] Conversely, the activation of muscle causes retraction of sprouting, localization of AChR, and reduction in the number of nerve terminals abutting on embryonic muscle (although this effect is observed after cell death).[63] Some of these characteristics have been analyzed in duck embryo muscles where, although the total number of endplates becomes reduced, the size of each endplate increases in area and is innervated by more nerve terminals than in control muscle fibers.[56]

Observations on neuronal survival after neuromuscular blockade have now been extended to the chick ciliary ganglion system. These experiments are novel because the target includes striated, as well as smooth, muscles and both nicotinic and muscarinic AChRs mediate synaptic responses. In turn, a complication in this system arises from the ontogeny of target receptor pharmacology in the iris. A brief explanation of AChR development in this system is needed for interpretation of the data presented below. Ciliary neurons project to the striated muscles of the iris and ciliary body, where early synaptic transmission (St 35-38) in the iris is mediated by muscarinic AChR. Later (St 38), newly inserted junctional nicotinic AChRs gradually predominate.[64] Muscarinic AChRs, however, remain in the extrasynaptic regions of the muscle. While neuromuscular pharmacology is known to be nicotinic in the ciliary body by St 40, it is unclear whether muscarinic AChRs are ever present in the ciliary body muscle. Choroid neurons project to smooth muscle of the choroid coat and make varicose synapses where transmission is mediated entirely by muscarinic AChR. In this system, neuronal death can be examined after differential pharmacological blockade of muscarinic (by atropine) and nicotinic (by dTB or α-BTX) AChR during the normal neuronal death period. Total neuron counts were made of both ciliary and choroid populations and compared with differential cell counts of ciliary neurons alone after their selective labeling with HRP.[65]

Because some of the drugs used in these studies act at both neuromuscular and ganglionic synapses, it is important to define the effects of ganglionic blockade alone. Chlorasondamine, a ganglionic blocker, decreases ciliary neuron survival by 25%.[66] In contrast, α-BTX increases ciliary cell survival by 90%. Since muscarinic AChRs predominate at ciliary neuromuscular junctions during the early portion of the neuronal death period, it is not surprising that atropine increases ciliary cell survival by 71%. Interestingly, dTB, which blocks both ganglionic and neuromuscular synapses, increases neuronal survival by 36%. This increase is smaller than that seen after atropine or α-BTX, perhaps due to the opposing consequences of both ganglionic and neuromuscular blockade.

Blockade of ganglionic activity during the cell death period influences survival in the choroid cell population, consistent with the hypothesis that ganglionic transmission blockade decreases cell survival. Chlorasondamine and dTB both block only ganglionic transmission in the choroid cell-smooth muscle system, and both decrease cell survival by about 25%. However, muscarinic AChR blockade with atropine administered daily decreases choroid cell survival by about 25%. Atropine does not block ganglionic transmission. Furthermore, daily nicotinic AChR blockade with α-BTX increases choroid cell survival by about 90% without affecting either ganglionic or neuromuscular transmission at the smooth muscle junction.

These surprising results raise several possibilities. Smooth muscles may not respond to neuromuscular blockade by supporting more neuronal survival, as striated muscles do. It is known, for example, that smooth muscles do not undergo atrophy after denervation,[67] but muscarinic AChR have been shown to up- or down-regulate in response to antagonists or agonists, respectively.[68,69] The possibility that atropine is directly toxic to choroid cells is unlikely, since it is used at concentrations that are not toxic to cultured cells.[70] In choroid neurons, there is a correlation between blockade of smooth muscle junctions and decreased neuronal survival. Functional activity at choroid neuromuscular junctions certainly has an opposite role, if any, to its role in regulating cell survival at striated neuromuscular synapses. Inasmuch as we are not aware of any interactions between muscarinic receptors and α-BTX, the successful rescue of choroid cells with α-BTX is even more intriguing. Tentatively we consider that the drugs may be influencing local cells (e.g., macrophages) that remove neural elements at strategically localized portions of the nerve terminals; and, alternatively, atropine acting on presynaptic receptors may inhibit axonal sprouting, whereas α-BTX induces more axonal sprouting of choroid nerves. In any event, the dissociation of activity and survival is clear.

Moreover, experiments in the trochlear nucleus suggest an activity-independent regulation of cell survival. Sohal et al.[71] used myasthenia gravis immunoglobulin (MGI) to delay trochlear motoneuron cell death in duck embryos. Interestingly, this treatment applied chronically did not produce visible muscle paralysis when MGI was injected into the egg. Another set of experiments with drugs that block action potential propagation has suggested that behavioral inactivity is not a sufficient condition for prevention of cell death. In *Xenopus* frog, where neuromuscular blockade effectively prevents cell death in the spinal cord,[72] anesthetic treatment with chlorbutanol (larval immersion) or procaine (silastic chip insertion into axial musculature) has no effect on cell death.[73] Why these treatments failed to prevent cell death is unclear. However, when Harris and McCaig used tetrodotoxin (TTX) to block activity in rat embryos a 30% increase in thoracic motoneuron survival was observed.[74] Therefore, the aforementioned experiments suggest that peripheral muscular or neuromuscular activity blockade may not always be sufficient to delay normal cell death. Also, while we have limited our discussion to consequences of activity blockade (with TTX or neuromuscular blockers) on neuromuscular interactions, it is also possible for the above-mentioned treatments to directly influence central activation of motoneurons. In fact, Landmesser and Szente[75] investigated the effects of daily neuromuscular blockade sufficient to prevent normal cell death on the activation patterns of hindlimb motoneuron pools. Both dTB and α-BTX blocked patterned bursting activity of spinal nerves. Several other manipulations with pharmacological agents that interfere with neuromuscular function have been performed that argue against activity as a sufficient condition for the delay of normal cell death. Application of eserine or carbachol to chick embryos produces neuromuscular paralysis, but decreases spinal motoneuron survival.[76] Hemicholinium-3, which inhibits ACh synthesis and consequently blocks neuromuscular transmission, prevents motoneuron cell death in the spinal cord,[76] but decreases ciliary ganglion cell survival.[77]

In summary, what was originally thought to be a simple hypothesis has turned out to be fraught with contradictory results and complications. Apparently, this has distracted our efforts to explain the mechanism of cell death by forcing us to try to understand the effects of various drugs on developing synaptic structures. Even if we were able to answer all questions concerning such pharmacological actions, we may still not be able to unravel the mechanisms underlying normal cell death. In spite of these complications, there appears to be a common denominator. All treatments which increase the survival of motoneurons also enhance the number or size of motor nerve terminals. Synaptic inactivity induces sprouting after muscle or nerve paralysis. Conversely, activity decreases polyneuronal innervation in embryonic muscle. Although MGI has been shown to increase cell survival without causing muscle paralysis,[71] an increase in nerve terminal sprouting has been observed after MGI treatment.[78] Pharmacological agents that cause muscle paralysis (*Botulinum* toxin and TTX) have also been shown to increase nerve terminal sprouting.[79,80] It would be interesting if choroid nerves of the ciliary ganglion respond to α-BTX treatment with increased sprouting and to atropine with decreased sprouting. Motor nerve terminal sprouting could then be a common denominator in the various treatments enhancing cell survival.[63] Terminal growth could be induced in a variety of activity-dependent and activity-independent manners.

There is good evidence that substances produced by the muscle can influence neuronal survival. For example, Manthorpe et al.[81,82] have purified a substance from the eye tissues which supports ciliary ganglion survival in vitro. Chronically paralyzed muscle fibers seem to be able to provide support to more spinal motoneurons than normal muscle fibers. Limb muscle homogenates from chronically paralyzed muscles (α-BTX) maintain more motoneuron survival in vitro than do control limb muscle homogenates.[83] Activity has also been shown to regulate other proteins in muscle cells. Denervation increases NGF production by the rat iris[84] and N-CAM (the neural cell adhesion molecule) expression on chick hindlimb muscle surfaces.[85,86] Therefore, it would not be surprising if activity regulates the production of a trophic substance(s) by the muscle cell as well (see Chapter 11). While there is no controversy over the existence of these trophic substances, all of the assays for cell survival have been performed using cultured neurons. It is imperative to assay purified trophic substances for influences on cell survival in the intact embryo. Some experiments along this line have provided promising results. For example, there has been a preliminary report of increased spinal motoneuron survival in the chick embryo after administration of crude muscle homogenates.[87]

B. Neuronal Messengers

1. Cyclic GMP

Motoneuron survival may require the production of an intracellular second messenger triggered by successful competition for trophic support in the target region. Weill[88] reported a four- to fivefold increase in cGMP, but not in cAMP, levels in spinal neurons both in vivo and in vitro after association with skeletal muscle. This increase coincided with neuromuscular junction formation and was specific for spinal cord neurons co-cultured with skeletal muscle, as glial cells could not be substituted for spinal cord neurons and fibroblasts could not replace muscle cells. The increase in cGMP persisted even when spinal cord neurons were co-cultured with muscle in the presence of TTX or dTB. Since neuromuscular contacts develop in the presence of both TTX and dTB,[54,56] these findings may indicate the production of a second messenger in response to synapse formation. Additionally, Weill showed that the neuronal increase in cGMP can be produced by muscle-conditioned media, presumably mediated by a diffusible muscle substance.[88] This does not necessarily imply that a diffusible substance from muscle occurs in vivo. Muscle cells in culture are known to secrete proteins which are associated in vivo with the muscle basal lamina.[89] It is possible that the substance secreted in vitro is associated with the basal lamina of the muscle in vivo; therefore, neuromuscular contact may be required for the response in vivo (see Chapter 8).

Since neuromuscular junction formation in spinal motoneurons coincides with both the onset of cell death and cGMP production, Weill and Greene[90] treated chick embryos with a membrane-permeable cGMP analogue (dibutyryl cGMP) in an attempt to influence normal motoneuron death. Daily injections of dibutyryl cGMP during the cell death period prevented about 58% of the normal cell death in the LMC of the spinal cord (Figure 2). Segmental analysis of cell survival revealed a uniform increase across all lumbosacral segments. This is in contrast to the unevenly distributed effect (described above) observed after neuromuscular junction blockade.[16] Neuronal cGMP may act as a second messenger for neuronal survival in an as yet undetermined manner. Presumably, dibutyryl cGMP circumvents the requirements for the muscle substance(s).

These observations could not be confirmed by Oppenheim,[91] who was unable to show any effect on cell survival following a similar protocol. The only difference between Oppenheim's procedure and that of Weill and Greene was the strain of chicks used. Weill claims that different strains may have different dose-response curves for the effect of dibutyryl cGMP on cell survival. Recently, Weill repeated the experiment using White Leghorn embryos, the same strain Oppenheim used.[92] At the dose previously used in the Harco ×

Partner embryo experiments (0.91 mg/day) dibutyryl cGMP had no effect on motoneuron survival. However, when the dose was decreased to 0.864 mg/day, a 53.5% increase in survival was observed. Weill claims that the dose reponse curve is biphasic, with the range of effective doses uncommonly narrow. Recently, dibutyryl cGMP administration has been shown to increase cell survival in the ciliary cell population of the ciliary ganglion.[93] Although the effect of different doses was not examined, this observation supports the hypothesis that cGMP can influence motoneuron survival.

2. Neuropeptides

There has recently been a surge in the discovery of neuropeptides. Some of these oligomeric molecules have been shown to act in many systems to regulate transmitter release.[94-96] They have also been implicated in a number of phenomena, including neuronal differentiation.[97] An interesting feature of neuroactive peptides is that they are derived from large precursor proteins and in many cases are produced by posttranscriptional processing involving proteolytic cleavages.[98] In some cases, peptides have been detected early during neural development, even before transmitter function is apparent.[99,100] In the avian ciliary ganglion, opiate peptides are present early, and their immunoreactivity is temporally correlated with the neuronal death period.[101] These observations prompted us to investigate whether opiates could mediate cell survival, possibly by activating an intracellular second messenger such as cGMP. It has been shown that opiates elevate cGMP levels in striatal slices and neuroblastoma.[102,103] Previously, Erichsen et al.[104] reported enkephalin-like immunoreactivity in presynaptic terminals of both ciliary and choroid neurons. Cultures of ciliary ganglion cells which lack preganglionic terminals also show enkephalin-like immunoreactivity.[105] These observations suggest that opiate peptides exist endogenously in vivo both pre- and postganglionically. To examine the possibility that opiates can influence cell death in the ciliary ganglion, Meriney et al.[106] administered morphine to developing chick embryo daily during the normal neuronal death period. This treatment dramatically increased neuronal survival in both populations of ciliary ganglion neurons (by about 90%), and the effect was completely blocked by naloxone (a specific blocker of opiate receptors). When morphine treatment was discontinued after the cell death period, neuronal degeneration increased toward control levels. Therefore, morphine appears to delay cell death only until the treatment is discontinued. This effect is similar to the ability of neuromuscular blocking agents to delay cell death in the spinal cord until discontinued. Interestingly, morphine does not affect spinal motoneuron survival during the normal cell death period. The influence of morphine on cell survival may be restricted to cells with endogenous opiate receptors.

To test the hypothesis that endogenous opiates normally regulate neuronal death in the ciliary ganglion, naloxone was administered daily during the cell death period.[106] Naloxone alone was unable to significantly influence cell number. This argues against the possibility that endogenous opiates regulate cell death. However, the hypothesis is still viable, since neither the complete identity of the pre- and postganglionic peptides nor the nature and ontogeny of opiate receptor appearance in this tissue have been determined. It is possible that morphine influences cell survival by modulating neurotransmission, as has been shown in many systems.[94-96] However, this seems unlikely, because iris muscle contractions elicited by nerve stimulation are not blocked in chicks treated chronically with morphine.[107] Choroid smooth muscle activation is only slightly blocked by morphine. Morphine has more pronounced effects on ganglionic transmission, as choroid ganglionic synapses are completely blocked by acute administration of this drug.[108] Pharmacological blockade of these synapses is thought to decrease neuronal survival. Therefore, it does not appear that neuromuscular blockade of the target muscles plays any role in the morphine-induced rescue of neurons in the ciliary ganglion. Whatever the role of opiates in the regulation of neuronal cell death, it is clear that morphine can dramatically influence cell survival in the ciliary ganglion.

IV. FACTORS INFLUENCING CHOLINERGIC METABOLISM IN VITRO

A. Cell Survival In Vitro

The search for substances that mediate motoneuron survival and differentiation is progressing at a rapid rate. Thus far, however, only NGF has been shown to be responsible for the epigenetic determination of survival of any neuron (in this case, sympathetic and sensory neurons). For cholinergic neurons it has not been possible to identify a source with a sufficient amount of survival-promoting activity. Recently, Barbin et al.[82] have purified a fairly potent agent from embryonic chicken eye, ciliary neuronotrophic factor (CNTF), a protein which supports the survival of cholinergic ciliary neurons in culture. Several other groups have identified similar proteins from various sources.[109-113] Since the topic of survival substances has been well reviewed elsewhere,[114-120] it is not surveyed here. Interestingly, CNTF-like activity has been shown to increase during the periods in which the innervating neurons are most dependent upon the presence of target for survival in vivo.[121]

Survival is not the only parameter that may be important in assessing the influence of target tissues upon neurons. These influences may be divided into many categories: survival dependent, growth promoting, neurite extending and promoting, CAT promoting, or more complex parameters such as neuronal chemosensitivity or characteristics of transmitter release. This classification is somewhat arbitrary, and some target-derived factors may belong to more than one of these categories.

Since motoneurons of both the autonomic and central nervous system (CNS) are predominantly cholinergic, CAT activity and ACh synthesis have been used as assays for motoneuron growth and differentiation. ACh is synthesized by the esterification of choline (usually derived from extracellular sources through a specific uptake system) with an acetyl group (derived from acetyl CoA). In addition, most cholinergic terminals possess a high-affinity choline uptake system and AChE, although neither is exclusively limited to cholinergic cells. To distinguish general growth from differentiation and amplification of cholinergic properties, parameters are often reported in terms of specific activity and in comparison with measures of general metabolic enzymes such as lactate dehydrogenase. The remainder of this chapter will concentrate on target properties that influence the differentiation of cholinergic maturation.

B. Cholinergic Metabolism in Sympathetic Postganglionic Cells

In any discussion of environmental effects upon cholinergic metabolism, the dramatic plasticity of cultured sympathetic neurons must be considered. The addition of conditioned medium from certain nonneuronal cells (e.g., embryonic heart cells) to cultured sympathetic neurons results in development of the cholinergic synthetic machinery in adrenergic neurons.[122-124] In this system perhaps a single substance from conditioned medium may be responsbile for the switch of predominant transmitter phenotypes from adrenergic to cholinergic. This conditioned medium substance has been shown to be a glycoprotein with a molecular weight of approximately 45 kdaltons.[125-127] A controversy over the character of this adrenergic-to-cholinergic switch exists between Patterson's group, who propose an all-or-none transition,[123] and Bunge's group, who allow for the existence of permanent dual-function neurons.[124] This disagreement, however, does not extend to the observation that the conditioned medium substance causes a 1000-fold increase in ACh synthesis in cells which would normally be adrenergic. These observations support the model that a specific protein secreted by either target cells or cells in a target area can profoundly alter transmitter phenotype in sympathetic ganglion neurons. In a less dramatic fashion, contact of sympathetic neurons with fixed heart cell membranes can also induce a cholinergic phenotype, suggesting that membrane contact may also convey information similar to that of the diffusible substance.[128] No such substance, however, is known to be able to alter cholinergic parasympathetic to adrenergic phenotypes, although such cells have been shown to contain certain catecholamine-synthesizing enzymes.[129,130]

C. Retrograde Effects on Cholinergic Motoneurons

1. CAT-Promoting Interactions

Giller et al.[131] observed that CAT activity was increased more than tenfold in mouse spinal cord cells upon co-culture with muscle cells. This induction could also be produced by adding conditioned media from myotube (fused myoblast) or myoblast cultures.[132,133] The levels of AChE, neuronal muscarinic AChR, and ACh content also appear to increase after exposure to conditioned media,[134] but to a much lesser degree than does CAT. These increases are assumed to be specific for cholinergic neurons, since other transmitter enzymes such as glutamic acid decarboxylase are not affected.

A much smaller increase in CAT activity (50%) was reported by Berg[135] for co-cultures of chick spinal cord cells and myotubes. However, as Brookes et al.[134] observed, the neuronal density of Berg's cultures was relatively high (3×10^5 cells per cm^2), and high cell density has been known to mask inductive cholinergic effects due to production of trophic substances by the neurons themselves. Although cultures of CNS tissues contain many kinds of cells, these effects were assumed to occur in existing cholinergic neurons rather than in a new subpopulation of neurons, since the cell number remained constant. Recently, Calof and Reichardt[17] successfully cultured purified chick spinal motoneurons after retrograde transport of a fluorescent marker and cell sorting. Although they did not look for changes in cholinergic metabolism, they isolated two substances from the myotube-conditioned medium; a survival- and a neurite-promoting substance which may be a laminin-like peptide.

Parasympathetic ciliary ganglion cell cultures have an obvious experimental advantage over sympathetic ganglia and CNS tissues cultures, because they consist of only two populations of cholinergic motoneurons. Addition of extracts of homogenized muscle or eye tissue increases CAT activity.[136,137] In addition, Nishi and Berg[138] found a major component of chick eye extract that increased CAT levels per neuron but had no effect on general growth as measured by either protein values or lactate dehydrogenase activity. This substance has a molecular weight of about 50 kdaltons and has a greater effect on neurons grown in physiological levels of potassium (3 m*M*) than on neurons grown in medium containing a much higher potassium concentration (26 m*M*). In turn, Nishi and Berg[138] and Kato and Rey[139] reported little or no difference in CAT activity and ACh synthesis between ciliary ganglion neurons cultured either alone or with muscle. This is not the case, however, when cultures are grown with 10% chick embryo extract in the presence of potassium at physiological concentrations.[140,141] In fact, co-culture of ciliary neurons with chick pectoral myotubes for 7 days doubled CAT activity and basal ACh synthesis, and maintained a higher level of neuronal chemosensitivity to applied ACh. It is important to note that these results were obtained with a growth medium with embryo extract containing saturating levels of agents responsible for survival and growth of these neurons in vitro. Thus, the presence of live myotubes and/or functional nerve-muscle contacts stimulated cholinergic development in vitro over and above that observed with soluble substances added to the growth medium.

2. Synthetic Responsiveness in the Ciliary Ganglion

Although levels of CAT activities reflect potential transmitter synthetic capacity, regulation of ACh synthesis in intact cells in vivo is normally thought to be limited by the availability of the substrate, choline.[142-144] This theory is supported by the fact that synthesis rates in the ciliary ganglion are two to three orders of magnitude less than would be expected from existing CAT levels and are greatly affected by changes in exogenous choline levels.[145,146] Although CAT levels and basal ACh synthesis rates in ciliary ganglion terminals in the iris increase exponentially during embryonic development, adult synaptic reliability occurs rather abruptly at hatching.[46] Synaptic reliability is described as the maintenance of the same level of electrically recorded postsynaptic responses after continuous presynaptic stimulation for 30 sec and can be used as a discrete index of synaptic maturity. In fact, the period of synaptic reliability coincides with the ability of the nerve terminal to accelerate ACh synthesis immediately after significant amounts of transmitter release. We have used well-characterized cultures of ganglionic neurons plus or minus striated muscle in a routine assay for ACh synthesis that is very sensitive to alterations in presynaptic function.[140] Neurons are derived from 9-day ciliary ganglia and either plated alone or co-cultured for 7 days with 11-day chick embryo pectoral muscle. ACh synthesis in neurons is measured by incorporation of ^{3}H-choline (8.4 Ci/mmol) into ACh during a 30 min incubation at 37°C in normal Tyrodes solution. This synthesis may be stimulated by a 10-min preincubation in high potassium (55 m*M*), which causes transmitter release and a decrease in ACh internal storage.

Our cultures of E9 ciliary ganglion neurons grown in 10% horse serum and 10% chick embryo extract do not display this synthetic responsiveness to a high-potassium (55 m*M*) preincubation period until 2 weeks in vitro. However, if these neurons are co-cultured with chick pectoral myotubes, ACh synthesis rates are practically doubled by a 10-min high-potassium preincubation period.[140] Co-culture also increases basal ACh synthesis and CAT levels to a similar degree. Thus, myotubes can exert cholinergic influences on cultured neurons beyond that of chick embryo extract both quantitatively (in overall rates) and qualitatively (in the pattern of responsiveness).

3. Contact-Mediated Effects

Addition of conditioned media from ciliary neuron cultures, muscle cultures, or co-cultures did not affect the neuronal synthetic responsiveness or basal ACh synthesis. These results suggest that maturation of ACh synthesis may be mediated via neuron-muscle contact. Contact-mediated maturation of transmitter systems by neuronal targets has also been described in the CNS. Prochiantz et al.[145] measured the onset and rate of high-affinity dopamine uptake in cultures of rat mesencephalic (hippocampal) dopaminergic neurons. When these neurons were co-cultured with rat striatal cells, their innervation target in vivo, the high-affinity uptake rate more than doubled. Subsequently, these effects were also obtained when crude membrane preparations from rat striatal cells were added to cultures of hippocampal neurons.[146] The source of this trophic activity was specific to striatal regions of the CNS and appeared to be developmentally regulated. A similar cell-contact model is supported by recent observations in our culture system. In these experiments E11 chick pectoral muscle was dissociated and cultured for 5 days. These cultures were then osmotically lysed in sterile distilled water for 2 hr and washed three times in normal growth medium. Ciliary ganglion neurons (E9) were then dissociated and plated upon the substrate of lysed myotube membrane remnants. As shown in Figure 5, osmotically lysed membrane remnants of myotube cultures were able to induce synthetic responsiveness in ciliary ganglion neurons, although not to as great an extent as live myotubes.[147,148,160] The difference in effect between live and lysed myotubes may well be due to the absence of either a soluble substance produced and released by intact myotubes or a membrane-associated agent induced by neuron-muscle contact (see Chapter 8). However, it is also possible that mechanical factors, such as poor attachment of membrane remnants to the collagen substrate, may play a role. This influence, however, was specific in that substrates of lysed fibroblast remnants, which also provided a complex and adhesive surface, were not able to induce synthetic responsiveness.

Table 1 shows the effects of co-culture with myotubes or myotube remnants upon cholinergic chemosensitivity of ciliary ganglion neurons cultured either for 1 week or for more than 2 weeks.[141,149] In cultures of neurons alone, the number of neurons displaying a postsynaptic response to applied ACh decreases steadily over 12 days in culture. Co-culture of ciliary ganglion neurons with myotubes prevents this decrease, as does co-culture with

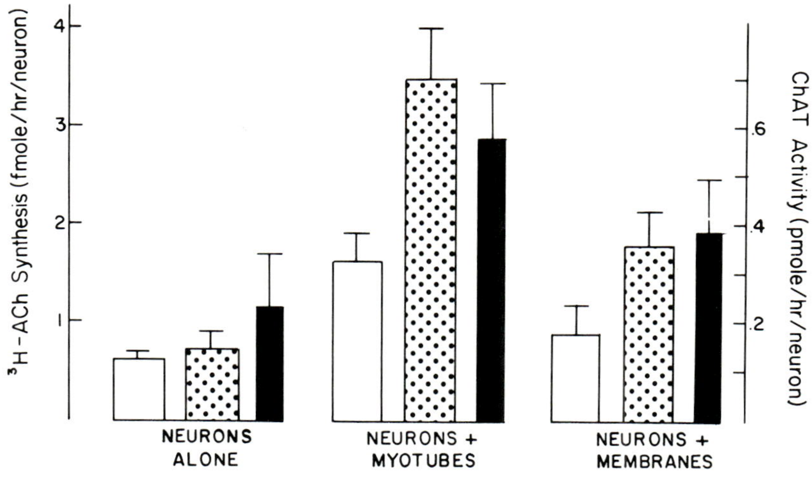

FIGURE 5. Effects of target on ACh synthesis in cultured ciliary ganglion neurons. Ciliary ganglia from 9-day chick embryos were dissociated and cultured, and ACh synthesis rates were determined as described in the text. Neurons were either cultured alone or co-cultured with myotubes or lysed myotube membrane remnants for 7 days. White bars represent basal ACh synthesis rates; stippled bars represent compensatory ACh synthesis following a 10-min princubation in high potassium (28 m*M*); black bars represent CAT levels (see scale at right). Values are the mean plus or minus the standard error of the mean (n = 6).

Table 1
EFFECTS OF TARGET AND AGE IN CULTURE ON CHOLINERGIC PROPERTIES OF CILIARY GANGLION NEURONS

Culture conditions	% Neurons responding to applied ACh (weeks)		% Increase in ACh synthesis after high-potassium preincubation (weeks)	
	1	2	1	2
Neurons alone	0	70	14	118
Neurons + myotubes	100	100	109	126
Neurons + myotube remnants	88	0	64	—
Neurons + fibroblast remnants	0	0	17	—

myotube remnants at 1 week in vitro. This is similar to the effect of myotube remnants on ACh synthetic responsiveness, which is also shown in Table 1. Membrane remnants of lysed fibroblasts or lysed myotubes were not able to maintain attachment with the collagen substrate after 8 to 10 days with regular feeding. As a result, estimates of neuronal survival became difficult during this period, and, therefore, measurements of ACh synthesis and chemosensitivity were not included. Cultures of neurons alone regain chemosensitivity and develop synthetic responsiveness if allowed to survive in culture for 2 weeks. Hence, the effect of target-derived tissue appears to be either an acceleration of maturation or maintenance of differentiation in which contact-mediated mechanisms play a prominent role during the first week in culture. These results do not completely rule out the possibility that diffusible substances, which may be secreted from live myotubes, bind to the myotube membrane upon lysis and mediate the remnant effect on synthetic responsiveness. However, no change in effectiveness was observed even after multiple washings of the membrane remnant substrate prior to neuron plating. The membrane-mediated effects are sensitive to extracellular trypsin digestion prior to lysis of myotubes, suggesting that the active components are exposed on the outer surface of the sarcolemma and contain a protein moiety.[150] Tuttle[151] has shown that a differential detergent extraction of myotube remnants can produce a soluble fraction which, when cross-linked to a collagen substrate, maintains neuronal ACh sensitivity in vitro.

The evidence cited above provides a strong case for a complex model of retrograde interactions between a developing neuron and its target environment. Not only are there separate factors for neurons with different transmitter phenotypes, but for each neuron there are multiple factors for survival, growth, and functional properties. These interactions appear to be mediated in some cases by both soluble substances and specific cell membrane contact. Although the mechanism by which these factors operate is unknown, much work has been directed toward elucidating how this interaction is regulated.

D. Role of Electrical Activity

As mentioned earlier, neuronal electrical activity and synaptic transmission with target tissues have been shown to affect neuronal development. Electrical excitability, synapse formation, and transmission have been reproduced in cultures of spinal cord and ciliary ganglion neurons.[152,153] Since such early synaptic transmission occurs between the developing neuron and target, the cell culture system provides an extremely useful model for testing the role of electrical activity in neuron-target interactions (see Chapter 4).

1. Spinal Cord

In dissociated spinal cord cultures, spontaneous electrical activity appears to be necessary for maturation and long-term survival.[154] In these experiments, chronic exposure to micromolar levels of TTX completely blocks spontaneous action potentials and synaptic activity in spinal cord cells in vitro. This inhibition results in decreased CAT levels and eventual death in spinal neurons. This effect is not seen in dissociated dorsal root ganglion cells or cerebral cortex explants.[155] The decreased spinal neuron survival in the presence of TTX occurred only if the entire media was exchanged during the culture feeding period. Cultures in which only 50% of growth media was withdrawn at feeding did not show sensitivity to chronic treatments of TTX.[156]

These results imply that a substance(s) secreted into the media somehow circumvents the requirements for spontaneous electrical activity. This hypothesis finds some support in recent experiments utilizing co-cultures of dissociated spinal cord neurons and dorsal root ganglion cells.[154] In this system, chronic addition of TTX also results in spinal neuron death and decreased CAT activity. Addition of conditioned media from nonpoisoned sister cultures blocks the effect of TTX addition on survival and CAT activity. This conditioned media, however, does not block the effect of TTX on electrical activity. Since conditioned media from TTX-poisoned sister cultures could not mimic this effect, it was concluded that substances were released into the medium from some source during electrical activity. Although this conclusion is somewhat speculative, it is apparent that something in the nonpoisoned conditioned media is able to affect the TTX poisoned neuron's ability to survive. This result is in contrast to models of trophic interaction (such as synapse repression)[157] where simultaneous activity of target and neuron is required.

In co-cultures of spinal neurons with skeletal muscle, chronic addition of TTX has also been shown to decrease neuronal survival.[132] This effect is apparently not due to a block of cholinergic transmission at the neuromuscular synapses that form in culture, since addition of α-BTX does not affect neuronal survival or muscle-induced increase in CAT activity. Although there may be alternative reasons for this result, an intriguing possibility is that a molecule other than ACh is released during transmission that either directly affects the neurons or causes a survival agent to be released by the muscle (see Chapter 6).

Table 2
EFFECTS OF CHRONIC EXPOSURE TO TTX ON ACh SYNTHESIS IN CULTURED CILIARY GANGLIA

Culture conditions	Basal ACh synthesis (fmol ACh/hr/neuron)	% Increase after potassium preincubation
	Neurons Alone	
Control	0.7 ± 0.12	10 ± 16
+ TTX (0.1 μ*M*)	0.8 ± 0.21	15 ± 9
+ 28 m*M* potassium	1.1 ± 0.28	16 ± 18
	Neurons + Myotubes	
Control	1.8 ± 0.31	115 ± 26
+ TTX (0.1 μ*M*)	2.2 ± 0.19	54 ± 34[a]
+ TTX (5 n*M*)	1.7 ± 0.27	65 ± 29[a]
+ 28 m*M* potassium	2.9 ± 0.49	105 ± 25
	Neurons + Remnants	
Control	1.1 ± 0.18	61 ± 18
+ TTX (0.1 μ*M*)	0.9 ± 0.15	53 ± 22

Note: Values are the mean plus or minus the standard error of the mean of at least six experiments for each culture condition.[a]

[a] Significantly different from respective controls ($p < 0.05$).

2. *Ciliary Ganglion*

Cultures of ciliary ganglion neurons either alone or with myotubes also display both spontaneous electrical activity and functional synaptic transmission.[138,141] In addition, these cultured neurons display membrane currents similar to those in developing and mature ganglia, including fast sodium channels sensitive to agents like TTX.[158] We were interested in examining the role TTX-sensitive activity played in mediating the effects of target on neuronal synthetic responsivness. Cultures of ciliary ganglion neurons alone and in co-culture with myotubes, or lysed myotube remnants, were chronically exposed to TTX at concentrations ranging from 1 n*M* to 1 μ*M*. Even at 1 μ*M* there was no effect of TTX on survival in either co-cultures or cultures of neurons alone.

In cultures of neurons alone, ACh synthesis rates under standard conditions and following high-potassium preincubation were not affected by chronic exposure to TTX (see Table 2). However, in muscle co-cultures, 30 to 50% of the stimulus-induced acceleration of ACh synthesis was sensitive to TTX at concentrations of at least 1 n*M*. Acute exposure of neurons to TTX during synthesis measurements has no effect on ACh synthesis or responsiveness. This is important because it may be argued that the co-culture effects were due to an effect of TTX on the synthesis assay. These experiments were repeated on cultures of neurons plated on membrane fragments from lysed myotubes. This substrate also supports the in vitro development of a compensatory ACh synthetic capacity following stimulation, although not as great as that in co-culture with live muscle as described above. Chronic exposure to TTX had no significant effect on ACh synthesis rates in these cultures.

Thus, TTX-sensitive electrical activity may be responsible for a portion of the compensatory increase in ACh synthesis trophically induced in ciliary ganglion neurons by co-culture with muscle cells. It is tempting to conclude that the other portion of this interaction is mediated by membrane-contact factors. However, other possibilities, such as the TTX-insensitive electrical activity, cannot be ruled out. Chronic exposure of cultures to a high-potassium (28 m*M* KCl) medium, which depolarizes cells and might be expected to increase inward calcium currents, increases ACh synthesis rates overall. In turn, this lowers the amount of compensatory ACh synthesis following stimulation when expressed as a percentage of basal rates.[159] This argues against a role of TTX-insensitive electrical activity or calcium flux in regulating ACh synthetic responsiveness.

V. SUMMARY AND CONCLUDING REMARKS

A general picture of the phenomena of normal cell death has emerged. It is now apparent where, why, and how these events occurs, i.e., in most motor nuclei, due to redundancy and by competition. It is also accepted that motoneuron survival is contingent upon retrograde influences from the target (muscle fibers). However, the specific nature of the target requirements and which aspects of this interaction are necessary for survival are still matters of debate. In both the LMC of the spinal cord and the ciliary ganglion, experimental evidence suggests that motoneurons are overproduced and decrease in number during synaptogenesis. Motoneuron pools appear to compete independently for trophic support from their target cells and survive to the extent of that support.

Experiments with blockers of striated neuromuscular synapses have led to the hypothesis that striated muscles chronically paralyzed during the normal cell death period provide trophic support to more motoneurons. Therefore, the extent of cell death during this period is reduced. It is believed that blockade of synaptic activity during this period allows muscle fibers to be receptive to more than one axonal innervation. However, MGI can prevent motoneuron death in the trochlear nucleus of the duck without blocking neuromuscular activity. To interpret these results, it is important to identify which factor(s) from peripheral target muscles are necessary for survival and how these influence neural development. We have presented several schematic possibilities by which target muscles might influence motoneuron survival during a critical period in development (also see Chapters 10 and 11). Activation of genes necessary to manufacture specific proteins important for neural development (such as secretory proteins or enzymes) may be initiated by intraneuronal signal(s). These hypothetical signal(s) may either be activated by a trophic substance (produced by the muscle, taken up, and retrogradely transported by the neuron) or by contact with the target muscle. A selective contact-mediated mechanism would necessitate a recognition between complementary membrane molecules of pre- and postsynaptic elements, which would result in a specific intraneuronal messenger. In this regard, it is worth noting that addition of cGMP can increase motoneuron survival when administered during the normal cell death period. This raises the possibility that cGMP is acting as a second messenger in response to synapse formation or cell contact. In support of this possibility are data showing that exposure of developing neurons to their targets increase neuronal cGMP independence of synaptic activity.

The work reviewed in the first part of this chapter attempts to identify the neuron-muscle interactions which may trigger intraneuronal signals important for cell survival. Whether these interactions are based on release of trophic substances or direct membrane-membrane recognition, the amount of neuronal contact with the target is important. More nerve terminals provide more surface area for uptake of a trophic substance, and more apposition of nerve and muscle membranes. Functional activity and the number of AChRs may not be essential for neuronal survival, but may regulate trophic or contact-mediated interactions between nerve and muscle. A number of variations on these mechanisms are viable alternatives. Most of the experiments described here can be explained using variations of these schemes.

These mechanisms, described in general terms, might regulate neuronal maturation as well as survival. The use of in vitro systems to analyze muscle-neuron trophic interactions has resulted in the identification of multiple factors which may regulate neuronal survival and development. Specific substances can be secreted by cultured target tissues which alter transmitter phenotype expression in sympathetic cells. In cultured neurons from spinal cord and ciliary ganglia, substances derived from target tissues can also stimulate specific cholinergic parameters such as CAT activity and basal ACh synthesis independent of general metabolic growth.

Although these interactions can be ascribed to soluble, releasable trophic substances, more complex cholinergic parameters, such as maintenance of neuronal chemosensitivity and ability to accelerate ACh synthesis on demand, may depend upon contact-mediated interactions between neurons and muscle membranes. These trophic effects appear to be regulated to some degree by TTX-sensitive electrical activity. Whether this regulation occurs via changes in membrane conductances or electrically stimulated release of a trophic substance(s), or is mediated transsynaptically is not clear.

Some extremely important questions which this chapter has not touched upon concern the precise intraneuronal signals or cellular machinery which is responsible for nerve cell survival and maturation as this cell competes for stable membership in a functioning nervous system.

ACKNOWLEDGMENTS

We thank L. Landmesser, L. Dahm, M. Ford, H. Fryer, C. Tate, and B. Wautlet for helpful discussions and critical review of the manuscript, and S. Putnam for assistance in the preparation of this manuscript. This work was supported by BNS 840581, NIH 10338, and the University of Connecticut Research Foundation.

REFERENCES

1. **Hume, R. I., Role, L. W., and Fischbach, G. D.,** Acetylcholine release from growth cones detected with patches of acetylcholine receptor-rich membrances, *Nature (London)*, 305, 632, 1983.
2. **Young, S. H. and Poo, M.-M.,** Spontaneous release of transmitter from growth cones of embryonic neurons, *Nature (London)*, 305, 634, 1983.
3. **Kelly, A. M. and Zacks, S. I.,** The fine structure of motor endplate morphogenesis, *J. Cell Biol.*, 42, 154, 1969.
4. **Lentz, T. L.,** Development of the neuromuscular junction. I. Cytological studies on the neuromuscular junction of differentiating muscle in the regenerating limb of the newt triturus, *J. Cell Biol.*, 42, 431, 1969.
5. **Chevallier, A.,** Role of the somitic mesoderm in the development of the thorax in bird embryos. II. Origin of thoracic and appendicular musculature, *J. Embryol. Exp. Morphol.* 49, 73, 1979.
6. **Whalen, R. G., Sell, S. M., Butler-Browne, G. S., Schwartz, K., Bouveret, P., and Pinset-Härström, I.,** Three myosine heavy-chain isozymes appear sequentially in rat muscle development, *Nature (London)*, 292, 805, 1981.
7. **Butler, J. and Cosmos, E.,** Differentiation of the avian latissimus dorsi primordium: analysis of fiber type expression using the myosin ATPase histochemistry reaction, *J. Exp. Zool.*, 218, 219, 1981.
8. **Cowan, W. M., Fawcett, J. W., O'Leary, D. D. M., and Stanfield, B. B.,** Regressive events in neurogenesis, *Science*, 225, 1258, 1984.
9. **Van Essen, D. C.,** Neuromuscular synapse elimination: structural, functional and mechanistic aspects, in *Neuronal Development*, Spitzer, N. C., Ed., Plenum Press, New York, 1982, 333.
10. **Landmesser, L. and Pilar, G.,** Synaptic transmission and cell death during normal ganglionic development, *J. Physiol. (London)*, 241, 737, 1974.
11. **Oppenheim, R. W. and Chu-Wang, I.-W.,** Spontaneous cell death of spinal motoneurons following peripheral innervation in the chick embryo, *Brain Res.*, 125, 154, 1977.
12. **Pittman, R. and Oppenheim, R. W.,** Cell death of motoneurons in the chick embryo spinal cord. IV. Evidence that a functional neuromuscular interaction is involved in the regulation of naturally occurring cell death and the stabilization of synapses, *J. Comp. Neurol.*, 187, 425, 1979.
13. **Truman, J. W.,** Cell death in invertebrate nervous systems, *Annu. Rev. Neurosci.*, 7, 171, 1984.
14. **Goodman, C. S. and Bate, M.,** Neuronal development in the grasshopper, *Trends Neurosci.*, 4, 163, 1981.

15. **Truman, J. W. and Schwartz, L. M.,** Programmed death in the nervous system of a moth, *Trends Neurosci.* 5, 270, 1982.
16. **Pittman, R. H. and Oppenheim, R. W.,** Neuromuscular blockage increases motoneuron survival during normal cell death in the chick embryo, *Nature (London)*, 271, 364, 1978.
17. **Calof, A. and Reichardt, L. F.,** Motoneurons purified by cell sorting respond to two distinct activities in myotube-conditioned medium, *Dev. Biol.*, 106, 194, 1984.
18. **Beaudoin, A. R.,** The development of lateral motor column cells in the lumbo-sacral cord in *Rana pipiens*. I. Normal development and development following unilateral limb ablations, *Anat. Rec.*, 121, 81, 1955.
19. **Prestige, M. C.,** The control of cell number in the lumbar ventral horns during the development of *Xenopus laevis* tadpoles, *J. Embryol. Exp. Morphol.*, 18, 359, 1967.
20. **Hamburger, V.,** The effects of wing bud extirpation on the development of the central nervous system in chick embryos, *J. Exp. Zool.*, 68, 449, 1934.
21. **Hamburger, V.,** Regression versus peripheral control of differentiation in motor hypoplasia, *Am. J. Anat.*, 102, 365, 1958.
22. **Oppenheim, R. W., Chu-Wang, I.-W., and Maderdrut, J. L.,** Cell death of motoneurons in the chick embryo spinal cord. III. The differentiation of motoneurons prior to their induced degeneration following limb bud removal, *J. Comp. Neurol.*, 177, 87, 1978.
23. **Tosney, K. and Landmesser, L.,** Pattern and specificity of axonal outgrowth following varying degrees of chick limb bud ablation, *J. Neurosci.* 4, 2518, 1984.
24. **Landmesser, L. and Pilar, G.,** Synapse formation during embryogenesis on ganglion cells lacking a periphery, *J. Physiol. (London)*, 241, 715, 1974.
25. **Landmesser, L. and Pilar, G.,** Fate of ganglionic synapses and ganglion cell axons during normal and induced cell death, *J. Cell Biol.*, 68, 357, 1976.
26. **Hughes, A. F. W.,** A quantitative study of the development of the nerves in the hindlimb of the *Eleutherodactylus martinicensus*, *J. Embryol. Exp. Morphol.*, 13, 9, 1965.
27. **Lamb, A. H.,** Selective bilateral motor innervation in *Xenopus* tadpoles with one hind limb, *J. Embryol. Exp. Morphol.*, 65, 149, 1981.
28. **Lamb, A. H.,** Motoneuron counts in *Xenopus* frogs reared with one bilaterally innervated hindlimb, *Nature (London)*, 284, 347, 1980.
29. **Denton, C. J., Lamb, A. H., Wilson, P., and Mark, R. F.,** Innervation patterns of muscles of one-legged *Xenopus laevis* supplied by motoneurons from both sides of the spinal cord, *Dev. Brain Res.*, 17, 85, 1985.
30. **Lance-Jones, C. and Landmesser, L.,** Motoneuron projection patterns in the embryonic chick limb following partial deletions of the spinal cord, *J. Physiol. (London)*, 302, 559, 1980.
31. **Landmesser, L.,** The generation of neuromuscular specificity, *Annu. Rev. Neurosci.*, 3, 279, 1980.
32. **Lance-Jones, C. and Landmesser, L.,** Motoneuron projection patterns in the chick hindlimb following early partial reversals of the spinal cord, *J. Physiol. (London)*, 302, 581, 1980.
33. **O'Leary, D. D. M. and Cowan, W. M.,** Further studies on the development of the isthmo-optic nucleus with special reference to the occurrence and fate of ectopic and ipsilaterally projecting neurons, *J. Comp. Neurol.*, 212, 399, 1982.
34. **McLennan, I. S.,** Size of motoneuron pool may be related to number of myotubes in developing muscle, *Dev. Biol.*, 92, 263, 1982.
35. **Tanaka, H. and Landmesser, L.,** Cell death of lumbosacral motoneurons in chick, quail, and chick-quail chimera embryos: a test of the quantitative matching hypothesis of neuronal cell death, *J. Neurosci.*, in press.
36. **Pilar, G., Landmesser, L., and Burstein, L.,** Competition for survival among developing ciliary ganglion cells, *J. Neurophysiol.*, 43, 233, 1980.
37. **Hollyday, M. and Hamburger, V.,** Reduction of the naturally occurring motor neuron loss by enlargement of the periphery, *J. Comp. Neurol.*, 170, 311, 1976.
38. **Morris, D. G.,** Development of functional motor innervation in supernumerary hindlimbs of the chick embryo, *J. Neurophysiol.*, 41, 1450, 1978.
39. **Hollyday, M., Hamburger, V., and Farris, J. M. G.,** Localization of motor neuron pools supplying identified muscles in normal and supernumerary legs of chick embryos, *Proc. Natl. Acad. Sci. U.S.A.*, 74, 3582, 1977.
40. **Hollyday, M. and Mendell, L.,** Analysis of moving supernumerary limbs of *Xenopus laevis*, *Exp. Neurol.*, 51, 316, 1976.
41. **Tanaka, H. and Landmesser, L.,** Interspecies selective motoneuron projection patterns in chick-quail chimeras, *J. Neurosci.*, in press.
42. **Lamb, A. H.,** Motoneuron death in the embryo, *Crit. Rev. Clin. Neurobiol.*, 1(2), 141, 1984.
43. **Landmesser, L. and Pilar, G.,** The onset and development of transmission in the chick ciliary ganglion, *J. Physiol. (London)*, 222, 691, 1972.

44. **Dahm, L. and Landmesser, L.,** The distribution of chick ciliary axon collaterals within individual ciliary nerves with respect to competition and cell death, in preparation, 1988.
45. **Chiapinelli, V., Giacobini, E., Pilar, G., and Uchimura, H.,** Induction of cholinergic enzymes in the chick ciliary ganglion and iris muscle cells during synapse formation, *J. Physiol. (London)*, 257, 749, 1976.
46. **Pilar, G., Tuttle, J., and Vacca, K.,** Functional maturation of motor nerve terminals in the avian iris: ultrastructure, transmitter metabolism and synaptic reliability, *J. Physiol. (London)*, 321, 175, 1981.
47. **Landmesser, L. and Pilar, G.,** Interactions between neurons and their targets during *in vivo* synaptogenesis, *Fed. Proc.*, 37, 2016, 1978.
48. **Pilar, G. and Landmesser, L.,** Ultrastructural differences during embryonic cell death in normal and peripherally deprived ciliary ganglia, *J. Cell Biol.*, 68, 339, 1976.
49. **Chu-Wang, I.-W. and Oppenheim, R. W.,** Cell death of motoneurons in the chick embryo spinal cord. I. A light and electron microscopic study of naturally occurring and induced cell loss during development, *J. Comp. Neurol.*, 177, 33, 1978.
50. **Lance-Jones, C. and Landmesser, L.,** unpublished, 1986.
51. **Levi-Montalcini, R. and Angeletti, P. V.,** Essential role of nerve growth factor in the survival and maintenance of dissociated sensory and sympathetic embryonic nerve cells *in vitro, Dev. Biol.*, 7, 653, 1963.
52. **Levi-Montalcini, R.,** The nerve growth factor. Its role in growth, differentiation and function of the sympathetic adrenergic neurons, *Prog. Brain Res.*, 45, 235, 1976.
53. **Hamburger, V. and Yip, J. W.,** Reduction of experimentally induced neuronal death in spinal ganglia of the chick embryo by nerve growth factor, *J. Neurosci.*, 4, 767, 1984.
54. **Freeman, S. S., Engle, A. G., and Drachman, D. B.,** Experimental acetylcholine blockade of the neuromuscular junction. Effects on endplate and muscle fiber ultrastructure, *Ann. N.Y. Acad. Sci.*, 274, 46, 1976.
55. **Anderson, M. J. and Cohen, M. W.,** Nerve-induced and spontaneous redistribution of acetylcholine receptors on cultured muscle cells, *J. Physiol. (London)*, 268, 757, 1977.
56. **Sohal, G. S., Creazzo, T. C., and Oblak, T. G.,** Effects of chronic paralysis with α-bungarotoxin on development of innervation, *Exp. Neurol.*, 66, 619, 1979.
57. **Landmesser, L.,** personal communication, 1986.
58. **Oppenheim, R. W. and Nuñez, R.,** Electrical stimulation of hindlimb increases neuronal cell death in chick embryo, *Nature (London)*, 295, 57, 1982.
59. **Roufa, D. and Martinosi, A. N.,** Effect of curare on the development of chick embryo skeletal muscle *in ovo, Biochem. Pharmacol.*, 30, 1501, 1981.
60. **Harris, A. J.,** Embryonic growth and innervation of rat skeletal muscles. I. Neural regulation of muscle fiber numbers, *Philos. Trans. R. Soc. London*, 293, 257, 1981.
61. **Sperry, D. G. and Grobstein, P.,** Regulation of neuron numbers in *Xenopus laevis:* effects of hormonal manipulation altering size at metamorphosis, *J. Comp. Neurol.*, 232, 287, 1985.
62. **Creazzo, T. L., and Sohal, G. S.,** Neural control of embryonic acetylcholine receptor and skeletal muscle, *Cell Tissue Res.*, 228, 1, 1983.
63. **Brown, M. C., Holland, R. L., and Hopkins, W. G.,** Motor nerve sprouting, *Annu. Rev. Neurosci.*, 4, 17, 1981.
64. **Pilar, G., Nuñez, R., McLennan, I., and Meriney, S. D.,** Muscarinic and nicotinic synaptic activation of the developing chicken iris, *J. Neurosci.*, in press.
65. **Meriney, S. D., Pilar, G., Ogawa, M., and Nuñez, R.,** Differential neuronal survival in the avian ciliary ganglion after chronic acetylcholine receptor blockade, *J. Neurosci.*, in press.
66. **Wright, L.,** Cell survival in chick embryo ciliary ganglia is reduced by chronic ganglionic blockade, *Dev. Brain Res.*, 1, 283, 1981.
67. **Sachs, D. I., Kloog, Y., Korczyn, A. D., Heron, D. S., and Sokolovsky, M.,** Denervation, supersensitivity and muscarinic receptors in the cat iris, *Biochem. Pharmacol.*, 28, 1513, 1979.
68. **Rush, R. A., Crouch, M. F., Morris, C. P., and Gannon, B. J.,** Neural regulation of muscarinic receptors in chick expansor secundariorum muscle, *Nature (London)*, 296, 569, 1982.
69. **Klein, W. L., Nathanson, N., and Nirenberg, M.,** Muscarinic acetylcholine receptor regulation by accelerated rate of receptor loss, *Biochem. Biophys. Res. Commun.*, 90, 506, 1979.
70. **Pilar, G.,** unpublished, 1986.
71. **Sohal, G. S., Leshner, R. T., and Swift, T. R.,** Myasthenia gravis immunoglobulin augments motor neuron survival without producing muscle paralysis, *Muscle Nerve*, 6, 122, 1983.
72. **Olek, A. J.,** Effects of alpha and beta bungarotoxin on motor neuron loss in *Xenopus laevis, Neuroscience*, 5, 1557, 1980.
73. **Olek, A. J. and Edwards, C.,** Effects of anesthetic treatment on motor neuron death in *Xenopus, Brain Res.*, 191, 483, 1980.

74. **Harris, A. J. and McCaig, C. D.,** Motoneuron death and motor unit size during embryonic development of the rat, *J. Neurosci.*, 4, 13, 1984.
75. **Landmesser, L. and Szente, M.,** Activation patterns of embryonic chick hindlimb muscles following blockade of activity and motoneuron cell death, *J. Physiol. (London)*, in press.
76. **Oppenheim, R. W. and Chu-Wang, I.-W.,** Aspects of naturally-occurring motoneuron death in the chick spinal cord during embryonic development, in *Somatic and Autonomic Nerve-Muscle Interactions*, Burnstock, G., O'Brien, R., and Vrbova, G., Eds., Elsevier, Amsterdam, 1983, 57.
77. **Nuñez, R. and Pilar, G.,** unpublished, 1983.
78. **Brownell, B., Oppenheimer, D. R., and Spalding, J. M. R.,** Neurogenic muscle atrophy in myasthenia gravis, *J. Neurol. Neurosurg. Psychiatry*, 35, 311, 1972.
79. **Duchen, L. W. and Stitch, S. J.,** The effects of botulinum toxin on the pattern of innervation of skeletal muscle in the mouse, *Q. J. Exp. Physiol.*, 53, 84, 1968.
80. **Brown, M. C. and Ironton, R.,** Motor neuron sprouting induced by prolonged tetrodotoxin block of nerve action potentials, *Nature (London)*, 265, 459, 1977.
81. **Manthorpe, M., Barbin, G., and Varon, S.,** Purification of ciliary neuronotrophic factor from chick embryo eye, *Trans. Am. Soc. Neurochem.*, 15, 180, 1984.
82. **Barbin, G., Manthorpe, M., and Varon, S.,** Purification of the chick eye ciliary neuronotrophic factor, *J. Neurochem.*, 43, 1468, 1984.
83. **Tanaka, H.,** personal communication, 1985.
84. **Ebendal, T., Olson, L., Seiger, A., and Hedlund, K.-O.,** Nerve growth factors in the rat iris, *Nature (London)*, 286, 25, 1980.
85. **Rieger, F., Grumet, M., and Edelman, G. M.,** N-CAM at the vertebrate neuromuscular junction, *J. Cell Biol.*, 101, 285, 1985.
86. **Couvalt, J. and Sanes, J.,** Neural cell adhesion molecule (N-CAM) accumulates in denervated and paralyzed skeletal muscle, *Proc. Natl. Acad. Sci. U.S.A.*, 82, 4544, 1985.
87. **Petruzelli, G. J. and Hughes, W. F.,** A reduction in motoneuron death in the chick lateral motor column *in vivo* by extracts of embryonic skeletal muscle, *Soc. Neurosci.*, 11, 258, 1985.
88. **Weill, C. L.,** Cyclic GMP accumulation by spinal cord neurons during the period of neuromuscular junction formation, *Neurosci. Lett.*, 30, 263, 1982.
89. **Wilson, B. W., Nieberg, P. S., Walker, C. R., Linkhart, T. A., and Fry, M. D.,** Production and release of acetylcholinesterase by cultured embryo muscle, *Dev. Biol.*, 33, 285, 1978.
90. **Weill, C. L. and Greene, D. P.,** Prevention of natural motoneuron cell death by dibutyryl cyclic GMP, *Nature (London)*, 308, 452, 1984.
91. **Oppenheim, R. W.,** Cyclic GMP and neuron death, *Nature (London)*, 313, 248, 1985.
92. **Weill, C. L.,** Cyclic GMP and neuron death, *Nature (London)*, 313, 248, 1985.
93. **Meriney, S. D., Pilar, G., and Nuñez, R.,** The effects of chronic neuromuscular blockade and cGMP administration on differential motoneuron survival in the avian ciliary ganglion, *Soc. Neurosci.*, 10, 640, 1984.
94. **Mulder, A. H., Wardeh, G., Hogenboom, F., and Frankhuyzen, A. L.,** Kappa and delta-opioid receptor agonists differentially inhibit striatal dopamine and acetylcholine release, *Nature (London)*, 308, 278, 1984.
95. **Bixby, J. L. and Spitzer, N. C.,** Enkephalin reduces quantal content at the frog neuromuscular junction, *Nature (London)*, 301, 431, 1983.
96. **Konishi, S., Tsunoo, A., and Otsuka, M.,** Enkephalin as a transmitter for presynaptic inhibition in sympathetic ganglia, *Nature (London)*, 294, 80, 1981.
97. **Haynes, L. W., Smith, M. E., and Smyth, D. G.,** Evidence for the neurotrophic regulation of collagen-tailed acetylcholinesterase in immature skeletal muscle by beta-endorphin, *J. Neurochem.*, 42, 1542, 1984.
98. **White, J. D., Stewart, K. D., Krause, J. E., and McKelvy, J. F.,** Biochemistry of peptide-secreting neurons, *Physiol. Rev.*, 65, 553, 1985.
99. **Zagon, I. S., Rhodes, R. E., and McLaughlin, P. J.,** Distribution of enkephalin immunoreactivity in germinative cells of the developing rat cerebellum, *Science*, 227, 1049, 1985.
100. **Zagon, I. S. and McLaughlin, P. J.,** Increased brain size and cellular content in infant rats treated with an opiate antagonist, *Science*, 221, 1179, 1983.
101. **White, J. D., Krause, J. E., Karten, H. J., and McKelvy, J. F.,** Presence and ontogeny of enkephalin and substance-P in the chick ciliary ganglion, *J. Neurochem.*, 45, 1319, 1985.
102. **Minneman, K. and Iverson, L.,** Enkephalin and opiate narcotics increase cyclic GMP accumulation in slices of rat neostriatum, *Nature (London)*, 262, 313, 1976.
103. **Gullis, R. J., Traber, J., and Hamprecht, B.,** Morphine elevates levels of cyclic GMP in a neuroblastoma glioma hybrid cell line, *Nature (London)*, 256, 57, 1975.
104. **Erichsen, J. T., Karten, H. J., Eldred, W. D., and Brecha, N. C.,** Localization of substance P-like and enkephalin-like immunoreactivity within preganglionic terminals of the avian ciliary ganglion: light and electron microscopy, *J. Neurosci.*, 2, 994, 1982.

105. **Ogawa, M., Crean, G., and Pilar, G.,** Survival and differentiation of ST-40 neurons isolated from chick ciliary ganglion in culture, *Soc. Neurosci.*, 10, 429, 1984.
106. **Meriney, S. D., Gray, D. B., and Pilar, G.,** Morphine-induced delay of normal cell death in the avian ciliary ganglion, *Science*, 228, 1451, 1985.
107. **Meriney, S. D. and Pilar, G.,** Chronic morphine administration delays normal cell death in the avian ciliary ganglion, *Soc. Neurosci.*, 11, 260, 1985.
108. **Meriney, S. D.,** unpublished, 1986.
109. **Nishi, R. and Berg, D. K.,** Two components from eye tissue that differentially stimulate the growth and development of ciliary ganglion neurons in cell culture, *J. Neurosci.*, 1, 505, 1981.
110. **Bonyhady, R. E., Hendry, I. A., and Hill, C. E.,** Reversible dissociation of a bovine cardiac factor that supports survival of avian ciliary ganglionic neurons, *J. Neurosci. Res.*, 7, 11, 1982.
111. **Manthorpe, M., Skaper, S. D., Williams, L. R., Fryer, H. J. L., and Varon, S.,** Isolation of mammalian peripheral nerve ciliary neuronotrophic factor, *Soc. Neurosci.*, 11, 1082, 1985.
112. **Collins, F.,** Electrophoretic similarity of the ciliary ganglion survival factors from different tissues and species, *Dev. Biol.*, 109, 255, 1985.
113. **Wallace, T. L. and Johnson, E. M.,** Characterization and partial purification from pig lung of survival-promoting and choline acetyltransferase-sustaining activities for parasympathetic neurons, *Soc. Neurosci.*, 11, 1083, 1985.
114. **Manthorpe, M. and Varon, S.,** Regulation of neuronal survival and neuritic growth in the avian ciliary ganglion by trophic factors, in *Growth and Maturation Factors*, Guroff, G., Ed., John Wiley & Sons, New York, 1983, 77.
115. **Manthorpe, M., Longo, F. M., and Varon, S.,** Comparative features of spinal neuronotrophic factors collected in fluids *in vitro* and *in vivo*, *J. Neurosci. Res.*, 8, 241, 1982.
116. **Berg, D. K.,** Cell death in neuronal development: Regulation by trophic factors, in *Neuronal Development*, Spitzer, N. C., Ed., Plenum Press, New York, 1982, 297.
117. **Berg, D. K.,** New neuronal growth factors, *Annu. Rev. Neurosci.*, 7, 149, 1984.
118. **Ebendal, T., Hedlund, K. O., and Norrgren, G.,** Nerve growth factors in chick tissues, *J. Neurosci. Res.*, 8, 153, 1982.
119. **Edgar, D. and Barde, Y. A.,** Neuronal growth factors, *Trends Neurosci.*, 6, 260, 1983.
120. **Thoenen, H. and Edgar, D.,** Neurotrophic factors, *Science*, 229, 238, 1985.
121. **Landa, K. B., Adler, R., Manthorpe, M., and Varon, S.,** Cholinergic neuronotrophic factors. III. Developmental increase of trophic activity of chick embryo ciliary ganglion neurons in their intraocular target tissues, *Dev. Biol.*, 74, 401, 1980.
122. **Patterson, P. H. and Chun, L.,** The induction of acetylcholine synthesis in primary cultures of dissociated rat sympathetic neurons. I. Effects of conditioned medium, *Dev. Biol.*, 56, 263, 1977.
123. **Patterson, P. H.,** Environmental determination of autonomic transmitter functions, *Annu. Rev. Neurosci.*, 1, 1, 1978.
124. **Bunge, R. B., Johnson, M., and Ross, C. D.,** Nature and nurture in the development of the autonomic neuron, *Science*, 199, 1409, 1978.
125. **Fukada, K.,** Purification of a glycoprotein which controls transmitter choice, *Soc. Neurosci.*, 11, 668, 1985.
126. **Giess, M. C. and Weber, M. J.,** Acetylcholine metabolism in rat spinal cord cultures: regulation by a factor involved in the determination of the neurotransmitter phenotype of sympathetic neurons, *J. Neurosci.*, 4, 1442, 1984.
127. **Weber, M. J.,** A diffusible factor responsible for the determination of cholinergic functions in cultured sympathetic neurons, *J. Biol. Chem.*, 256, 3447, 1981.
128. **Hawrot, E.,** Cultured sympathetic neurons: effects of cell-derived and synthetic substrata on survival and development, *Dev. Biol.*, 74, 136, 1980.
129. **Teitelman, G., Joh, T. H., Grayson, L., Park, D. H., Reis, D. J., and Iacovitti, L.,** Cholinergic neurons of the chick ciliary ganglia express adrenergic traits *in vivo* and *in vitro*, *J. Neurosci.*, 5, 29, 1985.
130. **Iacovitti, L., Joh, T. H., Albert, V. R., Park, D. H., Reis, D. J., and Teitelman, G.,** Partial expression of catecholaminergic traits in cholinergic chick ciliary ganglia: studies *in vivo* and *in vitro*, *Dev. Biol.*, 110, 402, 1985.
131. **Giller, E. L., Schrier, B. K., Shainberg, A., Fisk, H. R., and Nelson, P. G.,** Choline acetyltransferase activity is increased in combined cultures of spinal cord and muscle cells from mice, *Science*, 182, 588, 1973.
132. **Giller, E. L., Neale, J. N., Bullock, P. N., Schrier, B. K., and Nelson, P. G.,** Choline acetyltransferase activity of spinal cord cell cultures increased by co-culture with muscle and muscle-conditioned medium, *J. Cell Biol.*, 74, 16, 1977.
133. **Godfrey, E. W., Schrier, B. K., and Nelson, P. G.,** Source and target cell specificities of a conditioned medium factor that increases choline acetyltransferase activity in cultured spinal cord cells, *Dev. Biol.*, 77, 403, 1980.

134. **Brookes, N., Burt, D. R., Goldberg, A. M., and Bierkamper, G. G.,** The influence of muscle-conditioned medium on cholinergic maturation on spinal cord cell cultures, *Brain Res.*, 186, 474, 1980.
135. **Berg, D. K.,** Acetylcholine synthesis by chick spinal cord neurons in dissociated cell culture, *Dev. Biol.*, 66, 500, 1978.
136. **McLennan, I. S. and Hendry, I. A.,** Influence of cardiac extracts on cultured ciliary ganglia, *Dev. Neurosci.*, 3, 1, 1980.
137. **Varon, S., Manthorpe, M., and Adler, R.,** Cholinergic neuronotrophic factors. I. Survival, neurite outgrowth and choline acetyltransferase activity in monolayer cultures from chick embryo ciliary ganglia, *Brain Res.*, 173, 29, 1979.
138. **Nishi, R. and Berg, D. K.,** Survival and development of ciliary ganglion neurons grown alone in culture, *Nature (London)*, 277, 232, 1979.
139. **Kato, A. C. and Rey, M. J.,** Chick ciliary ganglion in dissociated cell culture. I. Cholinergic properties, *Dev. Biol.*, 94, 121, 1982.
140. **Tuttle, J. B., Vaca, K., and Pilar, G.,** Target influences on ^{3}H-ACh synthesis and release by ciliary ganglion *in vitro, Dev. Biol.*, 97, 255, 1983.
141. **Crean, G., Pilar, G., Tuttle, J. B., and Vaca, K.,** Enhanced chemosensitivity of chick parasympathetic neurones in co-culture with myotubes, *J. Physiol. (London)*, 331, 87, 1982.
142. **Haubrich, D. R. and Chippendale, T. J.,** Regulation of acetylcholine synthesis in nervous tissue, *Life Sci.*, 20, 1465, 1977.
143. **Tucek, S.,** *Acetylcholine Synthesis in Neurons*, Chapman & Hall, London, 1978, 1.
144. **Wetzel, G. T. and Brown, J. H.,** Relationships between choline uptake, acetylcholine synthesis and acetylcholine release in isolated rat atria, *J. Pharmacol. Exp. Ther.*, 226, 343, 1983.
145. **Prochiantz, A., DiPorzio, U., Kato, A., Berger, B., and Glowinski, J.,** *In vitro* maturation of mesencephalic dopaminergic neurons from mouse embryos is enhanced by the presence of their striatal target cells, *Proc. Natl. Acad. Sci. U.S.A.*, 76, 5387, 1979.
146. **Prochiantz, A., Daguet, M. C., Herbert, A., and Glowinski, J.,** Specific stimulation of *in vitro* maturation of mesencephalic dopaminergic neurons by striatal membranes, *Nature (London)*, 293, 570, 1981.
147. **Gray, D. B. and Tuttle, J. B.,** Target influences upon transmitter synthesis in cultured chick ciliary ganglion neurons, *Soc. Neurosci.*, 9, 846, 1983.
148. **Gray, D. B. and Tuttle, J. B.,** Target influences upon terminal morphology and ACh synthesis in cultured chick ciliary ganglion neurons, *Soc. Neurosci.*, 10, 1057, 1984.
149. **Tuttle, J. B.,** Interaction with membrane remnants of target myotubes maintains transmitter sensitivity of cultured neurons, *Science*, 220, 977, 1983.
150. **Gray, D. B.,** unpublished, 1986.
151. **Tuttle, J. B.,** personal communication, 1985.
152. **Fischbach, G. D.,** Synaptic potentials recorded in cell cultures of nerve and muscle, *Science*, 169, 1331, 1970.
153. **Betz, W.,** The formation of synapses between chick embryo skeletal muscle and ciliary ganglia grown *in vitro, J. Physiol. (London)*, 254, 63, 1976.
154. **Brenneman, D. E., Fitzgerald, S., and Nelson, P. G.,** Interaction between trophic action and electrical activity in spinal cord cultures, *Dev. Brain Res.*, 15, 211, 1984.
155. **Crain, S. M., Bornstein, M. B., and Peterson, E. R.,** Maturation of cultured embryonic CNS tissue during chronic exposure to agents which prevent bioelectric activity, *Brain Res.*, 8, 363, 1968.
156. **Brenneman, D.,** personal communication, 1986.
157. **Thompson, W. J.,** Activity and synapse elimination at the neuromuscular junction, *Cell Mol. Neurobiol.*, 5, 167, 1985.
158. **Bader, C. R., Bertrand, D., and Kato, A. C.,** Membrane currents in a developing parasympathetic ganglion, *Dev. Biol.*, 98, 515, 1983.
159. **Tuttle, J. B. and Gray, D. B.,** Activity-dependent regulation of ACh synthesis in cultured ciliary ganglia, *Soc. Neurosci.*, 11, 1146, 1985.
160. **Gray, D. B. and Tuttle, J. B.,** ^{3}H-ACh synthesis in cultured ciliary ganglion neurons: effects of myotube membrane, *Dev. Biol.*, 119, 290, 1987.

Chapter 10

PLASTICITY OF ADULT NEUROMUSCULAR JUNCTIONS

Norman Robbins

TABLE OF CONTENTS

I. INTRODUCTION

The mature neuromuscular junction (NMJ), once viewed as the epitome of synaptic stability, is now thought to undergo turnover and remodeling in adult life. This newer view, articulated by Young in 1952,[1] was first given experimental support by Barker and Ip,[2] who noted the presence of sprouts at mammalian endplates. More recently, it was found that the morphology of the NMJ subtly changes over the course of a lifetime,[3] and in the frog, from winter to summer.[4] Thus, synapse turnover is not simply a process of one-for-one replacement, but rather is subject to modulation, e.g., during disuse and aging.

At this juncture in our understanding of neuromuscular plasticity, a discussion of the modifiable morphologic and physiologic parameters may serve to highlight gaps in our understanding, raise questions about mechanism(s), and point to areas for future investigation. A central theme will be the life cycle of nerve terminals and its implications for the understanding of neuromuscular synaptic plasticity. This chapter deals mainly with the concepts of NMJ plasticity in relation to underlying cellular processes and mechanisms, rather than with the various experimental circumstances in which this phenomenon has been observed. Accordingly, citations have been selected only for the purpose of illustration. Also, unless otherwise specified, most of the information is drawn from literature on mammals. For an excellent review organized by type of experiment see Grinnell and Herrera,[5] and for a discussion of plasticity in a variety of neuronal systems see the wide-ranging review by Cotman et al.[6]

II. LIFE CYCLE OF THE NERVE TERMINAL

A. Background and Scheme

Although most regions of a mature NMJ presumably undergo molecular turnover without structural deviation from the classical picture, other regions appear to be engaged in a cycle of structural modifications. Thus, in vertebrates a percentage of ultrastructural profiles of mature junctions show postsynaptic folds free of nerve terminals (abandoned); nerve terminal branches that are partially or completely engulfed by the Schwann cell; myelin figures or other remnants of membranous (presumably axonal) material; small axonal branches with a low density of synaptic vesicles or mitochondria or with dense core vesicles (associated with regenerating axons); terminals occupied by microtubules and neurofilaments; and fine sprouts emanating from mature terminals.[2,7-11] In the absence of direct observation of the life cycle over time in any one NMJ, it is theoretically possible that each of the above-mentioned observations represents only a temporary deviation from the normal state or, alternatively, a stable structural variant of the idealized normal NMJ.

One indication that there is indeed a turnover of nerve terminals comes from the numerous sequential observations on living sensory nerve terminals in tadpole tails.[12] Degeneration, retraction, and regrowth were observed in parts of the terminal tree of a given axon, while other parts of the same tree remained static. Focal regrowth after degeneration did not invariably restore the original pattern of innervation, but often deviated in new directions or produced extra axonal branches. A key observation was that each branch of the terminal arborization of a single axon changed or reacted to local perturbations independently of its neighbors. One may question whether these observations in premetamorphic sensory terminals embedded in an actively renewing epithelium are relevant to the mature NMJ. Yet other findings[2,7-11] indicate that analogous processes almost certainly occur at the NMJ, although they may occur at a slower rate. Speidel's observations[12] are in agreement with the fact that motor nerve terminals at a given NMJ do not behave as an homogeneous entity, but show different events in different regions of the same junction.[9] It would seem that whether the determining influence is intrinsic or extrinsic, the structurally dynamic subunit

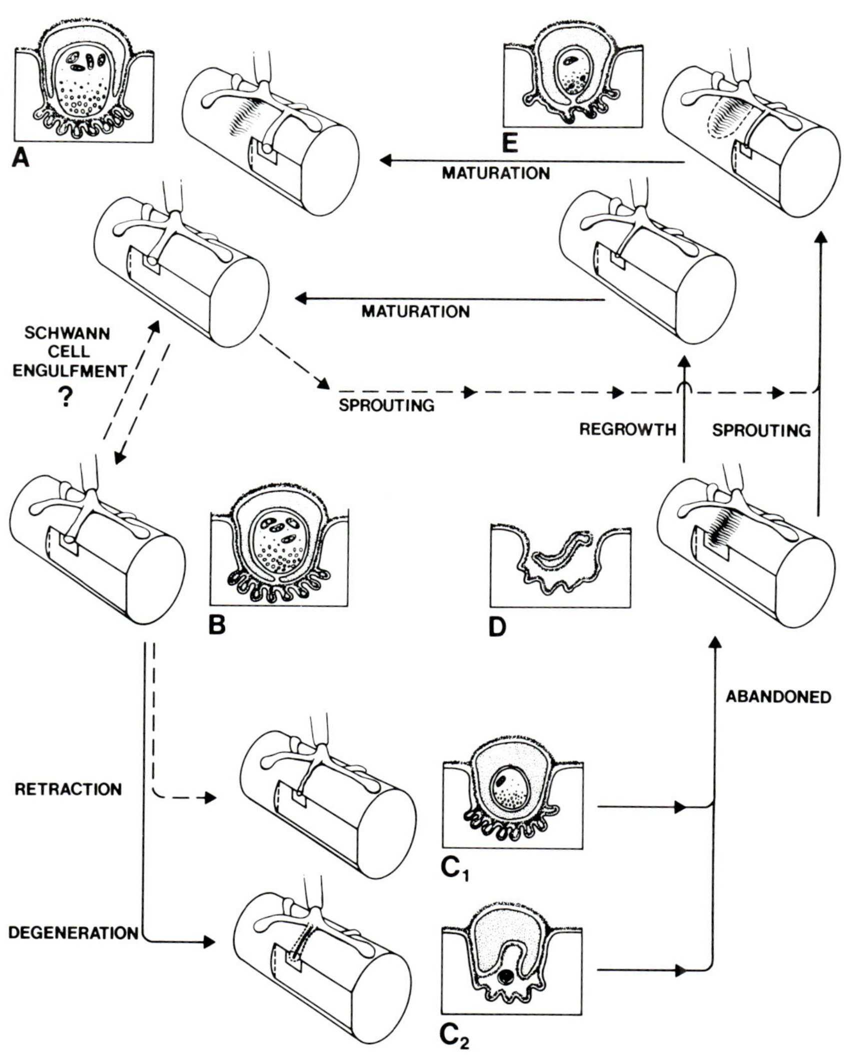

FIGURE 1. Scheme of the life cycle of nerve terminals at the mammalian NMJ, illustrated with diagrams of light and electron microscopic findings at each stage. For further description, see the text.

at the NMJ is often the individual terminal branches and not the entire terminal. Other evidence also supports the notion of heterogeneity and independence of individual branches at the nerve terminal. For instance, neurites kept in vitro after they have been experimentally separated from the cell body continue to grow and make directional choices for at least a number of hours.[13] Also, axotomized nerve terminals in adult frogs can sprout if the appropriate stimulus is applied before Wallerian degeneration intervenes.[14] Thus, in discussing the life cycle of motor nerve terminals, emphasis will be placed on the behavior of the individual branches, i.e., the turnover of the terminal as a whole representing the summation of events in individual branches.

By combining static data from the NMJ and dynamic data from other parts of the peripheral nervous system, one can propose a provisional scheme of the life cycle of nerve terminals at the NMJ. In this scheme (Figure 1), an early event in such a life cycle would be a partial engulfment of a nerve terminal branch by the Schwann cell (Figure 1B). If this engulfment is not reversed, it could lead to either withdrawal or degeneration of the terminal branch

(Figure 1 C1 and C2). Engulfment could also represent a sensitive secondary reaction to incipient retraction or degeneration. Completion of the latter two processes could either leave a permanently abandoned postsynaptic fold (Figure 1D) or be followed by reinnervation of the temporarily abandoned site (Figure 1D to E). In addition, either as an independent event or as one related to terminal withdrawal from a postsynaptic site, sprouts would grow to new sites, where either temporary or relatively long-lasting synaptic regions would be established (Figure 1, short circuit between A and E). In all cases, the cycle eventually concludes with a mature presynaptic bouton or branch (Figure 1A). Although this process of overt structural reorganization probably occurs very gradually through most of life and more extensively in aging, current work in our laboratory indicates that more subtle rearrangements are more common. These include transient emission and retraction of short protrusions and sprouts or partial longitudinal or radial retraction and reexpansion of terminal branches. These minor adjustments could be the structural manifestation of the fine-tuning of synaptic shape by mechanisms normally present in a dynamic balance until some event such as disuse or aging perturbs the equilibrium.

Unfortunately, the sequence of events proposed cannot be easily inferred from morphological data obtained by electron or light microscopy. For instance, up to 25% of mammalian NMJ may show sprouting by light microscopy,[3] but in some muscles active sprouting and regression may be occurring at 100% of the junctions as a rapid and transient event, as recent evidence from our laboratory indicates. On the other hand, the number of times (frequency) a particular stage of the cycle is observed under the electron microscope should be proportional to the volume fraction of the nerve terminal's arborization in that stage. If all regions of the NMJ went through the same life cycle described above in an asynchronous manner, then the percent of junctional regions showing a particular stage would be proportional to the percent of the cycle occupied by that stage. Finally, knowing the duration of a given stage and the percentage of the entire junction in that state, one could compute the average turnover time of a neuromuscular synapse.

The rationale for the scheme in Figure 1 derives from a variety of separate observations. The scheme provides a provisional framework in which synaptic remodeling under circumstances such as disuse or aging can be analyzed as a perturbation of the life cycle of nerve terminals in young mature animals. Each of the stages in the proposed life cycle of the nerve terminal is discussed in the following sections.

B. Schwann Cell Engulfment

Typically, the Schwann cell covers the nerve terminal and does not greatly intrude into the synaptic space, except for occasional partial or complete engulfment of terminals.[3,7,15] It has been suggested that partial engulfment acts as a mechanism to regulate effective synaptic area of transmitter release.[16] However, Schwann cells may completely engulf axon terminals which show stigmata of focal degeneration such as large numbers of lysosomes, swollen mitochondria, and/or acid phosphatase positive organelles.[17] Indeed, most of the information about Schwann cell engulfment comes from studies on muscle denervation in which the engulfment process is one of the earliest events.

The mechanism(s) of Schwann cell engulfment is unclear, but it does not seem to involve a simple response to nerve terminal atrophy, at least at short times after axotomy. In denervated rodent diaphragm, the process of engulfment begins within 6 hr,[18] and in other fast muscle fibers it is complete by 12 hr.[19] At 6 hr, however, nerve terminal area is reduced only by about 17%,[19] and obvious degenerative changes other than swollen mitochondria are not apparent.[18]

Winlow and Usherwood[18] made two interesting proposals to account for Schwann cell engulfment after denervation. The first was that postdenervation cessation of activity halts synaptic vesicle recycling, leading to an altered synaptic nerve terminal membrane which in turn elicits Schwann cell engulfment. This suggestion is not consistent with the finding that if denervation is performed leaving a long nerve stump, Schwann cell engulfment is greatly delayed even though evoked transmitter release is instantly halted in this case as well.[20] The second proposal was that some unspecified denervation change in the nerve terminal membrane induced Schwann cell engulfment. Indeed, in the frog, freeze-fracture of axotomized terminals showed that prior to Schwann cell engulfment, there was disintegration of active zones on the presynaptic terminal surface.[21] Therefore, one of the possible signals for Schwann cell engulfment of nonaxotomized nerve terminals could be given by changes in nerve terminal membrane proteins.

Schwann cell engulfment could also involve a hierarchy of adhesiveness between nerve terminal and synaptic matrix vs. that between nerve terminal and Schwann cell. Thus, at sites where the bonds between nerve terminals and matrix became relatively weak, the next lower adhesive preference (between nerve terminal and Schwann cell) would be expressed, leading to partial or complete engulfment. Alternatively, signals from the nerve could stimulate the Schwann cell to secrete basal lamina components[22] which would increase nerve-Schwann cell adhesion (see Chapter 8). Thus, a sequence of events and a mechanism of engulfment of the nerve terminal similar to those observed in normal development of unmyelinated axons[23,24] may be reenacted during nerve terminal retraction or regeneration. Finally, there is some indirect evidence for the degradation of the synaptic extracellular matrix after denervation (e.g., the rapid decrease of the externally located collagen-tailed asymmetric acetylcholinesterase),[25] which could reduce adhesivity between nerve terminal and matrix. Conversely, focal concentration of cholinesterase (ChE) at points of nerve-matrix contact in nerve terminal sprouts otherwise engulfed by Schwann cells[11] may imply that there is restoration of the matrix adhesivity to nerve terminals at the same points. The test of this proposal will depend on the identification of matrix, Schwann cell, and nerve terminal adhesive components (see Chapter 8).

Engulfment of terminals by the Schwann cell is often associated with an increased incidence of focal denervation, as evidenced by an increased number of abandoned synaptic grooves elsewhere in the same junctions.[15,26] However, it is not clear to what extent substantial engulfment of a nerve terminal implies retraction, degeneration, or regeneration. Neither is it clear to what extent such engulfment is a normal and reversible phenomenon in nondegenerating nerve terminals. Finally, the similarity between Schwann cell engulfment in the peripheral nervous system and glial engulfment of presynaptic boutons in the central nervous system[27] is very striking and may point to similar mechanisms.

C. Retraction and Degeneration

1. Prevalence

Both retraction and degeneration of nerve terminals are known to occur. Light microscopy studies by Speidel[12] have shown retraction of sensory nerve terminals in sequentially observed living specimens, while other terminals of the same parent axon showed focal degeneration (identified by a beading of the linear terminal element similar to that seen after experimental denervation). It also seems certain that there is occasional retraction or degeneration of mature motor nerve terminal branches. In fact, cytochemical (ChE staining) and ultrastructural studies have shown synaptic clefts without overlying nerve terminals.[4,9] Inasmuch as ChE activity and development of synaptic folds are normally induced by contact with growing nerve terminals,[28] the absence of a nerve terminal at such sites implies previous occupation by a terminal. However, it is likely that the percentage of abandoned postsynaptic folds at any NMJ does not represent the total cumulative accretion of abandoned sites, since many are reoccupied and some may eventually become unrecognizable because of dedifferentiation. In the frog, for instance, it can be deduced from the work of Wernig et al.[4] that if the average length of nerve terminals is 600 μm, about 4% of the terminal length is on average abandoned every summer and reoccupied in the winter with a probability of about 90%.[4]

The fact is that there are no established static morphologic hallmarks of retraction, and, as several authors have noted, small nerve terminals at the NMJ (free or engulfed by Schwann cells) could represent either retracting or regenerating elements. Another difficulty in establishing the existence and extent of retraction at the NMJ is the possibility that this process may be so rapid that the chances of detection are very low. For instance, retraction of sensory nerve terminals occurs at rates of up to 1 μm/min[12] and that of neurite growth cones after axotomy in a matter of minutes.[13] Finally, polyneuronal innervation of developing vertebrate NMJ is withdrawn with no signs of focal denervation.[29,30] The conclusion here has been that the retraction occurred too rapidly to be detected.

Focal degeneration is easier to detect than retraction because intact terminals occasionally show one or more changes similar to those seen shortly after denervation. These include swollen mitochondria, clumping or diminution of synaptic vesicles, increased numbers of lysosomal or multivesicular inclusions, filamentous whorls, shrinking of terminals, and Schwann cell engulfment.[17-19,31] If the same time course holds both for spontaneous focal terminal degeneration and changes following denervation, then, in contrast to retraction, the degenerative process is relatively slow (taking 1 to 2 days for 50% loss of nerve terminals after denervation).[19]

The prevalence of focal NMJ degeneration is difficult to determine because of the small sample size obtained in typical ultrastructural studies. However, more data are available for the autonomic innervation of smooth muscle. For instance, in the anterior ciliary muscle of the monkey's eye, about 2% of normal nonaxotomized terminal axons were either degenerating or regenerating.[17] These data are not at odds with quantitative studies of the young mature rat skeletal muscle NMJ. For instance, from Table 2 in the study by Cardasis[15] and assuming that each longitudinal section contained an average of four terminal profiles, it can be estimated that at least 6% of motor nerve terminal branches contained lysosomes and 2% (retracting or growing?) were completely engulfed by Schwann cells. In young mature mouse muscle, however, there was no evidence of frank degeneration, although 4% of the terminals showed partial Schwann cell engulfment.[32] If all these percentages (e.g., of profiles showing partial Schwann cell engulfment, or terminal retraction or degeneration) refer to the fraction of total nerve terminal length, then only a portion of any individual branch may be affected.

2. Mechanism

Why do mature nerve terminals degenerate or retract? This is a critical question because the cycle of synapse turnover may be initiated by focal retraction or degeneration. An interesting hypothesis is that degeneration or retraction results from exhaustion or active destruction of a focal support system. For instance, retraction or degeneration might occur depending on a lesser or greater adhesivity, respectively, between the affected terminal and the synaptic matrix. The initial trigger could be a change in adhesive forces between the synaptic matrix and foci at the end, middle, or base of a terminal branch or sprout. It appears that cytoskeletal forces inside axons and terminals are such that there is a tendency to pull away from a cut axon or disattached growth cone.[13] Thus, if the distal section of a nerve terminal branch lost adhesivity with the matrix, it would retract due to these standing internal forces. On the other hand, if the middle section of a branch lost adhesivity while the most distal region remained attached, the collapsing force toward the base of the branch would pull away the middle section and break off or greatly attenuate the distal part, leading to degeneration. An appealing aspect of this hypothesis, given the heterogeneity of terminal morphology at an individual NMJ, is that localized changes in adhesivity are intuitively easier to visualize than regional exhaustion of vital cell components.

Whatever the mechanism(s) of retraction or degeneration, it is not clear whether there is a clock counting the lifetime of any particular terminal branch or whether the process is random, with a Poisson distribution of hits with only a statistical mean lifetime per terminal. In the latter event, a recently renewed terminal branch would have the same probability of undergoing focal degeneration as any other branch in the junction.

D. Regrowth and Sprouting

1. Prevalence

Sprouts are recognized with light microscopy of zinc iodide-osmium (ZIO) or heavy-metal-stained preparations as fine neuritic filaments emerging from nodes of Ranvier (collateral), heminodes (preterminal), or terminal branches (terminal). Only the latter will be considered here because collateral sprouting is more closely associated with partial denervation and regeneration after denervation,[33] and preterminal sprouts are less common than terminal sprouts at the unperturbed NMJ.[34]

At the light microscopic level, terminal sprouts are noted at 2 to 20% of young adult rodent junctions, but some technical and semantic distinctions are necessary to interpret the literature. With silver stains, the most distal nerve terminal extensions are often ill-defined, and when silver is used in combination with ChE stains, only the ultraterminal sprouts extending beyond the limits of the ChE-stained area can be clearly measured.[35] Sprouts near or between adjacent primary grooves are often difficult to detect by these methods. On the other hand, there is good evidence that nerve terminals are accurately defined with ZIO, since the patterns conform well to vitally stained living terminals,[36] scanning pictures of nerve terminals,[37] and the corresponding postsynaptic folds.[38] In ZIO preparations, terminal sprouts can be readily detected and validated as such by electron microscopy, but one cannot establish (as with ChE stain) the degree of postsynaptic specialization beneath the sprouts. Thus, with silver-ChE stains the percentage of sprouts is likely to be underestimated, and with ZIO methods the distinction between a thin, established terminal branch and an outgrowing sprout cannot be ascertained.

At the elecron microscopic level, it is difficult to distinguish growing from retracting or just normally small nerve terminals. Nonetheless, in young rat soleus (SOL) muscles, dense-core vesicles (characteristic of developing and regenerating terminals)[39] were noted in 1% of the nerve endings,[15] suggesting regrowth of terminals. Since the persistence of dense-core vesicles after regeneration is unknown, these findings can be taken only as a qualitative indication of focal regeneration. In two ultrastructural studies on ectopic synapse formation in adult animals, no unique features of growing terminals were noted,[40,41] only relatively small nerve terminals with low density of synaptic vesicles were reported in the early stages. Another characteristic specialization of growing axons, the growth cone, is not often observed at the NMJ.[42] This may be because the leading end of terminal sprouts is unlike that seen during axonal regeneration,[43] and/or the duration of the growth-cone stage is too short to be easily detected. Apparently, there are a number of technical problems which make it difficult to quantitatively estimate actual terminal sprouting. Development of more specific markers of nerve terminal growth would be most useful in this area of research. Indeed, with a new staining method,[98] it now appears that intraterminal sprouts are present at almost 100% of soleus nerve terminals and also at a much higher percentage of fast muscle NMJs than previously believed, based on ZIO stains.

2. Rate of Growth

Sprouting of motor nerve terminals appears to be a relatively rapid process. In tetrodotoxin (TTX) -treated nerves, sprouts up to 30 μm long appear over 1 to 2 days after a latency of 3 days.[44] Similarly, after specific and total destruction of nerve terminals by intramuscular injection of black widow spider venom and a latency of 3 days, nerve terminal regrowth (longest individual branch length of about 60 μm) replaces most of the original terminal length in 1 day.[98] Although slower restoration of nerve terminal growth has been reported,[42] in a normal NMJ typical terminal branches are short and regrowth after spontaneous focal retraction of degeneration may occur quite rapidly (a few hours). In fact, apparent growth rates in the NMJ may be comparable to those of neurites in tissue culture (e.g., 84 μm/hr),[45] sensory terminals in vivo (e.g., 40 μm/hr),[12] axonal regeneration, and slow axoplasmic transport.[46]

3. Persistence

The net effect of ongoing sprouting and retraction on the total area of the nerve terminal may be minimal or gradual as the animal ages. For instance, mouse SOL junctions undergo little change in total nerve terminal area, whereas the extensor digitorum longus (EDL) shows a linear increase throughout life (35% increase from 5 to 31 months of age).[3] It seems that throughout life terminal sprouts are evanescent or leave only a residue of an increase in branches emanating from the major branches, especially evident in scanning micrographs of junctions of old mice.[38] Such transient sprouts, growing beyond the confines of the synaptic grooves, are probably withdrawn by retraction rather than degeneration. Some sprouts, however, establish new synaptic colonies or regions which increase in prevalence in very old mice.[3] Similarly, in the frog, jumping fibers establish punctate postsynaptic specializations along their path of growth.[11] Thus, not all sprouts are withdrawn, but it is unclear what fraction goes on to mature.

4. Association between Sprouting and Retraction

In some cases, sprouting in one region of a terminal's synaptic tree is associated with withdrawal of branches in other regions. Thus, experimental disuse produces not only sprouting,[33] but also increased abandonment of postsynaptic sites.[26] In aging NMJ, which show an increased number of such abandoned sites,[9,32] the number of smaller nerve terminal regions (new focal synaptic sites) increases, while the area of larger regions decreases, and in SOL muscle the total synaptic area remains unchanged.[3] The implication is that as one branch of the terminal withdraws another sprouts, with conservation of nerve terminal mass. This property has also been noted in sensory nerve terminals in vivo[12] and in growing neurites in tissue culture.[45] On a longer time scale, in frog NMJ there is a predominance of nerve terminal regression in the summer and of sprouting in the winter.[4]

5. Stimulus for Sprouting

Sprouting may be triggered by muscle disuse as caused by either immobilization, nerve paralysis, or postsynaptic receptor blockade.[33] Conversely, direct stimulation of paralyzed muscle prevents such sprouting.[47] How disuse brings about sprouting is controversial. Disuse-induced extrajunctional acetylcholine receptors (AChR) may play a role because AChR blockade by α-bungarotoxin (α-BTX) prevents sprouting.[48] However, in addition to AChR, other factors may be involved because sprouting of regenerating nerve terminals after denervation is not reduced by chronic α-BTX treatment.[48] Terminals innervating normal muscle fibers, which are close to disused fibers, also show sprouting.[49] In addition, unilateral denervation of muscle causes sprouting of contralateral nerve terminals.[50] In this context, the possibility must be considered that perijunctional AChR simply function as receptors for ACh released by newly sprouted terminals. For instance, part of the process of validation and consequent persistence of new sprouts could require a postsynaptic response to ACh, a suggestion consistent with the well-known matching of nerve terminal area and muscle size (see Chapter 9).

Sprout-inducing factors are apparently released by disused or denervated muscle. In fact, muscle-conditioned medium increases neuritic outgrowth in tissue culture,[51] and injection of antibodies against presumed sprouting factors in denervated muscle reduce sprouting in vivo.[52] Thus, one scenario for the unperturbed NMJ would be that spontaneous nerve terminal retraction would focally denervate the muscle, causing local release of sprouting factors which would induce either regrowth of the previously withdrawn branch or sprouting elsewhere in the terminal. Alternatively, perturbations such as disuse would initially lead to increased output of sprouting factors and then to withdrawal of terminals from some synaptic sites. This process would provide cytoplasmic materials for the newly elicited sprouts and for preservation of a constant presynaptic terminal volume.

Another suggestion, relating sprouting to nerve terminal adhesion, rests on observations of interactions between growing neurites and the substrate in vitro.[53] When sensory neurons are cultured on polyornithine-coated vs. uncoated dishes, the number of growth cones emitted per cell and their growth rate are twice as great in 24 hr. Furthermore, embryonic chick ciliary ganglion neurons are continually protruding filopodia, but these normally retract unless heart-conditioned medium is added to the substrate, in which case the filopodia adhere more strongly and neurites appear within minutes.[54] While these observations concern neurites in an artificial situation, there is evidence that some factor in the neuromuscular synaptic matrix is preferentially more adhesive for regenerating motor axons than the surrounding nonsynaptic muscle surface.[55] In this regard, fibronectin and laminin, which are major matrix components, are known to promote neurite outgrowth (see Chapter 8).[56] These data suggest that nerve terminal sprouts are continuously probing the muscle surface for adhesivity. Either a loss of synaptic adhesivity or an increase of perijunctional adhesivity would lead to sprouting. If the differential adhesivity of the perijunctional region was great enough, new synaptic regions would be established. In theory, altered adhesion could be a function of either the matrix or surface molecules at the nerve terminal, e.g., one adhesive molecule implicated in NMJ formation, neural cell adhesion molecule (N-CAM), is an integral membrane component of both nerve and muscle cells.[57]

6. Maturation and Redistribution of Nerve Terminal Sprouts

Information from studies on nerve terminal regeneration after denervation may be useful in understanding the events which follow spontaneous or experimentally induced focal sprouting. Maturation of sprouts occurs quickly after destruction of nerve terminals without axonal damage (and without subsequent preterminal axonal regeneration). Following selective destruction of terminals by intramuscular injection of black widow spider venom into mouse SOL muscle, full restoration of terminal size and quantal release occurred after 8 and 9 days, respectively.[58] However, functional and structural maturation are not perfectly synchronized, e.g., in reinnervated frog muscle quantal content returns to control values by 26 days,[59] but complete restoration of active-zone morphology takes many months.[60] Still, additional time for maturation may be required when a new synaptic region is established by a terminal sprout growing onto a perijunctional site. Ectopic synapses in adult muscle exhibit functional transmission, clustered AChR, and concentrated ChE within 6 to 7 days,[61,62] but the nerve terminals do not reach normal size or quantal content until after 50 to 60 days.[40,41,63] In turn, at the established mammalian NMJ, postsynaptic morphogenesis of a new region may occur more rapidly at the perijunctional surface than at a distant ectopic site (see Chapter 5). This assumption is based on certain similarities between perijunctional and junctional areas, e.g., presence of AChR and ChE, albeit at a lower perijunctional density or concentration, respectively.[98] On the presynaptic side, maturation of sprouts may involve diversion of existing or newly transported synaptic components, as opposed to the requirements for entirely new terminals after nerve regeneration. In the frog, however, immature postsynaptic structures underlying small nerve terminals[4] and nonfunctional sprouts[59] have been detected during early stages of regeneration.

After repeated synapse withdrawal and sprouting, the topography of nerve terminals becomes progressively more complex. By midlife, there are more branch points and a greater elaboration of terminal patterns than in young adulthood (3- to 5-month-old rats and mice).[8,34] In some species, long ultraterminal sprouts appear,[8,64] but short outpouchings of the main terminals may be the most common form of branching.[38] Although these residua of synapse turnover present the most dramatic changes, it is likely that even in aging NMJ, most of the retraction-degeneration-regrowth cycles simply restore the initial conditions.

7. Regulation of Nerve Terminal Growth

Sprouts which persist would steadily increase total nerve terminal length, but it is clear that the total length of adult vertebrate motor nerve terminals is highly regulated and correlated with muscle fiber diameter.[5] Yet this is not a static process that ceases after development. Hypertrophied fibers in dystrophic mice show larger nerve terminals.[65] In lobster claw muscle, nerve fibers branch and rearrange the locus of synaptic specializations as the muscle fibers enlarge through much of their mature lifetime.[66] Also, the correlation between terminal length and diameter is preserved (in mouse SOL muscle) even when, with advanced age, nerve terminals redistribute into smaller terminal regions[3] with no change in muscle fiber diameter. The mechanism(s) of this regulation is not understood, but the factors involved may be similar to those which stimulate sprouting. With repeated turnover of motor nerve terminals, a small difference between degeneration-retraction and regrowth-sprouting may gradually increase the ratio of terminal length: fiber diameter over the course of a lifetime.[3]

Factors intrinsic to the motor neuron also affect the aforementioned relationship: after partial muscle denervation, the remaining axons sprout toward denervated junctions, but form synapses with shorter terminal length in relation to fiber diameter than do control junctions.[10] In such instances, nerve terminal growth may be limited by the supply of metabolites or cytoskeletal components.

8. Turnover of Motor Nerve Terminals

Cotman et al.[6] cited the finding[17] that 3.8% of mainly parasympathetic nerve terminals supplying monkey ciliary muscle were in a state of degeneration or regeneration. Using data for the degeneration time of axotomized sympathetic nerve terminals in the rat, they estimated a half-life of 18.2 days for the ciliary terminals. However, in the rat, axotomized parasympathetic nerve terminals degenerate more slowly than sympathetic terminals.[67] From data on humans,[68] it appears that nerve terminals degenerate about five times more slowly than in the rat. Using 20 days for nerve terminal degeneration in primates, the computed half-life becomes 365 days, and if only the percentage of degenerating terminals is used, the half-life doubles. On the other hand, if rapid retraction and regrowth were continually occurring, then the actual half-life would be shorter than that estimated from the prevalence of degeneration. It is clear that computations of synapse turnover require a variety of data from the same preparation.

Data on abandoned postsynaptic regions at least set a lower limit on the time period required for one cycle, as illustrated in Figure 1. From the rate of appearance of such abandoned sites in mouse and rat SOL muscles,[15,32] it seems that one new abandoned site is generated about every 48 days. Also, from studies on mice,[34] it appears that one new branch point is added about every 38 days. Thus, in the mammalian soleus NMJ, it is likely that there is at least one local cycle every 38 to 48 days. However, given the potential speed of growth and retraction, which greatly reduces the probability of finding a transient abandoned site or added branch, it is likely that the cycle time is far shorter. Furthermore, the amount of synapse turnover may vary between fast- and slow-twitch muscle.

III. STABILITY VS. PLASTICITY

A. Role of Presynaptic Cytoskeletal and Membranous Specializations

Neurofilaments and microtubules in motor nerve terminals are mostly located furthest from the site of nerve-muscle apposition and/or toward the center of the terminal.[32,69] Mitochondria are more abundant in the central core of the terminal, while synaptic vesicles and active zones are closer to the site of nerve-muscle apposition. These and other morphological and physiological characteristics clearly indicate that the motor nerve terminal is a polarized structure indistinctly segregated into one region similar to that of an unmyelinated axon and another specialized for exocytosis. Axoplasmic transport and dissipation of transported components maintain this polarity. Neurofilaments move down the axon as insoluble polymers until they are broken down at the base of the synaptic terminals, whereas membranous components are conveyed both nonspecifically to all neuronal membrane and specifically to the exocytotic region and presynaptic membrane.[70] A similar polarization occurs in growing neurites between the more proximal axon and the most distal regions of the growth cone, where actin, but not neurofilaments, is a major cytoskeletal component.[71]

Analogies between the growth cone and the exocytotic region of nerve terminals[72] are consistent with findings showing that both structures are sites of rapid membrane insertion and turnover,[70,73] lack neurofilamentous cytoskeleton, are strongly associated with voltage-dependent calcium channels,[74-76,77] and can release ACh in response to depolarization.[78,79] Calcium channel blockers inhibit the major function of each of these structures, i.e., growth cone development and nerve terminal exocytosis.[80] An important difference is that synaptic terminals recycle membrane,[73] whereas growth cones primarily exhibit membrane insertion and expansion. However, in overstimulated nerve terminals, where the normal balance between membrane insertion and recycling is upset, expansion of the nerve terminal membrane occurs.[73,81] Similar results were observed in vivo after intensive exercise-induced neuromuscular activity.[82]

The view suggested is of a fairly stable axonal core structure supporting an actively renewing synaptic terminal membrane. Preterminal regions of the axon (at the nodes of Ranvier) can sprout in response to appropriate stimuli, but the threshold is higher than at the terminals. This may explain why collateral sprouting occurs in response to strong stimuli such as partial denervation, whereas terminal sprouting is produced with milder perturbations such as disuse.[33,83] In other words, the synaptic region of the terminal has potential growth-cone-like properties, i.e., the exocytotic terminal region, which is free of constraining Schwann cells and stabilizing polymeric cytoskeletal elements, serves not only as the site of transmitter release but also as a specialization for plasticity. One prediction is that sprouts would arise on this aspect of the nerve terminal membrane more rapidly than from the more axonal region (cf. Reference 66 for an example).

B. Role of the Synaptic Matrix

The mature NMJ matrix contains components to which motor nerve terminals selectively adhere,[55] thus guiding regenerating axons back to their original postsynaptic targets (see Chapter 8). Such selective adhesion also implies that the intercellular bonds formed by the matrix may both organize and stabilize neuromuscular microtopography. For instance, the matrix alone can induce postsynaptic grooves and AChR clustering,[84] as well as presynaptic active zones and membrane recycling.[85] This organizational capacity of the matrix also serves to inhibit outgrowth in an established synapse. It follows that if the adult nerve terminal is tightly bound to the synaptic matrix, it must first be released from these bonds, possibly by a collagenase or another protease, before it can undergo structural rearrangement. In this regard, it has been suggested that proteases play a role in NMJ development[86] and are released upon nerve stimulation.[87]

IV. RELATION OF THE NERVE TERMINAL LIFE CYCLE TO PHYSIOLOGICAL PLASTICITY

A wide variety of perturbations (e.g., disuse, aging, and disease) cause changes in transmitter release at the mature NMJ.[5] However, there is little information on the physiological mechanism(s) and structural basis of altered release. Accordingly, few discernible connections betwen the nerve terminal's life cycle and physiological changes are known. One functional outcome of the regulation of nerve terminal turnover is maintenance of the relationship between synaptic size and transmitter release.[5] However, this relationship is not always maintained, e.g., within NMJ subgroups in frog muscles with similar input resistance, there is an inverse relation between nerve terminal length and transmitter release per unit length.[88] Furthermore, in aging CBF-1 mouse SOL muscle there is increased transmitter release[89] with little or no corresponding increase in nerve terminal length.[3]

Another approach to evaluating alterations in quantal release is to measure the slope of the transmitter release-external calcium concentration curve. In most instances, the slope is unchanged,[90,91] but some exceptions are known.[68] While these results imply no change in the cooperative action of calcium in transmitter release, other mechanisms which might be related to the state of regeneration or maturation of the terminal are possible (e.g., nerve terminal action potential duration or amplitude). Also, changes in synaptic vesicle density do not always correlate with altered release, e.g., in the aging mouse NMJ, synaptic vesicle density is reduced,[32] but transmitter release is increased.[89]

Related to the nerve terminal cycle and relevant to transmitter release alterations are the exocytotic area of nerve terminals,[16,60] sites of new synapse formation within the junction, and the density of active zones.[92] Other relevant factors may include the location and density of sodium, potassium, and calcium channels along nerve terminals. Sodium channels are present near the heminode,[74,93] whereas potassium and calcium channels are located more distally.[74] In regenerating terminals, sodium and potassium channels are initially intermixed, but later attain their normal distribution.[94] Thus, channel location, a possible determinant of synaptic transmitter release, may depend on the amount of synapse turnover. While a few structural variables that may correlate with alterations in transmitter release have been noted above, little is known about the events intervening between calcium influx and final transmitter exocytosis.[95] Despite this important gap in our knowledge, one may attempt to relate changes in the nerve terminal life cycle to physiological plasticity during disuse and aging.

V. RELATION OF THE NERVE TERMINAL LIFE CYCLE TO CHANGES DURING DISUSE AND AGING

Muscle disuse caused either by TTX or immobilization leads to increased nerve terminal sprouting.[26,33] In fact, immobilization enhances various features of synapse turnover, namely, retraction or degeneration (e.g., myelin figures, debris in Schwann cells, and abandoned postsynaptic folds) and putative regenerative features (e.g., terminals of small diameter with Schwann cells engulfing small axonal branches).[26] These observations suggest that either more terminals undergo degeneration-renewal than in controls or that degenerative-regenerative stages of the cycle are much longer. In either case, disuse apparently affects the nerve terminal cycle in more ways than a simple induction of sprouting. In aging NMJ, increased transmitter release has been reported in the mouse and the rat,[89,96] but in the mouse, this could not be attributed to an enlargement of the nerve terminal area.[3] As developed below, this increased release may be a consequence of increased synapse turnover.

Studies on aged rodents have shown signs of increased synaptic turnover similar to those caused by disuse.[15,32] As noted earlier, these ultrastructural findings do not distinguish between longer duration of phases of the cycle and greater prevalance of turnover (i.e., shorter resting mature phase). Also, certain phases of the cycle, such as focal abandonment of junctional folds or a slight asymmetry in the loss-growth of terminals, would tend to leave a cumulative record over the lifetime of the individual simply as the outcome of many previous turns of the cycle. This record consists of progressively more abandoned synaptic folds[15] and, in some muscles (e.g., EDL), steady growth in the nerve terminal area.[3] Certain types of debris, notably lipofuchsin and other poorly degradable membranous components (e.g., small axonal profiles or terminals with dense core vesicles[15]), would also accumulate. The presence of regenerative components suggests that nerve terminal turnover is occurring even though sprouting is not increased.[3,97] At a light microscopic level, the resemblance between intact junctions from old animals and regenerating terminals in young animals may reflect accelerated turnover in both cases. However, in the aging NMJ, diminished withdrawal of sprouts or greater perijunctional adhesivity cannot be excluded as mechanisms of remodeling.

The similarity of some aspects of aging to those of disuse raises the question as to whether the findings in aged synapses only reflect chronic disuse. This is not the case in CBF-1 mice because they show no statistical change in daily activity with age despite the presence of morphologic alterations.[3] Muscle disuse increases synapse turnover, whereas aging either enhances turnover or delays completion of the nerve terminal cycle. The increase in transmitter release observed in both cases may be explained by a redistribution of ion channels similar to that found in early stages of regeneration.[94] An increase in distal sodium channels would ensure greater propagation of depolarization from the heminode to the terminal. On the other hand, a decrease in potassium channels at terminal branches would prolong depolarization. With either mechanism, greater or more sustained depolarization would produce more transmitter release. Alternatively, increased sprouting in response to disuse could indicate that more growth-cone-like calcium channels are present, allowing greater calcium influx and thereby an increase in transmitter release. Either type of change might produce the enhanced transmitter release in certain aging NMJ, a hypothesis which is currently under investigation.

In sum, interpretation of morphologic findings in aging or disuse as perturbations of the normal nerve terminal life cycle leads to some testable hypotheses. Naturally, with better understanding of mechanisms of transmitter release and their relation to the nerve terminal cycle, other explanations of findings in disuse and aging will emerge.

VI. SUMMARY AND CONCLUDING REMARKS

Long-term modulation and regulation of neuromuscular structure and function can be understood in part as a perturbation of the normal cycle of turnover of nerve terminals. There are several steps in this cycle, each of which may have a number of regulatory influences. In this context, it should eventually be possible to account for morphological and related physiological changes during aging and muscle disuse in terms of the rate constants or probabilities of initiation of the various steps in the cycle.

As a working hypothesis (Figure 1), it is proposed that neuromuscular cycle begins with focal Schwann cell engulfment, perhaps concomitant with focal nerve terminal retraction or degeneration denuding a synaptic region. The latter either remains abandoned or is focally reinnervated. Associated with nerve terminal withdrawal or degeneration from one region, sprouts emerge from other regions. Under some conditions, sprouting may be stimulated without withdrawal elsewhere. While many sprouts are transient, some of them go on to form mature terminal branches or branchlets which complete the cycle. Focal reinnervation of abandoned folds is probably the more frequent event, which is followed by a rapid process of structural and functional maturation. Only very small portions of the NMJ are probably involved in synapse turnover at any one time, unless the junction is perturbed, e.g., by disuse or aging.

A number of factors may affect the stability of a given terminal and serve to explain the important observation that even at a single NMJ different terminal branches are either in the mature state or undergoing turnover. It is proposed that the neurofilament-associated axonal component of nerve terminals lends stability, while the exocytotic compartment, normally undergoing membrane turnover in the course of synaptic activity and sharing several properties with growth cones, is more susceptible to structural modulation in the form of either withdrawal or sprouting. It is also likely that extrinsic factors, especially the relative strength of adhesion of the nerve terminal to the synaptic matrix or to the Schwann cell, influence the likelihood of Schwann cell interposition or outgrowth and subsequently the likelihood of withdrawal or degeneration.

Disuse and aging both appear to affect either the probability of terminal branches entering the cycle of turnover or the duration of the various phases of the cycle. With appropriate markers, this distinction may be resolved. Many of the findings at the aging NMJ (e.g., both increased branching and number of abandoned postsynaptic grooves) can be viewed simply as expected remnants of repeated turnover of nerve terminals. Also, some of the physiological changes associated with these conditions may result from altered membrane properties (e.g., changes in distribution of sodium, potassium, or calcium channels) associated with focal growth and regeneration of nerve terminals. In this view, changes with age or disuse are perturbations of the same nerve terminal life cycle that continues throughout development and maturity.

ACKNOWLEDGMENTS

This work was partly supported by NIH grants AG00795 and NS18694. The author thanks Dr. Sean S. Kelly for helpful suggestions.

REFERENCES

1. **Young, J. Z.,** Growth and plasticity in the nervous system, *Proc. R. Soc. London Ser. B,* 139, 18, 1952.
2. **Barker, D. and Ip, M. C.,** Sprouting and degeneration of mammalian motor axons in normal and de-afferented skeletal muscle, *Proc. R. Soc. London Ser. B,* 163, 538, 1965.
3. **Robbins, N. and Fahim, M. A.,** Progression of age changes in mature mouse motor nerve terminals and its relation to locomotor activity, *J. Neurocytol.,* 14, 1019, 1985.
4. **Wernig, A., Pécot-Dechavassine, M., and Stover, H.,** Sprouting and regression of the nerve at frog neuromuscular junction in normal conditions and after prolonged paralysis with curare, *J. Neurocytol.,* 9, 278, 1980.
5. **Grinnell, A. D. and Herrera, A.,** Specificity and plasticity of neuromuscular connections: long-term regulation of motoneuron function, *Prog. Neurobiol.,* 17, 203, 1981.
6. **Cotman, C. W., Nieto-Sampedro, M., and Harris, E. W.,** Synapse replacement in the nervous system of adult vertebrates, *Physiol. Rev.,* 61, 684, 1981.
7. **Engel, A. G. and Santa, T.,** Histometric analysis of the ultrastructure of the neuromuscular junction in myasthenia gravis and in the myasthenic syndrome, *Ann. N.Y. Acad. Sci.,* 183, 46, 1971.
8. **Pestronk, A., Drachman, D. B., and Griffin, J. W.,** Effects of aging on nerve sprouting and regeneration, *Exp. Neurol.,* 70, 65, 1980.
9. **Cardasis, C. A. and Padykula, H. A.,** Ultrastructural evidence indicating reorganization at the neuromuscular junction in the normal rat soleus muscle, *Anat. Rec.,* 200, 41, 1981.
10. **Haimann, C., Mallart, A., Ferre, J. T., and Zilber-Gachelin, N. F.,** Patterns of motor innervation in the pectoral muscle of adult *Xenopus laevis:* evidence for possible synaptic remodeling, *J. Physiol. (London),* 310, 241, 1981.
11. **Anzil, A. P., Bieser, A., and Wernig, A.,** Light and electron microscopic identification of nerve terminal sprouting and retraction in normal adult frog muscle, *J. Physiol. (London),* 350, 393, 1984.
12. **Speidel, C. C.,** Adjustment of nerve endings, *Harvey Lect.* 36, 126, 1941.
13. **Shaw, G. and Bray, D.,** Movement and extension of isolated growth cones, *Exp. Cell Res.,* 104, 55, 1977.
14. **Rotshenker, S.,** Sprouting and synapse formation by motor axons separated from their cell bodies, *Brain Res.,* 223 141, 1981.

15. **Cardasis, C. A.,** Ultrastructural evidence of continued reorganization at the aging (11—26 months) rat soleus neuromuscular junction, *Anat. Rec.*, 207, 399, 1983.
16. **Herrera, A. A., Grinnell, A. D., and Wolowske, B.,** Ultrastructural correlates of naturally occurring differences in transmitter release efficacy in frog motor nerve terminals. *J. Neurocytol.*, 14, 193, 1985.
17. **Townes-Anderson, E. and Raviola, G.,** Degeneration and regeneration of autonomic nerve endings in the anterior part of rhesus monkey ciliary muscle, *J. Neurocytol.*, 7, 583, 1978.
18. **Winlow, W. and Usherwood, P. N. R.,** Ultrastructural studies of normal and degenerating mouse neuromuscular junctions, *J. Neurocytol.*, 4, 377, 1975.
19. **Pulliam, D. L. and April, E. W.,** Degenerative changes at the neuromuscular junctions of red, white and intermediate muscle fibers. I. Response to short stump nerve section, *J. Neurol. Sci.*, 43, 205, 1979.
20. **Pulliam, D. L. and April, E. W.,** Degenerative changes at the neuromuscular junctions of red, white, and intermediate muscle fibers. II. Responses to long stump nerve section and colchicine treatment, *J. Neurol. Sci.*, 43, 223, 1979.
21. **Ko, C.-P.,** Electrophysiological and freeze-fracture studies of changes following denervation at frog neuromuscular junctions, *J. Physiol. (London)*, 321, 627, 1981.
22. **Bunge, M. B., Williams, A. K., Wood, P. M., Uitto, J., and Jeffrey, J. J.,** Comparison of nerve cell and nerve cell plus Schwann cell cultures, with particular emphasis on basal lamina and collagen formation, *J. Cell Biol.* 84, 184, 1980.
23. **Peters, A. and Muir, A. R.,** The relationship between axons and Schwann cells during development of peripheral nerves in the rat, *Q. J. Exp. Physiol.*, 44, 117, 1959.
24. **Bray, G. M., Rasminsky, M., and Aguayo, A. J.,** Interactions between axons and their sheath cells, *Annu. Rev. Neurosci.*, 4, 127, 1981.
25. **Fernandez, H. L., Duell, M. J., and Festoff, B. W.,** Neurotrophic control of 16S acetylcholinesterase at the vertebrate neuromuscular junction, *J. Neurobiol.*, 10, 441, 1979.
26. **Pachter, B. R. and Eberstein, A.,** Neuromuscular plasticity following limb immobilization, *J. Neurocytol.*, 13, 1013, 1984.
27. **Sumner, B. E. H.,** A quantitative analysis of the response of presynaptic boutons to postsynaptic motor neuron axotomy, *Exp. Neurol.*, 46, 605, 1975.
28. **Dennis, M. J.,** Development of the neuromuscular junction: inductive interactions between cells, *Annu. Rev. Neurosci.*, 4, 43, 1981.
29. **Van Essen, D.,** Neuromuscular synapse elimination: structural, functional, and mechanistic aspects, in *Neuronal Development*, Spitzer, N. C., Ed., Plenum Press, New York, 1982, 333.
30. **Bixby, J. L.,** Ultrastructural observations on synapse elimination in neonatal rabbit skeletal muscle, *J. Neurocytol.*, 10, 81, 1981.
31. **Miledi, R. and Slater, C. R.,** Electrophysiology and electronmicroscopy of rat neuromuscular junctions after nerve degeneration, *Proc. R. Soc. London Ser. B*, 169, 289, 1968.
32. **Fahim, M. A. and Robbins, N.,** Ultrastructural studies of young and old mouse neuromuscular junctions, *J. Neurocytol.*, 11, 641, 1982.
33. **Brown, M. C., Holland, R. L., and Hopkins, W. G.,** Motor nerve sprouting, *Annu. Rev. Neurosci.*, 4, 17, 1981.
34. **Wernig, A., Carmody, J. J., Anzil, A. P., Hansert, E., Marciniak, M., and Zucker, H.,** Persistence of nerve sprouting with features of synapse remodeling in soleus muscles of adult mice, *Neuroscience*, 11, 241, 1984.
35. **Pestronk, A. and Drachman, D. B.,** A new stain for quantitative measurement of sprouting at neuromuscular junctions, *Muscle Nerve*, 1, 70, 1978.
36. **Yoshikami, D. and Okun, L. M.,** Staining of living presynaptic nerve terminals with selective fluorescent dyes, *Nature (London)*, 310, 53, 1984.
37. **Desaki, J. and Uehara, Y.,** The overall morphology of neuromuscular junctions as revealed by scanning electron microscopy, *J. Neurocytol.*, 10, 101, 1981.
38. **Fahim, M. A., Holley, J. A., and Robbins, N.,** Scanning and light microscopic study of age changes at a neuromuscular junction in the mouse, *J. Neurocytol.*, 12, 13, 1983.
39. **Bennett, M. R., MacLachlan, E. M., and Taylor, R. S.,** The formation of synapses in reinnervated mammalian striated muscle, *J. Physiol. (London)*, 233, 481, 1973.
40. **Saito, A. and Zacks, S. I.,** Fine structure of neuromuscular junctions after nerve section and implantation of nerve in denervated muscle, *Exp. Mol. Pathol.*, 10, 256, 1969.
41. **Koenig, J.,** Morphogenesis of motor end-plates *"in vivo"* and *"in vitro"*, *Brain Res.*, 62, 361, 1973.
42. **Gorio, A., Carmingnoto, G., Finesso, M., Polato, P., and Nunzi, M. G.,** Muscle reinnervation. II. Sprouting, synapse formation and repression, *Neuroscience*, 8, 403, 1983.
43. **Mason, C. A.,** Growing tips of embryonic cerebellar axons *in vivo*, *J. Neurosci. Res.*, 13, 55, 1985.
44. **Brown, M. C. and Ironton, R.,** Motor neuron sprouting induced by prolonged tetrodotoxin block of nerve action potentials, *Nature (London)*, 265, 459, 1977.

45. **Gundersen, R. W. and Barrett, J. N.,** Characterization of the turning response of dorsal root neurites toward nerve growth factor, *J. Cell Biol.*, 87, 546, 1980.
46. **Wujek, J. R. and Lasek, R. J.,** Correlation of axonal regeneration and slow component b in two branches of a single axon, *J. Neurosci.*, 3, 243, 1983.
47. **Brown, M. C. and Holland, R. L.,** A central role for denervated tissues in causing nerve sprouting, *Nature (London)*, 282, 724, 1979.
48. **Pestronk, A. and Drachman, D. B.,** Motor nerve terminal outgrowth and acetylcholine receptors: inhibition of terminal outgrowth by alpha-bungarotoxin and antiacetylcholine receptor antibody, *J. Neurosci.*, 5, 751, 1985.
49. **Betz, W. J., Caldwell, J. H., and Ribchester, R. R.,** Sprouting of active nerve terminals in partially inactive muscles of the rat, *J. Physiol. (London)*, 303, 281, 1980.
50. **Rotshenker, S.,** Sprouting of intact motor neurons induced by neuronal lesion in the absence of denervated muscle fibers and degenerating axons, *Brain Res.*, 155, 354, 1978.
51. **Smith, R. G. and Appel, S. H.,** Extracts of skeletal muscle increase neurite outgrowth and cholinergic activity of fetal rat spinal motor neurons, *Science*, 219, 1079, 1983.
52. **Gurney, M. E.,** Suppression of sprouting at the neuromuscular junction by immune sera, *Nature (London)*, 307, 546, 1984.
53. **Letourneau, P. C.,** Nerve fiber growth and its regulation by extrinsic factors, in *Neuronal Development*, Spitzer, N. C., Ed., Plenum Press, New York, 1982, 213.
54. **Collins, F.,** Axon initiation by ciliary neurons in culture, *Dev. Biol.*, 65, 50, 1978.
55. **Sanes, J. R., Marshall, L. M., and McMahan, U. J.,** Reinnervation of muscle fiber basal lamina after removal of myofibers, *J. Cell Biol.*, 78, 176, 1978.
56. **Rogers, S. L., Letourneau, P. C., Palm, S. L., McCarthy, J., and Furcht, L. T.,** Neurite extension by peripheral and central nervous system neurons in response to substratum-bound fibronectin and laminin, *Dev. Biol.*, 98, 212, 1983.
57. **Rutishauser, U.,** Influences of a neural cell adhesion molecule on axon growth and guidance, *J. Neurosci. Res.*, 13, 123, 1985.
58. **Duchen, L. W., Gomez, S., and Queiroz, L. S.,** The neuromuscular junction of the mouse after black widow spider venom, *J. Physiol. (London)*, 316, 279, 1981.
59. **DeCino, P.,** Transmitter release properties along regenerated nerve processes at the frog neuromuscular junction, *J. Neurosci.*, 1, 308, 1981.
60. **Ko, C.-P.,** Regeneration of the active zone at the frog neuromuscular junction, *J. Cell Biol.*, 98, 1685, 1984.
61. **Lømo, T. and Slater, C. R.,** Acetylcholine sensitivity of developing ectopic nerve-muscle junctions in adult soleus muscles, *J. Physiol. (London)*, 303, 173, 1980.
62. **Lømo, T. and Slater, C. R.,** Control of junctional acetylcholinesterase by neural and muscular influences in the rat, *J. Physiol. (London)*, 303, 191, 1980.
63. **Tonge, D. A.,** Physiological characteristics of re-innervation of skeletal muscle in the mouse, *J. Physiol. (London)*, 241, 141, 1974.
64. **Fagg, G. E., Scheff, S. W., and Cotman, C. W.,** Axonal sprouting at the neuromuscular junction of adult and aged rats, *Exp. Neurol.*, 74, 847, 1981.
65. **Harris, J. B. and Ribchester, R. R.,** The relationship between endplate size and transmitter release in normal and dystrophic muscles of the mouse, *J. Physiol (London)*, 296, 245, 1979.
66. **Govind, C. K. and Pearce, J.,** Remodeling of multiterminal innervation by nerve terminal sprouting in an identifiable lobster motoneuron, *Science*, 212, 1522, 1981.
67. **Roth, C. D. and Richardson, K. C.,** Electron microscopical studies on axonal degeneration in the rat iris following ganglionectomy, *Am. J. Anat.*, 124, 341, 1969.
68. **Cull-Candy, S. G., Miledi, R., and Uchitel, O. D.,** Denervation changes in normal and myasthenia gravis human muscle fibers during organ culture, *J. Physiol. (London)*, 333, 227, 1982.
69. **Birks, R., Huxley, H. E., and Katz, B.,** The fine structure of the neuromuscular junction of the frog, *J. Physiol. (London)*, 150, 134, 1960.
70. **Lasek, R. J.,** Translocation of the neuronal cytoskeleton and axonal locomotion, *Philos. Trans. R. Soc. London Ser. B*, 299, 313, 1982.
71. **Shaw, G., Osborn, M., and Weber, K.,** Arrangement of neurofilaments, microtubules and microfilament-associated proteins in culture dorsal root ganglia cells, *Eur. J. Cell Biol.*, 24, 20, 1981.
72. **Llinas, R.,** The role of calcium in neuronal function, in *The Neurosciences, 4th Study Program*, Schmitt, F. O. and Worden, F. G., Eds., MIT Press, Cambridge, Mass., 1979, 555.
73. **Heuse, J. E. and Reese, T. E.,** Evidence for recycling of synaptic vesicle membrane during transmitter release at the frog neuromuscular junction, *J. Cell Biol.*, 57, 315, 1973.
74. **Brigant, J. L. and Mallart, A.,** Presynaptic currents in mouse motor endings, *J. Physiol. (London)*, 333, 619, 1982.

75. **Meiri, H., Parnas, I., and Spira, M.,** Membrane conductance and action potential of a regenerating axonal tip, *Science.* 211, 709, 1981.
76. **MacVicar, B. A. and Llinas, R.,** Barium action potentials in regenerating axons of the lamprey spinal cord, *J. Neurosci. Res.,* 13, 323, 1985.
77. **Cohan, C. S., Haydon, P. G., and Kater, S. B.,** Single channel activity differs in growing and nongrowing growth cones of isolated identified neurons of *Helisoma, J. Neurosci. Res.,* 13, 285, 1985.
78. **Hume, R. L., Role, L. W., and Fischbach, G. D.,** Acetylcholine release from growth cones detected with patches of acetylcholine receptor-rich membranes, *Nature (London),* 305, 632, 1983.
79. **Young, S. H. and Poo, M.-M.,** Spontaneous release of transmitter from growth cones of embryonic neurons, *Nature (London),* 305, 634, 1983.
80. **Anglister, L., Farber, I. C., Shahar, A., and Grinvald, A.,** Localization of voltage-sensitive calcium channels along developing neurites: their possible role in regulating neurite elongation, *Dev. Biol.* 94, 351, 1982.
81. **Frontali, N., Ceccarelli, B., Gorio, A., Mauro, A., Siekevitz, P., Tzeng, M.-C., and Hurlbut, W. P.,** Purification from black widow spider venom of a protein factor causing the depletion of synaptic vesicles at neuromuscular junctions, *J. Cell Biol.,* 68, 462, 1976.
82. **Palacios-Prü, E. L., Mendoza, R. V., Palacios, L., and Colasante, C.,** Morphological changes in neuromuscular junctions during exercise, *J. Neurosci. Res.* 9, 371, 1983.
83. **Hopkins, W. G., Brown, M. C., and Keynes, R. J.,** Nerve growth from nodes of Ranvier in inactive muscle, *Brain Res.,* 222, 125, 1981.
84. **Burden, S., Sargent, P., and McMahan, U. J.,** Acetylcholine receptors in regenerating muscle accumulate at original synaptic sites in the absence of the nerve, *J. Cell Biol.,* 82, 412, 1979.
85. **Glicksman, M. A. and Sanes, J. R.,** Differentiation of motor nerve terminals formed in the absence of muscle fibers, *J. Neurocytol.,* 12, 661, 1983.
86. **O'Brien, R. A. D., Ostberg, A. J., and Vrbova, G.,** The effect of acetylcholine on the function and structure of the developing mammalian neuromuscular junction, *Neuroscience,* 5, 1367, 1980.
87. **Poberai, M., Savay, G., and Csillik, B.,** Function-dependent proteinase activity in the neuromuscular synapse, *J. Neurobiol.,* 2, 1, 1972.
88. **Nudell, B. M. and Grinnell, A. D.,** Regulation of synaptic position, size, and strength in anuran skeletal muscle, *J. Neurosci.,* 3, 161, 1983.
89. **Banker, B. Q., Kelly, S. S., and Robbins, N.,** Neuromuscular transmission and correlative morphology in young and old mice, *J. Physiol. (London),* 339, 355, 1983.
90. **Cull-Candy, S. G., Miledi, R., Trautmann, A., and Uchitel, O. D.,** On the release of transmitter at normal, myasthenia gravis and myasthenic syndrome affected human endplates, *J. Physiol. (London),* 299, 621, 1980.
91. **Fahim, M. A. and Robbins, N.,** Transmitter release and ultrastructure of young and old mouse neuromuscular junctions in low calcium solutions, *Soc. Neurosci.,* 11, 733, 1985.
92. **Fukunaga, H., Engel, A. G., Osame, M., and Lambert, E. H.,** Paucity and disorganization of presynaptic membrane active zones in the Lambert-Eaton myasthenic syndrome, *Muscle Nerve,* 5, 686, 1982.
93. **Konishi, T. and Sears, T. A.,** Electrical activity of motor nerve terminals, *Proc. R. Soc. London Ser. B,* 222, 115, 1984.
94. **Angaut-Petit, D. and Mallart, A.,** Presynaptic currents in newly formed neuromuscular junctions in the mouse, *Soc. Neurosci.,* 9, 880, 1983.
95. **Duncan, C. J.,** Role of calcium in triggering the release of transmitters at the neuromuscular junction, *Cell Calcium,* 4, 171, 1983.
96. **Smith, D. O.,** Acetylcholine storage, release and leakage at the neuromuscular junction of mature adult and aged rats. *J. Physiol. (London),* 347, 161, 1984.
97. **Rosenheimer, J. L. and Smith, D. O.,** Differential changes in the end-plate architecture of functionally diverse muscles during aging, *J. Neurophysiol.,* 53, 1567, 1985.
98. **Robbins, N.,** unpublished.

Chapter 11

TROPHIC MECHANISMS UNDERLYING NERVE-MUSCLE INTERACTIONS

Hugo L. Fernandez and J. Alejandro Donoso

TABLE OF CONTENTS

I. INTRODUCTION

Nerve and skeletal muscle cells independently exhibit propriotrophic functions that are essential for the preservation of their individual structural and functional integrities. To ensure a dynamic control of their physiological activities, these cells also exert long-term exterotrophic effects on each other. In other words, throughout life propriotrophic functions are continuously modulated by exterotrophic influences (Chapter 1). Apparently such vital cellular interdependence involves a number of factors which acting either alone or in combination mediate the regulation of several neuromuscular properties (Chapter 2). The main factors (mediators) which have been implicated in anterograde exterotrophic effects (nerve to muscle) include the neurotransmitter acetylcholine (ACh) itself (Chapter 3); nerve-evoked muscle electromechanical activity (Chapters 4 and 5); and neurogenic substances (Chapters 6 and 7) conveyed by axoplasmic transport (Chapter 3). In turn, the primary agent(s) proposed as a mediator(s) of retrograde exterotrophic effects (muscle to nerve) is a myogenic substance(s) which is presumably taken up by nerve terminals and transported to the neuron's cell body (Chapter 9). In addition to the above-mentioned factors, it has been recently suggested that certain extracellular matrix synaptic components also participate in both antero- and retrograde interactions (Chapter 8), especially during development, regeneration, and aging of the neuromuscular system (Chapter 10).

From the preceding chapters it is clear that the mechanisms underlying trophic events are remarkably complex, and considerable controversy exists over which cellular activities and regulatory factors are directly responsible for the mutual dependency of nerve and muscle cells. For instance, regarding anterograde effects, it has not yet been possible to determine to what extent the aforementioned neural factors specifically participate in the maintenance of a single or several muscle properties. Partly contributing to this problem are the following: attempts to experimentally differentiate between certain trophic effects by selectively eliminating individual neural factors have not been free from serious technical difficulties; often less than adequate indicators have been used to evaluate the muscle's response to the presence or absence of trophic mediators; and analyses of anterograde influences have frequently been based on the currently unjustified premise that only one regulatory factor can account for the maintenance of most if not all muscle characteristics. Another obvious complication has been that the extrapolation of data from experiments involving diverse animal species and many different in vivo and in vitro neuromuscular preparations has not always been rigorous. In this connection, a fact which is usually overlooked is that trophic phenomena undergo important modifications with age, and therefore the mechanisms that operate during cellular development may not be identical to those which are effectual through adulthood and/or aging. Similarly, differences exist in the regulation of endplate vs. non-endplate muscle features, as well as of different types of muscle fibers (fast and slow).

Most of the experimental approaches applied to study long-term neuromuscular interactions have traditionally dealt with attempts to show the existence of a single trophic mediator. Unfortunately, this strategy has consistently led to certain conflicting conclusions, e.g., ACh has been proposed as the only agent responsible for the control of muscle resting membrane potentials (RMP) and extrajunctional ACh receptors (AChR),[1] but the same has also been claimed for other neurogenic substances.[2] These opposing views may be theoretically reconciled by considering that trophic maintenance of a particular muscle property can involve the overlapping effects of more than one neural factor. In practice, however, it has not been possible to determine the precise relationship between such multiple trophic factors. In this context, it may be worthwhile to systematically examine the pathways through which trophic information is communicated and, more important, to define the possible sequences of molecular events underlying trophic mechanisms. To this end, this chapter presents a summarized account of current views and hypotheses which will hopefully prove useful in designing future experiments.

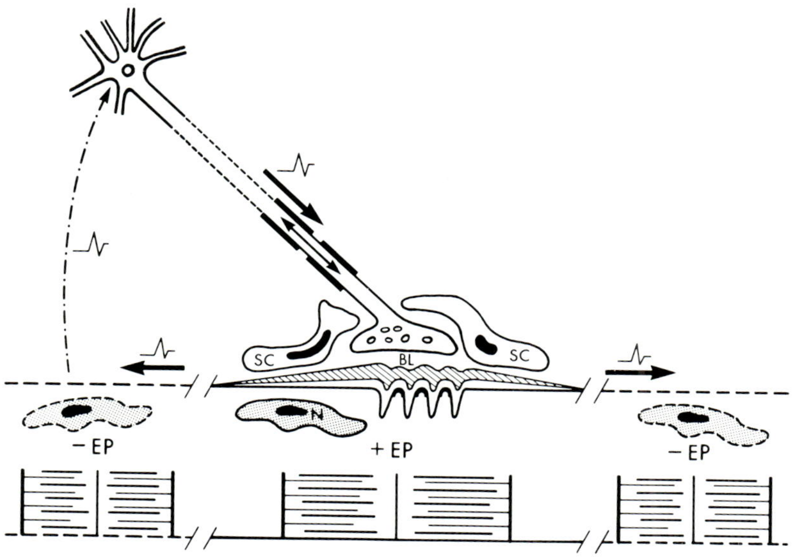

FIGURE 1. Schematic representation of a motorneuron and innervated extrafusal myofiber, illustrating the anatomical pathways for the bidirectional transfer of trophic signals. +EP, endplate region; −EP, nonendplate region; N, nucleus; BL, basal lamina; SC, Schwann cell; →, action potentials; ↔, axoplasmic transport; --→ , afferent activity.

II. PATHWAYS AND CELLULAR ACTIVITIES INVOLVED IN NERVE-MUSCLE INTERACTIONS

Prior to the establishment of synaptic contact, nerve and muscle cells are independently viable for a relatively short period of time. Once this contact takes place, however, both cells become interdependent and capable of long-term survival. This phenomenon certainly requires an efficient system of bidirectional transfer of trophic signals which are essential for the control of a wide range of cellular properties. In turn, the number of possible pathways through which such trophic information can be communicated is surprisingly small. As schematically shown in Figure 1, the anatomical relationship between nerve terminals and extrafusal myofibers is such that the transit of antero- and retrograde trophic signals must occur only by way of the neuromuscular junction, i.e., an ultrastructural specialization that accounts for no more than 0.1% of the entire muscle surface area. Obviously, any experimental and/or pathological alteration in the junction's organization will result in trophic imbalance and the consequent signs of neuromuscular atrophy. Aside from the aforementioned trophic pathway, it is conceivable for the afferent system to provide a parallel route through which additional retrograde information may be relayed to the motor neuron. It must be kept in mind, however, that the bioelectrical feedback signals conveyed by the muscle's sensory system may only subserve an indirect retrograde trophic function (Figure 1).

The above-mentioned pathways impose not only apparent anatomical restrictions, but also certain functional requirements on the operation of trophic mechanisms. Indeed, most of the cellular activities proposed as trophic mediators (e.g., axoplasmic transport, nerve and muscle action potential conduction, etc.) are interrupted at the neuromuscular junction, where they trigger a series of synaptic events. In this context, it must be recognized that the most important way by which such mediators are capable of initiating molecular modifications on the target cell is through the release (secretion) of neurogenic and myogenic substances encoding trophic information. Another important requirement is that processing of this

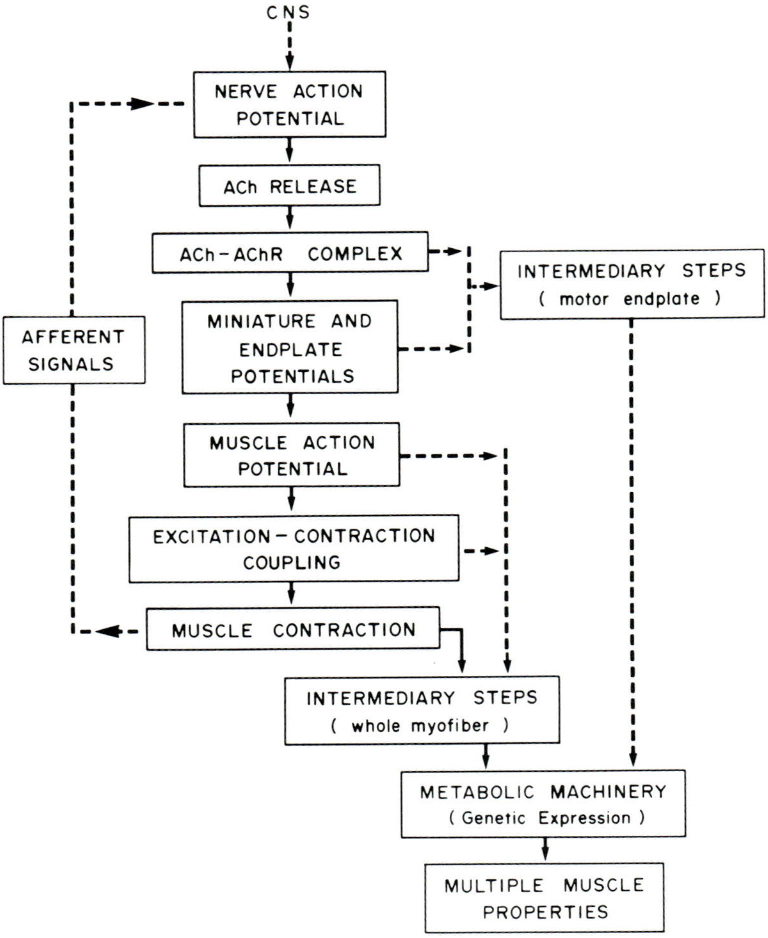

FIGURE 2. Sequence of anterograde trophic influences involving nerve action potentials.

information by target cells must involve the transfer of bioelectrical and/or molecular signals from the synaptic area to those cellular domains where trophic funtions are expressed. While in motor neurons such information transfer occurs from nerve terminals toward the cell body, in myofibers it takes place from the muscle's infoldings toward subsynaptic and extrasynaptic regions. Given these considerations, several possibilities emerge regarding the precise sequence of events implicated in neuromuscular trophic communication.

III. SEQUENCE OF ANTEROGRADE TROPHIC EFFECTS

Perhaps the simplest theoretical approach to elucidating the succession of actions underlying trophic mechanisms is to analyze separately the cellular activities (factors) involved and then attempt to determine their possible functional relationship(s). Following this line of thought, we initially assume that motor neurons exert regulatory influences on muscle cells mainly by means of those basic processes which are triggered by nerve action potentials. In general terms, part of the ACh that is released in response to nerve impulses binds to postsynaptic AChR and thereby sets in motion the well-known chain of events which ultimately lead to muscle contraction. As schematically illustrated in Figure 2, muscle contraction and certain of the intervening steps determining its onset finally control or modulate the myofiber's propriotrophic functions. Apparently, this is accomplished through specific effects on the genetic-metabolic machineries responsible for the maintenance of a large variety of muscle properties.

In particular, whether released spontaneously (quantal and nonquantal) or via nerve impulses (quantal), that ACh which reaches the muscle infoldings interacts with AChR. The resulting ACh-AChR complex then gives rise to nonpropagated miniature endplate potentials (MEPP; spontaneous quantal ACh release) and endplate potentials (EPP; evoked quantal ACh release). Judging from the operational confinement of these physiological processes within the myoneural junctional area, it is not unreasonable to think that they can be causally linked through certain intermediary steps to the trophic maintenance of such specialized muscle region (Figure 2). Among other possibilities the genetic expression of perijunctional nuclei may be trophically influenced by local ionic currents resulting from MEPP and/or EPP; a second messenger system activated by the ACh-AChR complex; or the physiological demands imposed by certain phenomena related to synaptic usage (e.g., AChR occupancy). Alternatively or additionally, EPP-induced propagated action potentials, as well as the consequent electrochemical events of excitation-contraction coupling and contraction, may either individually or in combination affect the metabolism of myofibers along their entire length. The possible intermediary steps involved under these conditions can be similar to those described above. Thus far, there are plenty of data supporting the important role played by muscle electromechanical activity (Chapter 4), but it has not yet been possible to define whether the ultimate trophic effect(s) is exerted via contraction itself or, rather, through those physiological steps leading to it (Figure 2). Regardless of the precise manner in which these mechanisms operate, it is essential to recall that their overall trophic outcome will depend on the pattern (frequency) of nerve impulses evoked in response to muscle afferent discharges and central nervous system (CNS) information received by peripheral motor neurons (Figure 2). Such a pattern apparently influences the phenotypic expression of myofibers in such a way that it defines at least in part the characteristic speed (fast or slow) of muscle contraction (Chapter 5).

At first glance, this reasoning appears to explain most, if not all, aspects of anterograde exterotrophic influences, including the control of both endplate and non-endplate muscle regions. Thus, as recently suggested,[1] there would seem to be no need to postulate any trophic mediator (factor) other than nerve-release ACh to account for the entire maintenance of skeletal muscle. At present, however, there is more than sufficient evidence supporting the participation of neurogenic substances other than ACh in the regulation of certain motor endplate properties (Chapter 2). Accordingly, we now consider some hypothetical ways in which such substances could affect myofibers. As represented in Figure 3, signals from the CNS and muscle afferent system not only determine the pattern of nerve impulse activity, but also modulate indirectly the motor neuron's genetic expression and metabolism. A significant proportion of the substances produced in the perikaryon is then transported along the axon (Chapter 3) to serve propriotrophic and presumably exterotrophic functions. Upon arrival at the nerve terminal, those substances capable of exerting exterotrophic influences (neurogenic trophic substances, or NTS) will be secreted in a nerve impulse-dependent and/or -independent manner. Whatever the case may be, the ultimate destination of some NTS is the synaptic cleft, where they partly contribute to the maintenance of myoneural contact through the replacement of extracellular matrix (ECM) components (Chapter 8). Indeed, certain unique ECM glycoproteins are known to be synthesized within the neuronal cell bodies; transported down the axons, probably in association with vesicles; and finally released into the cleft.[3] Interestingly, these glycoproteins might be related to those ECM constituents that induce the aggregation of junctional AChR[4] and AChE[5] during development and regeneration of the neuromuscular system.[6] Other NTS encoding exterotrophic instructions will reach the surface of the postsynaptic membrane, where reception and processing of molecular information are initiated. By analogy with the well-known mechanism of ACh-

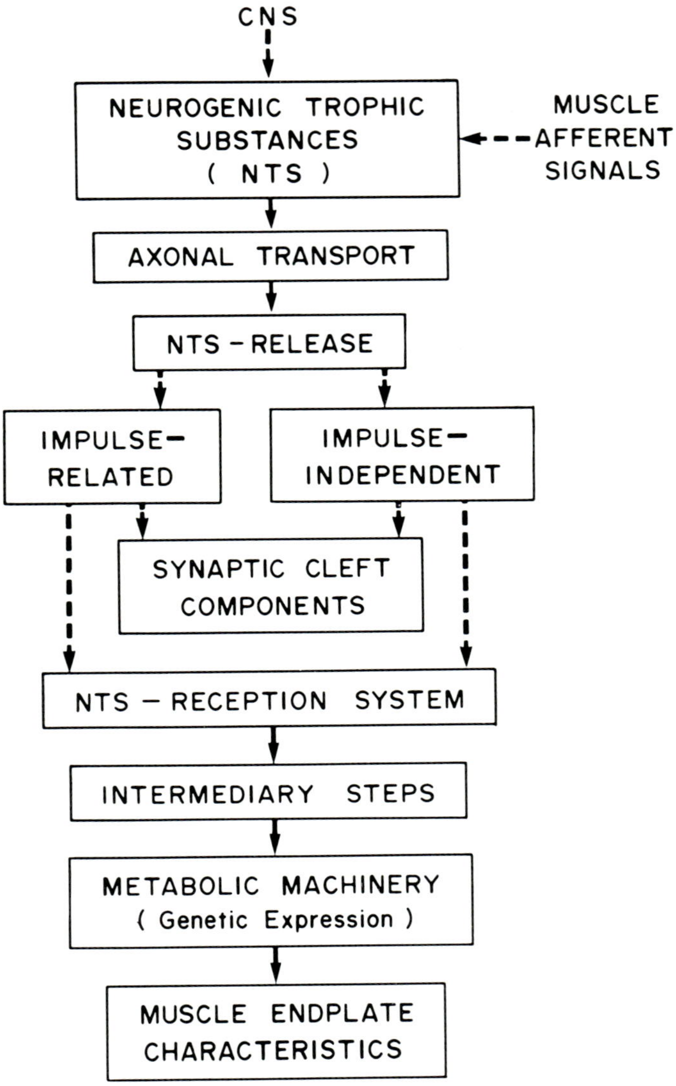

FIGURE 3. Sequence of anterograde trophic influences involving neurogenic substances (NTS).

AChR interactions, the phenomena mentioned above are likely to involve the binding of NTS to true receptors. Subsequently, the NTS-receptor complex may trigger the formation of a second-messenger and/or the allosteric activation of particular enzymes. Alternatively or additionally, NTS may interact with selective surface membrane binding sites and subsequentially undergo internalization via endocytosis. At this stage internalized NTS, second-messengers and/or allosterically activated enzymes must be somehow translocated to their final site(s) of action. Among a wide range of possibilities and through selective intermediary steps, these agents can regulate the genetic expression of perijuntional nuclei by affecting one or more of the multiple events involved (e.g., transcription, processing of RNA, translation, etc.).[7] Other operational levels at which trophic control by NTS may be accomplished include any of the various steps underlying those posttranslational mechanisms responsible for the formation of essential structural and functional molecular components of the motor endplate. Further elaboration of these as well as of other possibilities will be feasible when

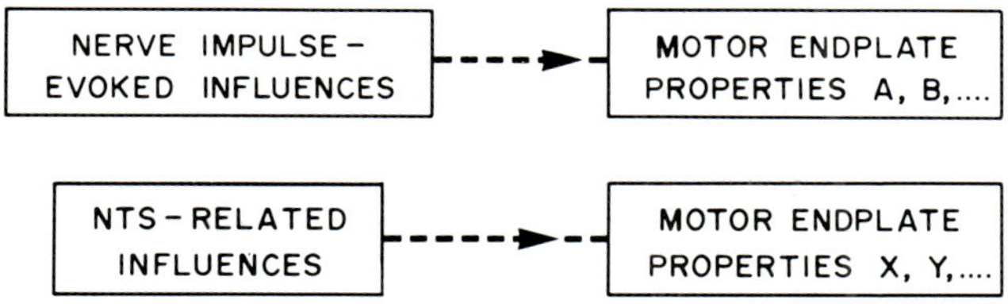

FIGURE 4. Independent trophic actions of nerve impulses and neurogenic substances (NTS) on distinct properties of skeletal muscle endplates.

more detailed isolation and biochemical characterization of NTS are achieved. Nonetheless, the important point to be stressed here is that the existence and essential role played by NTS in the maintenance of myoneural junctional properties can no longer be doubted (Chapters 6 and 7).

Current experimental evidence strongly suggests that separate mechanisms exist for the trophic regulation of endplate versus non-endplate muscle regions (Chapters 2 and 4). In particular, it has been concluded that the maintenance of non-endplate properties solely involves influences mediated by nerve-impulse-related events, whereas that of endplate features additionally requires the participation of specific neurogenic substances that carry trophic information (Chapter 5). In agreement with this concept, it is reasonably clear how propagated muscle action potentials can rapidly trigger trophic processes along the entire length of a multinucleated myofiber, influencing its metabolic-genetic machineries in a rather generalized manner (Figure 2). It is also fairly straightforward how NTS can modulate the myofiber's metabolism locally within the endplate region, but it is difficult to envision a system through which NTS could quickly reach more distant non-endplate areas to affect their operation (Figure 3). Accordingly, it is pertinent to analyze here anterograde exterotrophic effects on motor endplate characteristics in terms of the possible combined actions of those mediators previously considered alone and to attempt to determine whether their operation involves certain steps in common.

Close inspection of the anatomical pathway (Figure 1) and sequences of trophic events (Figures 2 and 3) presented earlier shows several cellular sites and operational steps at which individual actions elicited by nerve impulses and NTS appear superimposed. First, such actions sequentially take place along the same structural network (perikaryon, axon, nerve ending, etc.) extending at least up to the postsynaptic endplate region; i.e., NTS influences are most likely confined to this region, whereas those triggered by nerve-evoked muscle impulses certainly spread throughout the myofiber. Second, aside from this almost complete anatomical overlap and regarding the final outcome of both types of trophic processes, the following general possibilities come to mind:

1. The effects of nerve impulses and NTS on different motor endplate properties may be exerted in parallel, but in a completely independent manner (Figure 4). For instance, it can be argued that impulse-related and/or spontaneously released ACh is sufficient to control the myofiber's RMP, while NTS may regulate a certain junctional membrane constituent which does not participate in the RMP phenomena. Assuming that this is actually the case, a selective and temporary inhibition of ACh release or axoplasmic transport of NTS should cause changes only in those endplate properties separately controlled by the corresponding trophic agent. In other words, as schematized in Figure 4, experimental elimination of ACh or NTS should lead to specific alterations in A, B,... or X, Y,.., respectively.
2. Although both trophic agents can operate in parallel to influence distinct endplate properties separately, they may yet share one or more cellular events (Figure 5).

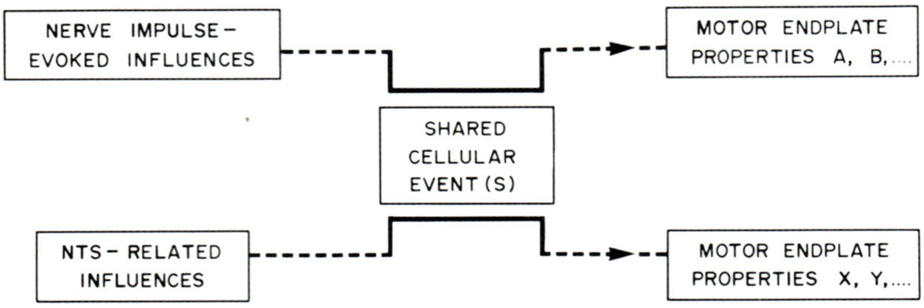

FIGURE 5. Independent trophic actions of nerve impulses and neurogenic substances (NTS) share one or more cellular events before effecting distinct properties of skeletal muscle endplates.

Considering the same example noted above, it is possible for both ACh and NTS to be released together from nerve endings (shared cellular event) via impulse-dependent and/or -independent secretory mechanisms. Thus, two separate modes of regulation underlying the maintenance of different muscle endplate features would be indirectly related to each other. Under these conditions, if nerve secretion is experimentally blocked, all properties subjected to such forms of trophic control (A, B,... and X, Y,...) should be affected.

3. Alternatively, the above-mentioned mechanisms may not necessarily operate in parallel, but, rather, in an interdependent manner, producing regulatory effects of a cooperative nature. In fact, those actions mediated by nerve impulses and NTS may be closely interrelated at certain of the multiple stages conforming the cascade of trophic events behind the regulation of one or more of the same junctional properties (Figure 6). These trophic agents may concomitantly or sequentially affect in a direct way a single or several of the intermediary steps described earlier, including the formation of a second messenger, allosteric activation of enzymes, internalization of NTS, pre- and posttranslational processes of gene expression, etc. (see text referring to Figures 2 and 3). In this context, one can envision that internalization of NTS may directly depend on local currents elicited by endplate potentials without affecting the further actions of NTS; local endplate currents invading perijunctional nuclei and NTS may simultaneously influence genetic expression (e.g., certain stages of RNA processing); endplate currents may promote the translocation of a second messenger from the muscle infoldings to its site(s) of action; etc. Interestingly, all these effects may occur regardless of the presence or absence of a shared cellular event, such as a possible coupled release of ACh and NTS (Figure 6).

Inasmuch as present experimental evidence supports the participation of nerve-impulse-mediated activities as well as NTS in the regulation of the same muscle endplate properties (e.g., RMP[1,2,8] and AChE[9,10] molecular forms) we are inclined to favor the last of these three possibilities. Although experimental testing of this, or a similar, hypothetical scheme based on the interdependent actions of multiple trophic mediators may prove unavoidably complex, it certainly deserves serious consideration. This becomes particularly apparent when one recognizes that the premise of a single nerve factor independently accounting for the entire trophic control of muscle is in general terms no longer tenable.[2,11-14]

IV. SEQUENCE OF RETROGRADE TROPHIC EFFECTS

Retrograde or antidromic trophic influences have been shown in a variety of neuromuscular preparations, including both in vivo and in vitro conditions. The resulting experimental

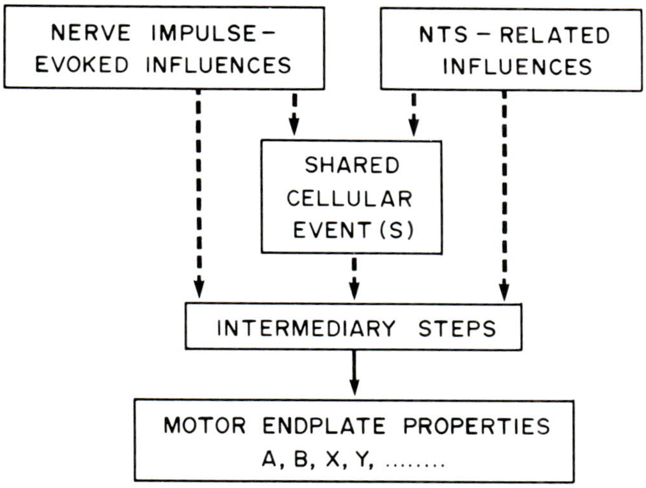

FIGURE 6. Cooperative trophic actions of nerve impulses and neurogenic substances (NTS) on the same properties of skeletal muscle endplates.

evidence has been extensively discussed in Chapter 9; therefore, we present here only a succinct view of the possible mechanisms involved. In general, at least three types of retrograde signals can be identified which may distinctively affect neuronal functions: (1) neurogenic substances conveyed by retrograde axoplasmic transport (turnaround after their arrival at nerve terminals via anterograde transport[15,16]) that essentially subserve a propriotrophic function (e.g., informational feedback system[17]); (2) afferent bioelectrical impulses triggered by the muscle's sensory system; and (3) myogenic trophic substances (MTS) which are released into the synaptic cleft. Regarding the latter, the precise subcellular origin of MTS is not certain; i.e., these substances could be produced solely within the endplate region of the myofiber and/or along its entire length. Regardless of their site(s) of origin, however, it seems clear that MTS must be released at and pass through the myoneural junction in order to exert their exterotrophic effects on both synaptic contact and motor neuron (see Section II). In this context, at least three final destinations can be envisioned: the synaptic cleft itself, the presynaptic nerve ending, and the nerve cell perikaryon. The last two destinations may involve the direct transfer of MTS themselves or, alternatively, that of the trophic information they contain.

Schematically shown in Figure 7 is a hypothetical sequence of events involved in MTS retrograde actions. This sequence is similar to that presented for NTS anterograde effects (see Figure 3) and essentially includes the following steps:

1. Release of MTS can be accomplished through mechanisms which may be either dependent or independent of ionic currents set up by local and/or propagated muscle electrical potentials. Most likely, this process occurs through the exocytotoxic secretion (see Chapter 9) of substances enclosed within certain types of vesicles.[18] Following their secretion, those substances which remain at the synaptic cleft serve to replace ECM components and other synaptic molecules. For example, released heparan sulfate proteoglycans appear to play an important role in the structural organization and function of the ECM.[19,20] Additionally, AChR[21] and most of the extracellular AChE molecular forms[22-24] at the neuromuscular junction also originate at and are secreted by myofibers (see Chapter 8).
2. Reception of those MTSs which traverse the synaptic cleft apparently takes place at the presynaptic nerve terminal membrane, where the processing of trophic information

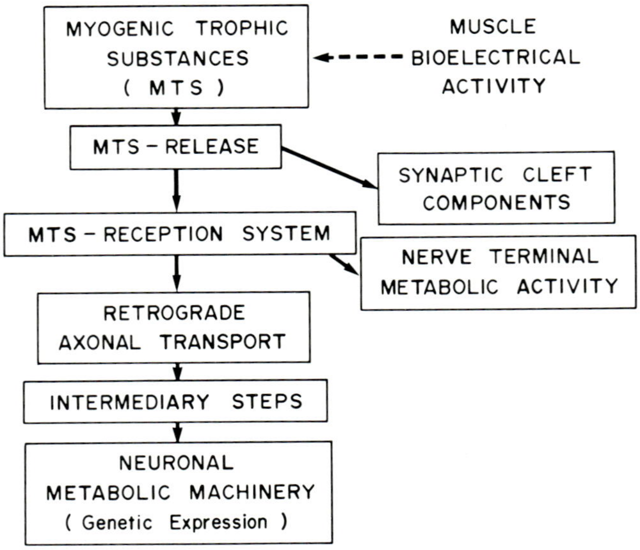

FIGURE 7. Sequence of events underlying retrograde trophic actions of myogenic substances (MTS).

is initiated. Such processing probably involves the interaction of MTS with MTS-receptors and/or with selective MTS-membrane binding sites. Subsequently, that trophic information encoded in the resulting MTS-molecular complexes or in an intermediary factor(s) induced by them (e.g., second messenger) must be internalized and sent to its final cellular destination(s).

3. Nerve terminal metabolic activities involving the assembly of synaptic vesicles, neurotransmitter synthesis, activation of proteases, etc. may be locally influenced by the internalized trophic information. Moreover, the MTS-molecular complexes or intermediary factors can be taken up by retrograde axoplasmic transport and thereby carried to the nerve cell body. At this site, perhaps via numerous intermediary steps, trophic signals effectively regulate the motor neuron's metabolic machinery and genetic expression.

V. NERVE-MUSCLE CELL TROPHIC LOOP

As illustrated in Figure 8, a single motor neuron sends outgoing anterograde exterotrophic signals to skeletal myofibers composing the motor unit, while it receives incoming information from the CNS and the innervated myofibers. That information which originates in myofibers is sent to the neuron as bioelectrical discharges along the afferent route and in the form of molecular signals which are transferred across the myoneural junction in the retrograde direction. This system, then, can be viewed as one forming a trophic loop in which the individual propriotrophic capabilities of neurons and myofibers are continuously being modulated by the other cell's exterotrophic influences.

At this point it is important to note that the anterograde flow of exterotrophic signals follows a divergent path, i.e., with one neuron affecting all myofibers within the corresponding motor unit. In contrast, the passage of retrograde exterotrophic information is convergent in the sense that many myofibers influence a single motor neuron. Perhaps the most remarkable characteristic of this particular cellular arrangement is that a considerable number of myofibers would have to malfunction before the innervating neuron undergoes any significant degenerative change, whereas impairment of a single axon is apparently

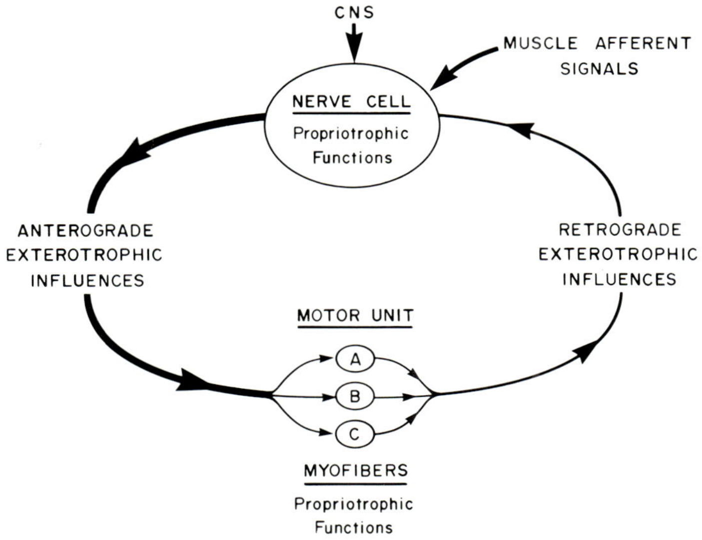

FIGURE 8. Nerve-muscle cell trophic loop involving propriotrophic and bidirectional exterotrophic influences.

sufficient to cause the atrophy of several myofibers. In this context, exterotrophic influences in the anterograde direction appear to be relatively more important than those exerted through the retrograde pathway. It follows that this arrangement provides a convenient functional and structural basis for the process of motor endplate regression-regeneration which takes place in normal muscle.[30] In fact, the uninterrupted operation of this trophic loop (Figure 8) is essential not only for cell survival, synaptogenesis, and cellular development (Chapters 8 and 9), but also for the maintenance of the aforementioned successive turnover and remodeling of myoneural junctions during adulthood and aging (Chapter 10). In addition, such a loop may provide an adequate model for similar cellular interactions throughout the nervous system, and elucidation of its mechanism(s) will most likely provide important insights into the general phenomena of plasticity and pathology of excitable cells.

VI. RESEARCH STRATEGIES

Considerable work has been directed toward arriving at a better understanding of those cellular actions which define the phenotype and maintain the functional integrity of nerve and muscle cells (Chapter 1 and 2). Based on this work, it can now be safely stated that neurogenic and myogenic substances, as well as neuromuscular bioelectrical activity, are all indispensable mediators of bidirectional trophic interactions (Chapters 3 to 7). Such mutual nerve-muscle cell interdependency is progressively built upon during development, adapted to new microenvironmental demands, impaired in pathological states, or totally renewed under certain physiological conditions. In agreement with these facts, trophic relationships have been studied for the most part under conditions when they are being established (development and synaptogenesis); undergoing extensive adaptation (enforced disuse and hyperuse); disrupted (denervation atrophy and degeneration); and renewed (reinnervation and regeneration).

Those studies include a large number of very ingenious approaches which have been grouped into two main classes: the elimination and replacement strategies.[25-27] Elimination refers to experimental paradigms in which trophic influences are suppressed either completely (surgical or chemical denervations) or selectively (e.g., partial blockade of axoplasmic transport). On the other hand, replacement involves attempts to effectively substitute either certain missing (e.g., neural substances added to noninnervated embryonic myofibers in tissue culture) or previously eliminated (e.g., contractile activity reestablished by electrical stimulation of denervated muscle) trophic mediators. Furthermore, numerous variations of these research strategies have been devised, including the multiple pharmacological approach (Chapter 2), in which relatively selective drugs are used to block certain intermediary steps of neuromuscular transmission and subsequently their effects are quantitatively compared with those of denervation;[1,27] and, most interesting, the ectopic reinnervation approach (Chapter 5), involving the transplantation of a foreign nerve onto a nonsynaptic region of a muscle which is made receptive to this treatment by the subsequent transection of its own nerve.[13,28,29] The usefulness of this approach, becomes evident when one considers that trophic signals may operate differently at different times throughout life. In this regard, it is also important to recall the organization of these strategies and available data presented in terms of the so-called developmental, aging, circadian, chemical, coincident, and intraneuronal times.[16] This classification establishes the significance of evaluating trophic phenomena in relation to multiple time-dependent roles and emphasizes the concept that trophic effects do not occur in isolation, but rather in an intimately connected fashion.

The long history of researching this field, together with the numerous diverse experimental approaches tried, understandably makes the formulation of new strategies quite difficult. Nonetheless, the increased knowledge gained on the phenomenology of nerve-muscle cell interactions, coupled with the advent of more sophisticated techniques, certainly permits application of the same or similar approaches to the study of trophic mechanisms within conceptual frameworks such as those previously described in this chapter. Illustrating this point are the following examples:

1. The traditional long-short nerve stump experimental protocol has been successfully employed to show that both the onset and maximal increase in the in vitro phosphorylation of a cAMP-modulated 56-kdalton soluble cytosolic muscle protein,[31] are directly related to the length of nerve that remains attached to rat soleus muscles after denervation.[12] The cytosolic protein was tentatively identified as the regulatory subunit of the cAMP-dependent protein kinase type II (RII[32]). Furthermore, its denervation-induced phosphorylation changes were ascribed to the gradual depletion of a neurogenic substance(s) which presumably continues to be transported and supplied to the muscle at least temporarily after nerve transection.[12] Although the nature of this substance(s) was not identified (e.g., ACh and/or other neurogenic molecules), such studies strongly suggest that nerve impulse-independent influences may be translated by muscle cells via cAMP-dependent protein kinases possibly involved in the operation of a second-messenger-type system.
2. Lending further support to the existence of separate intracellular intermediary steps such as those schematized in Figures 2, 3, and 6, it has been reported that TTX-treated rat myofibers in culture produce increased amounts of AChE and decreased levels of cell-surface AChR in the presence of the calcium ionophore A23187.[33] Inasmuch as the opposite was found for spontaneously contracting myofibers (not treated with TTX), these findings were taken to imply that muscle-contraction-induced or otherwise direct increases in cytoplasmic calcium partly modulate the amounts and forms of certain neurotransmission-related molecules.
3. It is also well documented that the precise distribution of synaptic molecules (i.e., clustering of AChR and AChE) is dictated initially by neurogenic substances[34,36] and eventually by basal lamina components.[35] Regarding the latter, AChR/AChE-aggregating substances extracted from ECM-enriched tissue fractions have been shown to be antigenically related to molecular constituents of the junctional basal lamina.[4,5] Aside from directing the clustering of AChR and AChE in regenerating myofibers,

whether those substances induce similar effects on developing muscle and/or maintain the particular distribution of synaptic macromolecules in normal adult muscles remains to be determined.[6] Whatever the case may be, it is apparent that substances which seem to have important macromolecular inductive-organizing functions at the synaptic cleft must be originally supplied and thereafter continuously renewed by nerve and/or muscle cells. While this supply-renewal process does not directly affect the propriotrophic operation of either neurons or myofibers, it subserves an exterotrophic macromolecular replacement function which effectively maintains synaptic contact and indirectly determines the localization of AChR and AChE.

4. The pharmacological approach involving TTX-treated cells in culture has been applied to explore the bioelectrical-activity-dependent mechanisms that partly control the levels of AChE,[38] basal lamina components,[39] certain enzymes that participate in energy metabolism,[40] and AChR.[40,41] More recently, this approach was used in investigations dealing with the role played by muscle electromechanical activity on AChR biosynthesis and, particularly, on the posttranslational processing of the receptor's α-subunit.[42,43] Results showed that muscle activity decreases the rate of synthesis of AChR α- and β-subunits without altering their degradation, does not affect α-subunits at the stage at which they acquire their ability to bind α-bungarotoxin, and diminishes the rate of assembly of the multiple receptor subunits into fully functional AChR. These findings strongly suggest that posttranslational changes in α-subunits may partly explain the activity-induced inhibition of AChR synthesis.[42] Other studies have shown that TTX-treated myotubes,[44] as well as denervated muscle, undergo significant increases in the levels of α-subunit mRNA, indicating that transcription and/or mRNA turnover are also implicated in the regulation of AChR.

Taken together these observations underscore the contention that at least one of the previously discussed trophic mediators (muscle electromechanical activity) affects the genetic expression of neuromuscular transmission-related molecules (e.g., AChR) probably at both pre- and posttranslational levels.

These examples, in conjunction with many others described in previous chapters, clearly reflect the complexities behind trophic phenomena by indirectly confirming the multifactorial nature of the underlying mechanisms. More important, they also emphasize that future advances in this field will require new modes of experimental analyses involving stepwise, focused approaches on the physiological interplay between such diverse trophic mediators. In this context, although our hypothetical schemes regarding trophic pathways (Figures 1 and 8) and possible sequences of operational events (Figures 6 and 7) go beyond those warranted by current data, they may be useful in leading to the identification of certain problems that can be approached experimentally. For instance, among several other related issues, there are the obvious necessities of distinguishing neurogenic and myogenic substances which serve a molecular replacement function from those that actually affect the target cell's genetic-metabolic machineries; establishing a relationship, if any, between motor endplate and non-endplate regulatory mechanisms; defining the precise modes of reception-translation of trophic instructions and the possible intermediary steps involved; studying more closely those cellular events that may be shared by bioelectrical-activity-dependent and -independent trophic influences (e.g., coupled spontaneous or evoked release of ACh and other substances); exploring the likely cooperative action of different trophic effects on second-messenger-type systems; etc. As mentioned earlier, resolution of some of these issues will be greatly facilitated once the purification and biochemical characterization of NTS and MTS are fully accomplished. By analogy with those studies on the mechanisms of action of nerve growth factor,[45] as well as on certain synaptic basal lamina components which appear to exhibit macromolecular inductive-organizing capabilities,[4-6,34-37] isolation of NTS and MTS should allow the use of sophisticated biochemical-immunochemical techniques in the exploration of their physiological role(s) and detailed mode(s) of operation.

Other general aspects which must be kept in mind when planning further experiments are discussed more extensively by individual contributing authors who have outlined a number of interesting avenues for future research. Most of them essentially agree in suggesting that significant advances in this area of research will be necessarily linked with the extent to which molecular genetic technology can be effectively applied to solve trophic problems. As noted in Chapter 1, which of the conceptual strategies and experimental methodologies considered will eventually prove the most successful remains as one of the most remarkable challenges in future studies on nerve-muscle cell trophic communication.

VII. SUMMARY AND CONCLUDING REMARKS

At present, it is well established that short-term survival of motor neurons and skeletal muscle cells separated from each other is possible because these cells individually express propriotrophic capabilities related to the preservation of their own structural and functional integrities. It is also apparent that their long-term survival critically depends not only on the formation and maintenance of a normal physiological contact, but also on the mutual exterotrophic influences they exert on each other throughout life. In fact, the muscle cell's propriotrophic operation is continuously controlled and modulated by the nerve cell's exterotrophic actions (anterograde trophic effects). Similarly, neuronal functions are effectively influenced by the corresponding innervated myofibers (retrograde trophic effects). Despite the extensive amount of data that has accumulated on the phenomenology of trophic phenomena, the underlying molecular mechanisms have yet to be fully resolved. Although it may seem obvious that the operation of trophic mechanisms can involve the combined effects of various interrelated regulatory factors, the traditional controversy over the participation of a single vs. multiple factors persists. This continuing controversy has historically been centered mainly around evidence supporting the individual involvement of either bioelectrical activity-dependent or -independent factors. Based on current experimental information, however, the multifactorial character of trophic mechanisms is progressively gaining more widespread recognition. Following this line of thought in this chapter, we present a summarized account of certain pathways and sequences of molecular events which are possibly involved in the communication of trophic signals in the neuromuscular system.

First, we consider that the number of possible pathways through which trophic information can be communicated is surprisingly small, i.e., the anatomical arrangement of nerve terminals and extrafusal myofibers is such that antero- and retrograde signals are essentially transferred by way of the motor endplate. In turn, the principal manner in which certain cellular activities (e.g., axoplasmic transport, nerve and muscle bioelectrical potentials, etc.) can initiate molecular modifications on target cells is via the release of neurogenic (ACh and NTS) and myogenic (MTS) substances that encode trophic information. Furthermore, processing of such information must involve its translation or direct transfer from the synaptic area to those intracellular domains of the target cell where the final trophic effects are to be exerted.

Regarding anterograde effects, we initially analyze nerve impulse transmission-related events separately from those involving NTS. Next, we combine all these events within hypothetical schemes in which the regulation of non-endplate muscle properties is primarily accomplished through impulse-related factors, whereas that of motor endplate features additionally requires the participation of NTS. In either case, processing of trophic information is thought to involve the following sequential operations: release, reception, internalization, translocation to subcellular sites of action, various intermediary steps, and eventual modulation of the myofiber's genetic-metabolic machineries. The control of non-endplate properties is assumed to include nerve-evoked propagated muscle action potentials, as well as the consequent electrochemical events of excitation-contraction coupling and contraction. These events would either individually or in combination and via intermediary steps (e.g., ionic currents invading extrajunctional nuclei, activation of a second-messenger system, etc.) affect multinucleated myofibers along their entire length. Muscle endplate characteristics are in turn trophically influenced by ionic currents that result from local potentials and reach perijunctional nuclei, a second-messenger system activated by ACh-AChR complexes, or the physiological demands imposed by certain parameters related to synaptic usage (e.g., AChR occupancy). Accordingly, not only nerve activity as such, but also the pattern of impulse activity affect the myofiber's phenotypic expression. In addition, after interacting with postsynaptic receptors or selective surface membrane binding sites, NTS can trigger the formation of a second messenger, trigger the allosteric activation of particular enzymes, or undergo internalization via endocytosis. Through selective intermediary steps these agents influence the genetic expression of perijunctional nuclei at one or more operational levels (e.g., transcription, processing of mRNA, etc.) and/or at any of the multiple posttranslational processes responsible for the formation of essential motor endplate components. While these mechanisms could conceivably operate in parallel to influence distinct endplate features separately, we favor the view that they function in an interdependent manner and exert trophic effects of a cooperative nature. Indeed, those actions mediated by nerve impulses and NTS may concomitantly or sequentially effect one or more of the multiple intermediary steps (e.g., a second-messenger system, enzyme activation, NTS internalization, pre- and posttranslational processes of gene expression, etc.) implicated in the regulation of a single or several junctional properties.

Retrograde effects are assumed to be mainly induced by MTS assembled within the myofiber's subsynaptic region and secreted via exocytosis. Following their release, some of these substances remain in the synaptic cleft to replace extracellular matrix components, whereas others (or the trophic information they contain) are taken up by nerve terminals. While certain MTS remain within the terminals where they can influence local metabolic activities, others are carried by retrograde axoplasmic transport to the nerve cell body. The hypothetical sequence of events proposed for MTS retrograde influences is similar to that presented for the anterograde actions of NTS. These sequences essentially include the same operational stages (release, reception, internalization, etc.), but subtle differences may certainly exist, particularly in some of the numerous intermediary steps involved. Both anterograde and retrograde regulatory effects are viewed as part of a trophic loop in which the flow of exterotrophic information toward the muscle follows a divergent path (one neuron controlling all myofibers of the motor unit), whereas that toward the nerve cell occurs along a convergent route (all motor unit myofibers affecting a single neuron). Apparently, the uninterrupted operation of this loop is essential for cell survival, synaptogenesis, and cellular development, as well as for the maintenance of turnover and remodeling of myoneural junctions during adulthood and aging.

A short review is presented regarding those research strategies which have been traditionally used in studies on trophic phenomena. Some examples are also given to illustrate how such strategies, together with modern methodologies, can be applied to further explore trophic mechanisms within conceptual frameworks such as those described in this chapter. In this context, we emphasize that future advances in this field of research will require new modes of analyses involving stepwise approaches focused on further exploration of the physiological interplay among multiple trophic mediators and the sequences of related operational events.

ACKNOWLEDGMENTS

We thank C. Hodges-Savola, S. Rummel, and C. Nyquist-Battie for helpful comments on the manuscript. This work was partly supported by the Medical Research Service of the Veterans Administration.

REFERENCES

1. **Drachman, D. B., Stanley, E. F., Pestronk, A., Griffin, J. W., and Price, L. P.,** Neurotrophic regulation of two properties of skeletal muscle by impulse-dependent and spontaneous acetylcholine transmission, *J. Neurosci.*, 2, 232, 1982.
2. **Guth, L. and Albuquerque, E. X.,** The neurotrophic regulation of resting membrane potential and extrajunctional acetylcholine sensitivity in mammalian skeletal muscle, in *Muscle Regeneration*, Mauro, A., Ed., Raven Press, New York, 1979, 405.
3. **Caroni, P., Carlson, S. S., Schweitzer, E., and Kelly, R. B.,** Presynaptic neurones may contribute a unique glycoprotein to the extracellular matrix at the synapse, *Nature (London)*, 314, 441, 1985.
4. **Fallon, J. R., Nitkin, R. M., Reist, N. E., Wallace, B. G., and McMahan, U. J.,** Acetylcholine receptor-aggregating factor is similar to molecules concentrated at neuromuscular junctions, *Nature (London)*, 315, 571, 1985.
5. **Wallace, B. G., Nitkin, R. M., Reist, N. E., Fallon, J. R., Moagery, N. N., and McMahan, U. J.,** Aggregates of acetylcholinesterase induced by acetylcholine receptor-aggregating factor, *Nature (London)*, 315, 574, 1985.
6. **Slater, C. R.,** Orchestration at the synapse, *Nature (London)*, 315, 543, 1985.
7. **Schmitt, F. O.,** A protocol for molecular genetic neuroscience, in *Molecular Genetic Neuroscience*, Schmitt, F. O., Bird, S. J., and Blum, F. E., Eds., Raven Press, New York, 1982, 1.
8. **Kotsias, B. A., Muchnik, S., Arrizurieta, E. E., Losario, A. S., and Sosa, M.,** Influence of trophic substances in the regulation of resting membrane potential and ionic concentration in skeletal muscle, *Exp. Neurol.*, 88, 56, 1985.
9. **Fernandez, H. L., Patterson, M. R., and Duell, M. J.,** Neurotrophic control of 16S acetylcholinesterase from mammalian skeletal muscle in organ culture, *J. Neurobiol.*, 11, 557, 1980.
10. **Lømo, T., Massoulié, J., and Vigny, M.,** Stimulation of denervated rat soleus muscle with fast and slow activity patterns induces different expression of acetylcholinesterase molecular forms, *J. Neurosci.*, 5, 1180, 1985.
11. **McArdle, J. J.,** Molecular aspects of the trophic influence of nerve on muscle, *Prog. Neurobiol.*, 21, 135, 1983.
12. **Held, I. R., Squinto, S. P., and McLane, J. A.,** Neurotrophic regulation of the phosphorylation of a soluble cytosolic protein in skeletal muscle, *J. Neurosci.*, 3, 2054, 1983.
13. **Lømo, T.,** Trophic factors and postsynaptic activity in synapse formation, *Nature (London)*, 305, 576, 1983.
14. **Max, S. R. and Markelonis, G. J.,** Neural control of muscle, *Neurochem. Int.*, 5, 675, 1983.
15. **Bisby, M. A.,** Retrograde axonal transport, *Adv. Cell Neurobiol.*, 1, 69, 1980.
16. **Ochs, S.,** *Axoplasmic Transport and Its Relation to Other Nerve Functions*, John Wiley & Sons, New York, 1982, 145.
17. **Kristensson, K.,** Retrograde transport of macromolecules in axons, *Annu. Rev. Pharmacol. Toxicol.*, 18, 97, 1978.
18. **Bursztajn, S.,** Coated vesicles are associated with acetylcholine receptors at nerve-muscle contacts, *J. Neurocytol.*, 13, 503, 1984.
19. **Yamada, K. M.,** Cell surface interactions with extracellular materials, *Annu. Rev. Biochem.*, 52, 761, 1983.
20. **Hook, M., Kjellen, L., Johansson, S., and Robinson, J.,** Cell-surface glycosaminoglycans, *Annu. Rev. Biochem.*, 53, 847, 1984.
21. **Rotundo, R. L. and Fambrough, D. M.,** Secretion of acetylcholinesterase: relation to acetylcholine receptor metabolism, *Cell*, 22, 595, 1980.
22. **McMahan, U. J., Sanes, J. R., and Marshall, L. M.,** Cholinesterase is associated with the basal lamina at the neuromuscular junction, *Nature (London)*, 271, 172, 1978.
23. **Fernandez, H. L. and Stiles, J. R.,** Intra- versus extracellular recovery of 16S acetylcholinesterase following organophosphate inactivation in the rat, *Neurosci. Lett.*, 49, 117, 1984.
24. **Rotundo, R. L. and Fambrough, D. M.,** Synthesis, transport, and fate of acetylcholinesterase in cultured chick embryo muscle cells, *Cell*, 22, 583, 1980.
25. **Guth, L.,** "Trophic" effects of vertebrate neurons, *Neurosci. Res. Prog. Bull.*, 7, 1, 1969.

26. **Smith, B. H. and Kreutzberg, G.,** Neuron-target cell interations, *Neurosci. Res. Prog. Bull.*, 14, 211, 1976.
27. **Drachman, D. B.,** Trophic interactions between nerves and muscles: the role of cholinergic transmission (including usage) and other factors, in *Biology of Cholinergic Function*, Goldberg, A. M. and Hanin, I., Eds., Raven Press, New York, 1976, 161.
28. **Lømo, T. and Slater, C. R.,** Acetylcholine sensitivity of developing ectopic nerve-muscle junctions in adult rat soleus muscles, *J. Physiol. (London)*, 303, 173, 1980.
29. **Weinberg, C. B. and Hall, Z. W.,** Junctional form of acetylcholinesterase restored at nerve-free endplates, *Dev. Biol.*, 68, 631, 1979.
30. **Eccles, J. C.,** Trophic interactions in the mammalian central nervous system, *Ann. N.Y. Acad. Sci.*, 228, 406, 1974.
31. **Squinto, S. P., McLane, J. A., and Held, I. R.,** Increased phosphorylation *in vitro* of a cytosolic polypeptide resolved from denervated skeletal muscle, *Neurochem. Res.*, 6, 203, 1981.
32. **Zoller, M. J., Kerlavage, A. R., and Taylor, S. S.,** Structural comparisons of cAMP-dependent proteinkinases I and II from porcine skeletal muscle, *J. Biol. Chem.*, 254, 2408, 1979.
33. **Rubin, L.,** Increases in muscle Ca^{++} mediate changes in acetylcholinesterase and acetylcholine receptors caused by muscle contraction, *Proc. Natl. Acad, Sci. U.S.A.*, 82, 7121, 1985.
34. **Anderson, M. J. and Cohen, M. W.,** Nerve-induced and spontaneous redistribution of acetylcholine receptors on cultured muscle cells, *J. Physiol. (London)*, 268, 757, 1977.
35. **Anderson, M. J.,** Nerve-induced remodeling of muscle basal lamina during synaptogenesis, *J. Cell Biol.*, 102, 863, 1986.
36. **Podleski, T. R., Axelrod, D., Ravdin, P., Greenberg, I., Johnson, M. M., and Salpeter, M. M.,** Nerve extract induces increase and redistribution of acetylcholine receptors on cloned muscle cells, *Proc. Natl. Acad. Sci. U.S.A.*, 75, 2035, 1978.
37. **Burden, S. J., Sargent, P. B., and McMahan, U. J.,** Acetylcholine receptors in regenerating muscle accumulate at original synaptic sites in the absence of nerve, *J. Cell Biol.*, 82, 412, 1979.
38. **Inestrosa, N. C., Miller, J. B., Silberstein, L., Ziskind-Conhaim, L., and Hall, Z. W.,** Developmental regulation of 16S acetylcholinesterase and acetylcholine receptors in a mouse cell line, *Exp. Cell Res.*, 147, 393, 1983.
39. **Sanes, J. R. and Lawrence, J. C.,** Activity-dependent accumulation of basal lamina by cultured rat myotubules, *Dev. Biol.*, 97, 123, 1983.
40. **Lawrence, J. C. and Solsgiver, W. J.,** Levels of enzymes of energy metabolism are controlled by activity of cultured rat myotubes, *Am. J. Physiol.*, 244, C348, 1983.
41. **Changeaux, J. P.,** Activity regulates the levels of acetylcholine receptor alpha-subunit mRNA in cultured chick myotubes, *Proc. Natl. Acad. Sci. U.S.A.*, 82, 4558, 1985.
42. **Carlin, B. E., Lawrence, J. C., Lindstrom, J. M., and Merlie, J. P.,** Inhibition of acetylcholine receptor assembly by activity in primary cultures of embryonic rat muscle cells, *J. Biol. Chem.*, 261, 5180, 1986.
43. **Carlin, B. E., Lawrence, J. C., Lindstrom, J. M., and Merlie, J. P.,** An acetylcholine receptor precursor α-subunit that binds α-bungarotoxin but not *d*-tubocurare, *Proc. Natl. Acad. Sci. U.S.A.*, 83, 498, 1986.
44. **Merlie, J. P., Isenberg, K. E., Russel, S. D., and Sanes, J. R.,** Denervation supersensitivity in skeletal muscle: analysis with a cloned cDNA probe, *J. Cell Biol.*, 99, 332, 1984.
45. **Shooter, E. M., Frey, P., Gunning, P. W., Landreth, G. E., Sutter, A., and Yankner, B. A.,** Modulation of neuronal function by nerve growth factor, in *Molecular Genetic Neuroscience*, Schmitt, F. O., Bird, S. J., and Bloom, F. E., Eds., Raven Press, New York, 1982, 425.

INDEX

A

D

G

H

I

J

K

L

O

P

T

U

V

W

X

Z